Hacking the Atom

Hacking the Atom

EXPLORATIONS IN NUCLEAR RESEARCH, VOL. 1

The new science of low-energy nuclear reactions (1990-2015)

Steven B. Krivit

Edited by Michael J. Ravnitzky

Pacific Oaks
Press
SAN RAFAEL, CALIFORNIA

Hacking the Atom: Explorations in Nuclear Research, Vol. 1
Copyright © 2016 by Steven B. Krivit

Pacific Oaks Press / *New Energy Times* / www.newenergytimes.com
369-B 3rd St. #556, San Rafael, CA 94901
Library of Congress Control Number: 2016902974

Krivit, Steven B., author.
    Hacking the atom / Steven B. Krivit ; editors,
  Michael Ravnitzky, Cynthia Goldstein, Mat Nieuwenhoven.
    pages cm -- (Explorations in nuclear research ; vol. 1)
    Includes bibliographical references and index.
    LCCN 2016902974
    ISBN 978-0-996886444 (hbk.)
    ISBN 978-0-996886451 (pbk.)
    ISBN 978-0-996886468 (Kindle)
    ISBN 978-0-996886475 (ePUB)

    1. Low-energy nuclear reactions--Research--History.
    2. Electroweak interactions--Research--History.
    3. Science--Social aspects.   I. Ravnitzky, Michael, editor. II. Goldstein,
  Cynthia, editor.   III. Nieuwenhoven, Mat, editor. IV. Title. V. Series:
    Krivit, Steven B. Explorations in nuclear research ; v. 1.

    QC794.8.L69K754 2016   539.7'5
    QBI16-600060

Cover design: Lucien G. Frisch (Photograph: © Jahoo | Dreamstime.com)
Interior design template: Book Design Templates Inc.
Typeset in Crimson 11 pt., designed by Sebastian Kosch
Editors: Michael Ravnitzky (Developmental Editor), Cynthia Goldstein (Copy Editor), Mat Nieuwenhoven (Technical Editor)
Index: Laura Shelly

Also by Steven B. Krivit

Fusion Fiasco: Explorations in Nuclear Research, Vol. 2 (2016)

Lost History: Explorations in Nuclear Research, Vol. 3 (2016)

Nuclear Energy Encyclopedia: Science, Technology, and Applications,
Steven B. Krivit, Editor-in-Chief; Jay H. Lehr, Series Editor,
Wiley Series on Energy (2011)

*American Chemical Society Symposium Series: Low-Energy Nuclear Reactions
and New Energy Technologies Sourcebook (Vol. 2),*
Jan Marwan, Steven B. Krivit, editors (2009)

*American Chemical Society Symposium Series: Low-Energy Nuclear Reactions
Sourcebook (Vol. 1),*
Jan Marwan, Steven B. Krivit, editors (2008)

The Rebirth of Cold Fusion: Real Science, Real Hope, Real Energy,
by Steven B. Krivit and Nadine Winocur (2004)

www.NewEnergyTimes.com (since 2000)

"Let us hope that in a decade or two, or, at least, just before the beginning of the 21st-century, the present meager years of theoretical physics will come to an end in a burst of entirely new revolutionary ideas similar to those which heralded the beginning of the 20th century."

George Gamow, Russian-American physicist —1966 (1904-1968)

To Jess
Steven B. Krivit

To my sons Nathan and Max Ravnitzky
Michael J. Ravnitzky

Acknowledgments

Acknowledgments for the Series

Several key people have assisted me in the past few years on these books. They are my personal heroes. First is my copy editor, Cynthia Goldstein, who has been part of my team since 2004. Cynthia is my secret weapon who, year after year, helps make my writing intelligible.

Three other people composed the core team that made these books possible. Initially, they responded to my request to critique each chapter as Cynthia and I produced them. All of them exceeded my expectations in each of their unique contributions. None of them is a scientist, but all of them have some scientific background or technical training. I asked them to find all possible flaws and errors, and indeed they brought much to my attention. Of course, if any errors remain, they are my fault alone.

Michael Ravnitzky, in Maryland, accepted my invitation to critically review the draft chapters. After I convinced him that I really did welcome every critique he had to offer, he provided an invaluable outpouring that contributed immensely to the development of the books. Michael was a dream to work with. I am forever in his debt for his insights, wisdom, and his relentless, constructive and diplomatic suggestions, and so many other thoughtful contributions.

Mat Nieuwenhoven, in the Netherlands, has an eye for details like I have not imagined possible. His dedication to this project and helping make it as technically accurate as possible was heroic.

Lucien G. Frisch, in Germany, could see and understand the larger vision of these books from the very beginning, as soon as he read the first completed chapter. Sometimes, he could see the purpose of these books more clearly than I did, as I was on occasion too close to it to see it myself. I thank him not only for his visionary guidance but also for his brilliant artistry that graces the covers of these books.

I wish to extend my heartfelt appreciation and gratitude to the following reference librarians, who have provided me with so much support for all three volumes: my hero Randy Souther, at the Gleeson Library, Geschke Center, University of San Francisco; Madelyn Wilson, FOIA Officer at the Department of Energy's Office of Scientific and

Technical Information; Lorna Whyte, Diane Delara and Pam Klein at the San Rafael Public Library; and Lorna Lippes, an independent researcher. I owe much appreciation to Libby Dechman, at the Library of Congress, for her help in establishing the new cataloging subject heading "low-energy nuclear reactions." Thanks to all of you; I have a newfound and deep appreciation for reference librarians and public libraries.

I wish to thank Larry, Dee, Mirabai and Tracy for their assistance in helping me see the path on which I was travelling. Thank you, Flori, Jessica, Christine, Al and Sean, for your encouragement and support. Last but not least, many thanks to Sophie Wilson for crucial conversations that have helped me understand how to tell these stories.

Acknowledgments for This Volume

I owe immense gratitude to several people who have since died. They diligently recorded and preserved this history: Jerry Bishop (*Wall Street Journal*), Hal Fox (*New Energy News, Journal of New Energy*), Eugene Mallove (*Cold Fusion* magazine and *Infinite Energy* magazine), and author Charles Beaudette (*Excess Heat: Why Cold Fusion Research Prevailed*). I wish to thank Christy Frazier, who has done and continues to do a heroic job of preserving and archiving the legacy of Gene's work. I would also like to thank cybrarian Jed Rothwell for his efforts to provide easy access to LENR papers.

Thanks to library clerk Diane Smith in the Carnegie History Center in Bryan, Texas, for help with archive requests and Shamil Khedgikar for help searching the Cornell University Library archives.

I owe much gratitude also to Dieter Britz for his keen sense to keep track of and meticulously index the published scientific papers that contain the facts of this history.

I wish to offer my deep appreciation to Marin Pichler for reading a draft of this manuscript and providing a thoughtful and helpful critique.

I also wish to thank the many scientists with whom I have spoken in these last 16 years. Every one of you has been a teacher for me. It has been a privilege and an honor to report and record your stories, discoveries and, at times, your anguish. In particular, I am most grateful to Lewis Larsen for many instructive and insightful conversations.

Table of Contents

Nullius in Verba, the motto of the Royal Society of Great Britain

Introduction

Nullius in Verba: "Take nobody's word for it – do the experiment."
Motto of the Royal Society of Great Britain

In 100 years of chemistry and physics, most scientists thought nuclear reactions could occur only in high-energy physics experiments and in massive nuclear reactors. But new research shows otherwise: Nuclear reactions can also occur in small, benchtop experiments.

Research shows that, unlike fusion or fission, these low-energy nuclear reactions can release their energy without emitting harmful radiation or greenhouse gases or causing nuclear chain reactions.

Few scientific topics in the last 100 years have created more conflict than this one. Changes in scientific thinking rarely take place without a fight, and this one is occurring right now. This book offers readers a ringside seat.

Perhaps the biggest surprise is that the research, labeled incorrectly in 1989 as "cold fusion," never stopped, although it had been pronounced dead again and again. Despite confusing data, highly irreproducible results, and more than a few hot tempers, significant, valid science took place, much of it unrecognized at the time. In some cases, the data were buried for many years.

This book depicts the development of a new field of science, giving readers a look behind the scenes. It also illuminates the important distinctions between science and pathological science. It shows once again that real science is based on observation, not belief.

Three Books

This is the first book in a three-book series. Each book stands alone, covering a distinct period of scientific exploration. The three are being published in reverse chronological order.

- Hacking the Atom: Explorations in Nuclear Research, Vol. 1 (1990-2015)
- Fusion Fiasco: Explorations in Nuclear Research, Vol. 2 (1989-1990)
- Lost History: Explorations in Nuclear Research, Vol. 3 (1912-1927)

It's Not Fusion

I have found no experimental evidence to support the "cold fusion" idea that deuterium nuclei (a form of hydrogen) fuse at room temperature at high rates. Nor have I found a viable theory that explains how deuterium-deuterium (D+D) fusion might occur in electrochemical cells.

Experimental Nuclear Data

However, I have found an abundance of experimental data, some of it from U.S. Department of Energy (DOE) national laboratories, including Oak Ridge, Lawrence Livermore and Los Alamos, that provides well-measured evidence of previously unrecognized nuclear phenomena.

Viable Theory

One theory, discussed in this volume, appears to explain most of the anomalous phenomena reported in the field. It has nothing to do with D+D fusion. It does not involve few-body, strong-interaction fission or fusion. Instead, it involves many-body, collective electroweak interactions that can enable high rates of nuclear transmutation processes, under moderate conditions, in electrochemical cells and other types of systems.

Nuclear Evidence

The most convincing evidence for this new nuclear science is not the measurement of excess heat but the measurement of nuclear products. These include isotopic shifts, elemental transmutations, tritium production and, sometimes, production of tritium and neutrons from the same experiments. For many years, the early proponents of this new science argued for its validity on the basis of excess-heat measurements. This series of books makes no such argument; instead, it reveals the evidence of direct nuclear products.

Volume 1: Hacking the Atom

Volume 1, *Hacking the Atom*, is the story of how the science initially and erroneously called "cold fusion" continued to progress slowly but incrementally after its near-death in 1989.

The most significant early advances were heavy- and light-water electrolysis experiments, performed at Hokkaido University in Japan and at the University of Illinois, respectively. The data revealed that a variety of nuclear transmutations were occurring in the low-energy nuclear reaction experiments, providing crucial insights into the new science.

Beginning in 1999, a new method of gas-loading experiments performed at Mitsubishi Heavy Industries in Japan revealed even more convincing evidence of low-energy nuclear transmutations.

In 2004, the U.S. DOE, responding to a request from five "cold fusion" researchers, sponsored a second review of the subject. The review did not change the position of the U.S. government, but it did reawaken worldwide interest in the topic.

In 2005, a preprint of a promising theory was released on arXiv by theoreticians Allan Widom and Lewis G. Larsen, and in 2006, it was published in the *European Physical Journal C — Particles and Fields*. This theory may turn out to be correct, incorrect, or somewhere in between. Regardless, it has served a useful purpose in showing at least one plausible, logical explanation — that does not violate laws of physics — to explain most of the reported experimental phenomena.

Since 2005, many scientists who had observed unexplained nuclear phenomena defended their belief in the D+D "cold fusion" idea, even when all evidence was to the contrary. *Hacking the Atom* explains these events.

Note on Terminology:

Until July 10, 2006, I had identified the subject as "cold fusion." At that time, as a result of my initial understanding of the Widom-Larsen theory, and my conversations with Larsen, I stopped using the term "cold fusion" except for historical purposes and instead adopted the term "low-energy nuclear reactions" (LENRs). By 2012, most people writing on the subject had shifted to the new term, LENR.

The word "low" in "low-energy nuclear reactions" refers to the magnitude of input energies that are required to trigger LENR reactions. Researchers chose the term to distinguish it from the field of high-energy particle physics, which uses very high temperatures or particle accelerators to trigger nuclear reactions.

In order to be historically accurate, and because I and most people called the research "cold fusion" until 2006, I will generally use that term until we arrive at 2006 in this narrative. Here are concise distinctions for the two terms: LENRs — non-fusion-based nuclear reactions that occur at or near room temperature; "cold fusion" — the incorrect hypothesis of nuclear fusion reactions that occur at or near room temperature. The glossary contains more-detailed descriptions of each term.

Volume 2: *Fusion Fiasco*

Volume 2, *Fusion Fiasco,* focuses on the 1989 "cold fusion" history. This science conflict began when electrochemist Martin Fleischmann (1927-2012), retired from the University of Southampton, England, and his colleague Stanley Pons (b. 1943), chairman of the University of Utah Chemistry Department, announced at a press conference that they had created a sustained fusion reaction in a modified test tube. (Fleischmann, 1989)

When I began writing this book in 2012, I didn't think another book was needed to tell the old 1989 story. I was certain that I and other authors had covered it thoroughly. I was wrong.

Two things happened. First, I found related chemistry-based transmutation research that took place in the 1910s and 1920s. That material became its own book, *Lost History.* That research is a remarkable precursor to the research that came 60 years later.

Newly Uncovered Facts

The second thing was that, as I was drafting what was to be a brief chapter for the 1989 "cold fusion" history, I began checking facts with some scientists who were involved at the outset. One of them was physicist Richard Garwin

Garwin shared hundreds of documents, most of which were internal documents used in the 1989 DOE-sponsored "cold fusion" review and which had remained out of sight for more than two decades.

These documents reveal the behind-the-scenes activity and the real story of this crucial event. The documents include reports from researchers at DOE laboratories, like Lawrence Berkeley Laboratory, who observed nuclear phenomena that they described as "false-positive, up to eight times background."

Garwin also sent me audio recordings of the first "cold fusion" workshop in 1989, which took place in Erice, Italy. No detailed accounts of the Erice workshop seem ever to have been published. These recordings reveal another side to this otherwise-bitter debate: friendly, collaborative relationships between physicists and chemists.

The DOE-sponsored review has been discussed in every historical account. However, another little-known review took place in October 1989, at a Washington, D.C., workshop. Data presented at this workshop does not appear to have been reported in any other book.

Renowned physicist Edward Teller participated in this workshop and, after hearing about isotopic shifts observed by scientists at two independent national laboratories, concluded that nuclear effects were taking place. (Teller, 1989) The open-minded Teller also had a hunch about an explanation for the mechanism. The facts concerning these two government-sponsored reviews have been buried for a quarter of a century.

Jerrold Footlick, an author and former editor at *Newsweek,* sent me audio tapes of his interviews with former staff members at the University of Utah. They reveal the special interests behind the infamous University of Utah fusion press conference.

Letters sent by Fleischmann to his good friend and electrochemistry colleague John Bockris reveal Fleischmann's actual motive for attempting electrolytic fusion.

After I had many extensive conversations and a meeting at Oxford University with theoretical particle physicist Frank Close, he provided me with several documents that shed new light on his accusations that Fleischmann and Pons had manipulated a gamma-ray graph.

Close also helped me sort out two other sensitive matters in this history: Pons' accusation that 1) Steven Jones had pirated his and Fleischmann's ideas and that 2) Pons' graduate student, Marvin Hawkins, had stolen Pons and Fleischmann's lab books.

Volume 3: *Lost History*

Lost History, tells the story of research that took place 100 years ago, a story that is surprisingly similar to events in the modern era reported in Vols. 1 and 2. It has been obscured and omitted from history books for nearly a century.

In the 1910s and 1920s, this research was known both in scientific circles and by the general public. It was reported in popular newspapers and magazines, such as the *New York Times* and *Scientific American*

Papers were published in the top scientific journals of the day, including *Physical Review, Science* and *Nature*. Prominent scientists in the U.S., Europe and Japan and even Nobel Prize recipients participated in this research. In the 1930s, however, it was all dismissed as error, primarily because the theory was not understood and the experimental results were difficult to repeat. Here are some highlights from *Lost History*:

Anomalous Production of Noble Gases

From 1912 to 1914, several independent researchers detected the production of the gases helium-4, neon, argon, and an as-yet-unidentified element of mass-3, which we now identify as tritium. Two of these researchers were Nobel laureates.

Wendt and Irion's Synthesis of Helium

In 1922, two chemists at the University of Chicago, Gerald L. Wendt and Clarence E. Irion, synthesized helium using the exploding electrical conductor method. Despite doubts and criticism, no one unambiguously identified any error in their 21 successful experiments.

Nuclear evidence from exploding electrical conductor experiments was confirmed 80 years later by researchers at the Kurchatov Institute in Russia. (Urutskoev, 2002)

Anomalous Production of Gold, Platinum and Thallium

In 1924, a German scientist accidently found trace amounts of gold and possibly platinum in the residue of mercury vapor lamps that he had been using for photography. A year later, scientists in Amsterdam carried out a similar experiment, but starting with lead, and observed the production of mercury and the rare element thallium. The same year, a prominent Japanese scientist, in a different kind of experiment, reported observing the production of gold and something that had the appearance of platinum. Newspapers reported that he toured the world showing people the gold he had made in the laboratory. No reports of challenges to his claim appear to exist, at least in English-language references.

Paneth and Peters' Hydrogen-to-Helium Transmutation

In 1926, German chemists Friedrich Adolf (Fritz) Paneth and Kurt Gustav Karl Peters pumped hydrogen gas into a chamber with finely divided palladium powder and reported the transmutation of hydrogen into helium. Paneth was at first very proud of his and Peters' achievement, claiming that they were the first scientists to perform a nuclear transmutation. Paneth dismissed the earlier 1912-1914 transmutation reports without thoroughly examining them and without clearly identifying any errors.

A year later, Paneth did an about-face: He worked hard to find explanations to dismiss his and Peters' helium-production claims. He was unable to completely explain away their results.

Correction to a Milestone in Scientific History

While doing research on this early transmutation era, I came across facts that contradict the depiction of an important milestone in scientific history. World-famous physicist Ernest Rutherford has been credited incorrectly with the first nuclear transmutation. Some historians, even Rutherford scholars, call it his greatest achievement.

Not only was he preceded by other researchers in the 1912-1914 era, but also the experiment that has been attributed to him, transmuting nitrogen to oxygen, was in fact performed by Patrick Maynard Stewart Blackett, a research fellow who was working under Rutherford.

All but a few historians — and all known Internet references as of 2015 — incorrectly credit this discovery to Rutherford.

Preparation for a Paradigm Shift

No Practical Devices Yet

LENRs may someday lead to practical energy or heating devices. Research shows that LENRs can reach local surface temperatures of 4,000-6,000 K and boil metals (palladium, nickel and tungsten) in small numbers of randomly scattered microscopic LENR-active hot-spot sites on the surfaces of laboratory devices. To date, routine production of excess heat in laboratory apparatus at levels greater than 1 Watt has been more difficult.

Although some people seem to understand the basic science of LENRs, much engineering research and development is needed for the science to evolve into practical device design and reproducible fabrication. Today, there are no commercially practical LENR reactor devices, even though some people and organizations in and associated with the field have episodically made such claims since 1989. I have investigated many such claims of commercially viable devices in recent years and found them to be unsubstantiated.

Nevertheless, the body of scientific data suggests that someone or some companies will eventually commercialize the technology for thermal power generation applications.

Welcome to the Journey

I have independently investigated and reported on this subject for 16 years. I invite scientists and non-scientists alike to join me on this journey of scientific exploration and discovery. It is my pleasure to share this adventure with you now.

Steven B. Krivit
San Rafael, California
Sept. 1, 2016

CHAPTER 1

The King of Cold Fusion

In March 1989, two chemists in Utah made an extraordinary announcement. They said they had discovered the solution to the world's energy problems using a test tube and water as fuel. In hours, they appeared on television and then in news stories around the world.

However, things quickly fell apart. Within weeks, excitement turned to skepticism as the experiments were not easily repeated. Their scientific peers became angry, and the pair was broadly denounced. This developed into one of the most divisive scientific conflicts in recent history. "Cold fusion," as it was known in the media, became known as a quintessential example of bad science.

But that account is incomplete. The real story has remained mostly hidden, buried in archives and behind closed doors, and scattered among the memories of participants. The events indicate the presence of a paradigm shift in nuclear science and suggest radical changes in energy technology.

For a century, low-energy nuclear phenomena have produced inexplicable experimental results, and there was no coherent theory. What really occurred is more astonishing and more fascinating than has ever been revealed. The consequences are potentially more important than anyone has ever imagined.

Fifteen years later, on Aug. 23, 2004, six other scientists stood before a panel of 11 peers at a meeting sponsored by the U.S. Department of Energy (DOE). The six scientists thought that their research might lead to a solution for the world's energy problems; clean energy, free of dangerous radiation, waste and greenhouse gases. It was a golden opportunity to make their case.

Two of the proposers — electrochemist Michael Charles McKubre (b. 1948) and theorist Peter L. Hagelstein (b. 1954) — had done the lion's share of gathering graphs, tables and data that, they asserted, provided evidence for room-temperature nuclear fusion. The two colleagues and business partners were treated with deference by other scientists in the low-energy nuclear reaction (LENR) field.

Behind the scenes, Michael Melich (b. 1940), a U.S. Navy physicist with a long, shadowy history in the research that he and his colleagues called "cold fusion," assisted them. McKubre had first introduced me to him in 2003. "Talk to that man," McKubre said, "he's a spook."

For years, Melich went out of his way to tell people that he was specifically tasked by the U.S. government to gather information about "cold fusion." He certainly acted the part of a wanna-be spook. At a conference in 2011, he walked around the room openly videotaping the faces of everyone in the audience. But it was, for the most part, an act. His evangelism about "cold fusion" and his ownership stake in a private company seeking commercialization of the science signaled his actual and more personal interests.

McKubre, a New Zealander by birth, the father of three, had worked at the prestigious SRI International laboratory in Menlo Park, California, since 1979. He was an outstanding spokesman for the field: He was intelligent, eloquent, and quick-witted, and to top it off, he spoke with a proper English (New Zealand) accent.

Michael McKubre (2007), Peter Hagelstein (2004) Photos: S.B. Krivit; Michael Melich (2008) Photo: D. Tran

At the end of an international "cold fusion" conference in Beijing, China, McKubre was the person who was asked to present the closing comments to summarize research progress. He was the person who, during a cultural outing during a conference in Rome, Italy, was invited by the local conference organizers to meet the Pope.

Among friends, who included researchers from around the world, he was known to drop his formality occasionally. While at the podium during a conference in Marseille, France, McKubre introduced Hagelstein, an associate professor in the Department of Electrical Engineering and Computer Science at the Massachusetts Institute of Technology (MIT). McKubre said that Boston and San Francisco were the only two places in the United States with intelligent people.

The audience, which included a number of U.S. citizens, responded with mixed reactions. When challenged by a Texan and asked who appointed him king, McKubre leaned back in his chair, put his hands behind his head, and exclaimed, "It's good to be the king!" Audience members familiar with the Mel Brooks movie *History of the World, Part I*, enjoyed a moment of laughter. Nevertheless, the title, at least in the field, fit him well.

Starting in the early 1990s, he managed an SRI laboratory called the Energy Research Center and held the title of director. The lab was funded with several million dollars, mostly from the Defense Advanced Research Projects Agency (DARPA) and the Electric Power Research Institute (EPRI), a respected nonprofit organization funded by the electric utility industry.

When the money ran out around 1998, control of the lab was given to another scientist at SRI, but McKubre was allowed to retain the title, according to Francis Tanzella (b. 1953), an SRI electrochemist who worked with McKubre. During that decade, McKubre and his team of 20 researchers performed experiments that provided some of the best-measured evidence of unexplained heat production in electrolytic cells.

But McKubre's pursuit of answers to the mystery of the science had been long and hard. Not only did he bear the brunt of hostilities from other scientists, but an explosion in the SRI lab on Jan. 2, 1992, took the life of his colleague Andy Riley. McKubre survived but continued to carry some of the shrapnel in his body.

Hagelstein had spent most of his adult life at MIT, in the same department. He had earned his bachelor's degree, master's degree and doctorate there. For a few years, he worked in California on weapons development. His Ph.D. dissertation, on the physics of X-ray laser design, led him to a job at Lawrence Livermore National Laboratory. There, along with scientists Edward Teller, George Chapline Jr. and Lowell Wood, he designed an X-ray laser-based weapon for ballistic missile defense. After five years, Hagelstein came back to MIT to teach in 1986. Work on the X-ray laser weapon was abandoned in 1992.

Unlike McKubre, Hagelstein preferred to stay away from the spotlight. In April 1989, Hagelstein was among the first scientists, if not the first one, to suggest that he could explain Martin Fleischmann and Stanley Pons' claims of room-temperature fusion. Days later, Ronald R. Parker, the head of the MIT Plasma Fusion Center, gave Hagelstein an indirect whack in the *Washington Post.* "In fusion research," Parker said, "there are always crackpot claims to produce fusion in a simple way." Later that year, Hagelstein was opposed for tenure by MIT faculty who thought that he embarrassed the institute by his "cold fusion" work.

When I interviewed McKubre in August 2003, he was frustrated that he no longer had financial support to do the research he was most inspired to do. "We have taken steps to lock up the intellectual property," McKubre said, "and we're in an unbelievably strong position with respect to the science. Yet we still can't get anybody to fund it."

Preparation for the DOE review began in March 2004. A month later, McKubre and Hagelstein formed Spindletop Corp., hoping that DOE would provide funding for the research and that the two of them would be among the recipients. Their associate was Matthew Trevithick, an advisor to Silicon Valley venture capitalists.

When Hagelstein stood before the DOE reviewers on Aug. 23, 2004, he presented those ideas, which, he said, were the best theoretical explanations for "cold fusion." His ideas were based on McKubre's experimental data, which, as McKubre told the reviewers, was the best experimental evidence of LENR research.

For 10 years, researchers had trumpeted McKubre's data as the best evidence for tabletop fusion. It wasn't. McKubre had fabricated and manipulated it. I uncovered and reported the story in 2010 (Chapter 33).

Moreover, in their presentations to the DOE reviewers in August 2004, McKubre and Hagelstein omitted a substantial body of legitimate

LENR research performed by other scientists. Specifically, they failed to mention experimental evidence of nuclear transmutations and isotopic shifts in the experiments.

Why the omissions? Because those data didn't support the idea of "cold fusion." The omitted data not only provided direct evidence of non-fusion nuclear reactions but also shed light on one of the most tumultuous and confusing periods of science in modern history.

A Difficult Birth

The public first heard about the idea of room-temperature "fusion" on March 23, 1989, when electrochemists Fleischmann and Pons spoke at the University of Utah press conference. They claimed that they had demonstrated the unprecedented feat of "sustained nuclear fusion," at room temperature no less, and even more remarkably, in a simple apparatus that resembled a test tube. The notion was that nuclei of deuterium — a form of hydrogen — could overcome their mutual electrostatic repulsion and undergo high rates of nuclear fusion at room temperature in benchtop experiments. Normally, deuterium-deuterium (D+D) fusion occurs only at millions of degrees within mammoth-size experiments. The Utah claim was shocking, to say the least.

Forty days later, angry physicists and one angry chemist lashed out at Fleischmann and Pons, accused them of fraud, incompetence, and delusion and claimed that the Utah chemists had led them on a wild-goose chase. Those other scientists had tried to repeat the experiment, but, instead of finding excess heat and nuclear evidence such as tritium, neutrons, and gamma rays, they found absolutely nothing.

The science discovery of the century had turned into the science disaster of the century, as physicists at an American Physical Society meeting literally cast their vote: The dream of limitless clean energy from room-temperature fusion and water in a simple glass apparatus was dead. The researchers were quickly assigned to the crackpot category. Vol. 2, *Fusion Fiasco*, tells the behind-the-scenes story of what really happened in 1989-1990. But it didn't end there.

Although most scientists dismissed the fusion fiasco in 1989, a few hundred scientists across the world stuck with the research despite the derision, insults and criticism they received from other scientists.

In 1989, many laboratories around the world, including some national laboratories, reported confirmatory data, not of fusion but of unexplained heat beyond the levels possible by known chemistry, low levels of neutron emission, and production of tritium.

However, reproducibility was a significant problem, as David H. Worledge, a program manager at the Electric Power Research Institute, said at the First Annual Conference on Cold Fusion in 1990. "Despite these advances, we have not yet succeeded in producing a recipe that can be handed to independent research groups that will lead to reproducible results," Worledge said. (Worledge, 1990)

At the end of that conference, in light of the many repeated observations of a wide range of experimental phenomena, Fleischmann objected to critics' complaints. Critics, he said, had attempted to explain away the results by an equally wide range of imaginary systematic errors, or resorted to cynicism, and depicted the meeting as a "séance of true believers." (Fleischmann, March 1990)

Theorists Jumping the Gun

Most scientists in 1989 who had claimed confirmation of some aspect of the Fleischmann-Pons deuterium-palladium electrolysis experiment or a variant, such as a deuterium gas experiment, steered clear of claims that they had observed fusion.

Virtually all of them limited their theoretical interpretations; they reported their observations as they had measured them and ascribed the results simply to some as-yet-unexplained nuclear process.

Hagelstein, however, jumped the gun. Between April 5 and 12, 1989, less than three weeks after Fleischmann and Pons announced that they had produced "nuclear fusion," Hagelstein submitted four papers for peer review to *Physical Review Letters*, according to MIT press releases dated April 12 and 21, 1989.

In them, and in the absence of supporting measured data, Hagelstein proposed a conceptual mechanism for "cold fusion," expressed as $d + d$

—> *Helium-4 + 23.8 MeV (heat)*. Specifically, Hagelstein's idea was that, for every atom of helium-4 produced, the reaction should release approximately 24 MeV of heat. However, nobody had reported observing any helium-4 in "cold fusion" experiments yet.

Hagelstein's equation is presumptive; it does not exist in established physics. It is a hypothetical variant of an actual equation describing the well-understood deuterium-deuterium nuclear fusion reaction $d + d$ —> *Helium-4 + 23.8 MeV gamma ray*.

A variant of the hypothetical "cold fusion" mechanism had been informally proposed five days earlier, on March 31, 1989, by Douglas Morrison (1929-2001), a high-energy particle physicist at CERN:

> Maybe the dominant reaction is fusion, $d + d$ —> *He-4*, but we need something else to share the energy and momentum produced — this could be the close neighboring structure of the lattice. Thus, the dominant reaction is to produce heat!

Morrison wrote this in the first of his prolific series of newsletters on the contentious subject. His first newsletter was the only one that was objective and accurately reflected the facts. Within five weeks, Morrison had denounced the entire topic as pathological science. Nevertheless, he became obsessed with it. He continued to publish his newsletters and attend conferences almost until his death, on April 29, 2001.

In August 1990, Morrison came to the Bhabha Atomic Research Centre (BARC) in Trombay, India's largest and most prestigious nuclear research facility to look at experimental evidence. Mahadeva Srinivasan (b. 1937), the former director of the Neutron Physics Division and assistant director of the Physics Group at BARC, showed him a "massive beta spectrum of tritium generated in titanium chips subjected to deuterium gas." Morrison, Srinivasan recalled, was speechless. Despite his attempts to discredit the research, Morrison was very happy at the conferences. "He honestly thought he was saving physics from some nuts," Srinivasan wrote.

Closely following Hagelstein on the "cold fusion" theory idea were two professors in the University of Utah Chemistry Department,

Cheves Walling and John Simons, who submitted a manuscript to the *Journal of Physical Chemistry* by April 14, 1989. (Walling and Simons, 1989) They, too, imagined that pairs of deuterons (deuterium nuclei) were fusing to form helium-4 at room temperature.

On April 17, Walling told reporter Lee Dye, at the *Los Angeles Times*, that he and Simons used one of the same experiments that produced heat for Fleischmann and Pons and that he (Walling) and Simons had detected helium-4 when they put the experiment into a mass spectrometer. Walling and Simons were the first researchers to (informally) report experimentally measured helium-4 production from the experiments.

Unlike excess heat, helium-4, as long it was measured in greater concentrations than atmospheric background, was definitive evidence of a nuclear reaction. Dye interviewed Walling — one of the most respected chemists at the time — by phone. "The amount of helium-4 corresponds to the amount that should have been there if the heat was coming from nuclear fusion," Walling said.

Mahadeva Srinivasan, Eugene Mallove, Thomas Passell, Douglas Morrison (2000) Photo: Barbara DelloRusso

But Walling and Simons did not publish their helium-4 data. The first results of helium-4 production from a benchtop electrolytic experiment weren't published until 1991, by Benjamin Bush (University of Texas) and Melvin Miles (b. 1937) (Naval Air Weapons Station — China Lake). (Bush, Benjamin, et al., 1991; Miles et al., 1991) Bush and Miles effectively wrote the same thing as Walling and Simons, that the amount of helium-4 observed in the gaseous products of the experiments corresponded approximately to the amount of excess heat.

The "cold fusion" idea that deuterons could overcome their mutual electrostatic repulsion at high rates at room temperature was, however, fatally flawed, not just because of disagreement with existing theory but because of contradictory experimental data, as explained in later chapters.

Alternate Helium-4 History

Electrochemist Melvin Miles' group published the first observation of helium-4 data in electrolytic experiments, but he denies that his group was the first. Instead, he gives that credit to Fleischmann and Pons. In a paper Miles presented in 2015, he wrote, "Fleischmann and Pons were actually the first to observe that helium-4 was produced in the Pd/D system." The historical record does not support this. Miles' only evidence is a Sept. 21, 1993, letter from Fleischmann stating, "We had our first indication of helium-4 in December of 1988!" (Miles, 2015)

Melvin Miles (Photo: S.B. Krivit)

Fleischmann and Pons made limited attempts to look for helium. Despite some unsubstantiated accounts, the two never published any observation of helium-4 in their deuterium-palladium experiments. In a memoir, Fleischmann wrote that their helium data were "unpublishable." (Fleischmann, 2000)

Some of Fleischmann and Pons' followers, including Hagelstein, have often written that the "cold fusion" idea of $d + d \longrightarrow Helium\text{-}4 +23.8\ MeV$ *(heat)* came from Fleischmann and Pons. (Hagelstein, 1998) The general idea of fusion did originate with Fleischmann and Pons, but the pair never claimed that helium-4 was the dominant nuclear product in "cold fusion," let alone propose that the amount of helium-4 was directly associated with excess heat.

Two Diverging Philosophical Schools

By March 1990, however, Hagelstein had backtracked. (Hagelstein, 1990) After considering the experimental evidence that had been reported in the previous 12 months, he abandoned his fusion idea, instead proposing a neutron-based idea.

Other theorists in 1990 stayed with the heat-and-helium-4 "cold fusion" concept, specifically Giuliano Preparata in Italy, and the team of Talbot A. Chubb and Scott R. Chubb (uncle and nephew) in the U.S.

By the end of March 1990, as Italian physicist Francesco Scaramuzzi explained, two diverging schools in the field emerged. One group comprised researchers who believed only in direct nuclear evidence such as neutrons. The other group believed in excess heat. Neutrons from benchtop experiments were barely accepted by the scientific community, Scaramuzzi wrote, and excess heat was outright rejected.

"I must confess," Scaramuzzi wrote, "that I belonged to the first school, being quite skeptical about heat production." The factionalism was so strong that, at one point, both groups were planning separate conferences for what eventually turned out to be the Second International Conference on Cold Fusion. (Scaramuzzi, 2000)

The fusion-believing scientists jumped to a conclusion and took the wrong fork in the road. It took 25 years — and heated battles — for the science to get back on the right path.

CHAPTER 2

Forbidden Research

In early 1990, while some theorists struggled to explain "cold fusion" as an actual fusion process, MIT's Peter Hagelstein thought it was better explained as a neutron-based idea.

Experimentally, three types of direct evidence of nuclear reactions were widely reported in 1989 and 1990: excess heat, low levels of neutrons, and tritium. Yet the clearest evidence for low-energy nuclear reactions (LENRs) at room temperature in 1989 was the isotopic shifts (changes to the ordinary mix of naturally occurring isotopes of elements) that had taken place at the Naval Research Laboratory (NRL) and the Lawrence Livermore National Laboratory (LLNL). (Vol. 2, *Fusion Fiasco*)

The strongest evidence for nuclear reactions is large isotopic shifts and the appearance of elements not present at the beginning of an experiment. Such results cannot be ascribed to chemical fractionation or contamination. There is no other explanation for such changes. Isotopic shifts, however, do not conjure images of alchemy, do elemental transmutations do.

Because tritium and helium-4 are well-known products of deuterium-deuterium (D+D) thermonuclear fusion, some early researchers assumed that the tritium and helium-4 observed in the experiments had to be produced by fusion. In hindsight, these products are more likely the result of neutron-driven nuclear transmutation, or, more precisely, nucleosynthesis.

One particular reported transmutation has been discussed for centuries, and its lore is embedded deeply in social consciousness and mythology: the alchemical transmutation of mercury to gold, which would require the reduction of a single proton from element number 80 to get to element number 79. Element number 79, gold, has no equal. It possesses visual beauty, resistance to corrosion, durability, and outstanding electrical and thermal conductivity. It can be used to draw wires of extreme thinness (ductility) and to hammer out sheets of extreme thinness such as for ceiling covers (malleability). Further, it is one of the most significant stores of wealth.

For centuries, people have schemed and dreamed of ways to manipulate Mother Nature to convert base (cheap) metals into gold. For most of this time, such efforts have been the object of charlatans who tried to defraud people; a modest fee to the alchemist, according to folklore, produced a quantity of gold of significantly greater value than the dross from which it was produced. Unsurprisingly, the alchemy accounts are devoid of stories of alchemists who used their skills to fill their own vaults with gold.

In modern times, physicists have demonstrated transmutations, including the creation of artificial gold by using nuclear reactors and particle accelerators. Yet this is not controversial. There are two reasons for this. First, people know that the cost of producing artificial gold from high-energy physics far exceeds the market value of such gold. Second, well-accepted nuclear physics can explain the process.

1920s Transmutations in Japan and Germany

In the last few decades, surprising evidence, spectroscopically measured, has accumulated that microscopic amounts of gold can be created by means other than nuclear reactors and accelerators. The creation of heavy elements, including gold, through low-energy nuclear transmutation reactions, however, conflicts with current scientific understanding. The observation of such anomalies, in fact, dates back to the 1910s and 1920s, as discussed in Vol. 3, *Lost History*.

Reports of the production of artificial gold by prominent scientists were published in the top scientific journals, such as *Nature* and *Naturwissenschaften*. The results made no sense according to theory at the time. But the people reporting these findings were legitimate scientists and performed their work like scientists, not medieval alchemists. Amounts of gold they produced in their experiments were miniscule, yet it was sufficiently above background levels to be scientifically significant.

But the experiments were not always easy to reproduce, and no conceptual theoretical framework was available at the time to explain the results. When activity in high-energy physics developed in the 1930s, it took precedence because that research was easy to reproduce and it had a well-defined theoretical basis. Consequently, the transmutation work from the previous two decades was relegated to the dustbin of scientific history, the results assumed to be wrong and generally forgotten.

1989 Low-Energy Transmutations in California

The stigma of alchemy cast a shadow over science in 1989. When Joseph C. Farmer, 34, and his colleagues at LLNL performed an electrolytic experiment and detected elements that hadn't been there before, it was so far beyond what they could conceive that they effectively ignored a valid result.

CHARACTERIZATION STRUCTURAL CHEMICAL BEFORE OR AFTER USE METHODS RESULTS	EDX of starting material — only Pd detected. Auger, SIMS, & XPS of surface after use. Elements detected included Si, S, Cl, C, Ca, Pd, O, Fe, and possibly Cu & Mg by Auger. SIMS detected H, Li, C, Na, Al, K, Ca, Fe, Pd, and possibly Ti, Cr, & Cu. XPS showed Fe, O, Ca, Pd, C, Si, and a trace of Al.
NOTABLE OBSERVATIONS	No evidence of nuclear fusion.
D / METAL RATIO ATTAINED	

EXPERIMENT YIELDED	yes	no
HEAT	_____yes	__X__no
NEUTRONS	_____yes	__X__no
TRITIUM	_____yes	__X__no
HELIUM	_____yes	__X__no

Portion of survey response sent from Lawrence Livermore National Laboratory researchers to the ERAB panel

As requested by the Department of Energy (DOE) for all of its national laboratories in 1989, Farmer and his colleagues at Livermore filled out a survey form around Aug. 15, 1989. The researchers wrote that, with energy-dispersive X-ray spectroscopy (EDX), they found palladium only in the starting material. On their survey form, the LLNL researchers wrote simply that they had seen "no evidence of nuclear fusion." In a more detailed report on Sept. 14, 1989, they told DOE that they had observed significant shifts in lithium-6 and lithium-7 isotopes.

1989 Low-Energy Transmutations in Texas

John O'Mara Bockris (1923-2013), at Texas A&M University, spearheaded the research involving transmutations of elements heavier than helium and tritium in "cold fusion." Bockris was a fascinating and prolific scientist, and his activities provided ample material for controversy. Many people who knew him called him a genius; others called him an egotist and a charlatan.

I met Bockris at his home in Texas in 2004, seven years after he had retired from Texas A&M at 81. Bockris had sustained more than his share of sharp criticism; nevertheless he maintained his sense of humor.

In LENRs, Bockris said that he and his colleagues had priority for three discoveries: 1) the synthesis of tritium; 2) the synthesis of helium-4; and 3) the synthesis of heavy elements.

A year before he died, he withdrew his claim of detecting helium-4; his reasoning is still unclear to me. His claim of priority for heavy-element nucleosynthesis in LENRs holds in the U.S., but a group in Russia beat him to publication by a few months.

Disruption of a Paradigm

The discovery of tritium produced in electrolytic cells in April 1989 at Texas A&M University was the first direct and unambiguous nuclear evidence for LENRs in the U.S.

The tritium report became well-known after Kevin Wolf (1942-1997), a Texas A&M nuclear chemist, spoke about it at the DOE-sponsored "Workshop on Cold Fusion Phenomena," in Santa Fe, New Mexico, May 23-25, 1989. (Vol. 2, *Fusion Fiasco*)

Unlike the claims of tritium, the claims of excess heat were merely indirect evidence of nuclear reactions and therefore subject to dispute as proof. The tritium evidence showed that nuclear reactions could occur in small, benchtop experiments.

The tritium was independently confirmed in India that year by dozens of researchers at the Bhabha Atomic Research Centre (BARC).

Padmanabha Krishnagopala Iyengar (1931-2011), the director of that laboratory, presented their results in Germany in July 1989, at the Fifth International Conference on Emerging Nuclear Energy Systems. (Iyengar, 1989)

Iyengar was a modest, unimposing man. He was passionate about science and open to new ideas. A dozen experiments performed by 10 groups in Iyengar's lab provided him with the evidence he needed — tritium and neutrons — to recognize the existence of the new phenomenon.

The level of reproducibility was low; nevertheless, the experiments that gave null results didn't bother him. He accepted the well-measured data despite the fact that science authorities like Richard Garwin, an IBM physicist and key member of the 1989 DOE cold fusion review panel, demanded that valid experiments must be highly reproducible.

Trouble With Tritium

The idea that nuclear reactions could take place in small, benchtop experiments and do so without emitting dangerous radiation was (and still is) shocking to most scientists. Even at Bockris' own university, some colleagues thought he might be falsifying the data.

Some professors at Texas A&M University began spreading rumors that Bockris or his students were spiking the electrolytic cells with tritium. The accusers never identified a plausible motive for Bockris and his colleagues to do this.

Eventually, the rumors in Texas attracted the curiosity of freelance writer Gary Taubes. He ignored the fact that tritium synthesis by electrolysis had also been observed at Los Alamos National Laboratory (LANL), at Oak Ridge National Laboratory and at BARC.

In June 1990, Taubes published an article in *Science* magazine about the Texas A&M tritium. (Taubes, 1990) The article wrongly implied that Bockris and his graduate student Nigel Packham had spiked their cells.

By his own admission, Taubes had "no smoking gun." His article provided no direct scientific evidence, no written records, no confessions, and no witnesses to support research misconduct, let alone the more serious charge of fraud. Yet most people who read Taubes' story interpreted Bockris and Packham's data as the result of fraud. Unfortunately, fraud made much more sense than the idea that nuclear reactions could take place in simple benchtop experiments. Despite the lack of an obvious motive for committing fraud, the prominent magazine *Science* seemed far more credible than a chemist claiming he could trigger nuclear reactions in a test tube.

It was easy for Taubes to depict Bockris and Packham as tritium adulterers because the experiments that produced tritium at BARC in India were not widely known in the West in 1990. The tritium counts reported by radiochemist Edmund Storms (b. 1931) at LANL were much lower and easier to dismiss than those of the Bockris group. The tritium counts observed by Thomas Claytor, another researcher at LANL, were also not as strong as the Texas A&M results. Claytor also kept a low profile and avoided publicizing his results.

For people who were willing to consider the Texas A&M tritium real, the findings provided the first substantive evidence for "cold fusion" as *bona fide* nuclear reactions. The effect was twofold. First, it contributed to the confidence of other researchers in the subject. Second, it exacerbated the condemnation from scientists who couldn't believe that nuclear reactions at room temperature were possible and that the tritium must have been a hoax.

Tritium Fallout

The fallout from the Taubes article in *Science* brought additional professional and personal hardship for Bockris and Packham. Follow-up articles nationwide forced the university administration to conduct a research integrity inquiry into Bockris' work.

In early July 1990, Provost E. Dean Gage appointed three faculty members to a panel to review the possibility of fraud, as depicted in the Taubes article. The panel was chaired by John Poston, a nuclear engineer, and included Joseph Natowitz, a nuclear chemist, and Ed Fry, a physicist. The panel reported nothing until November 1990.

Meanwhile, Packham, who would have been the first student to earn a Ph.D., in part based on his tritium experiments with Bockris, still had to complete his oral defense. The other part of his dissertation research was on an esoteric but non-controversial study of the biological production of hydrogen.

His oral defense took place on Aug. 16, 1990. Normally, these academic proceedings attract a sparse audience; Packham's final oral examination was standing-room-only. Fifty students and faculty members gathered to witness the event, according to Jennifer Nagorka, a writer at the *Dallas Morning News.*

The trouble started even before the oral defense. Texas A&M chemistry professor Manuel P. Soriaga had requested that Bockris include a physicist or a nuclear chemist on the graduate committee. Bockris refused, asserting, somewhat justifiably, that physicists were biased against such controversial research.

Under the dark shadow cast by *Science* magazine and Taubes — not to mention the perceived taint of "cold fusion" — some of the faculty were understandably tense. For example, Texas A&M nuclear chemist Kevin Wolf had, for nearly a year until the Taubes article published, claimed that he too had confirmed the production of tritium in his cells. Throughout this time, Wolf had been under pressure from his peers to recant his well-measured tritium results.

During the time when Taubes was gathering information for his article, Wolf spoke with Taubes and other reporters and suddenly came up with a story that his tritium had been a mistake, the result of previously undetected contamination. Days before the Taubes article published, Wolf did an about-face and renounced his data, even though he had no definitive evidence of any such contamination. (Vol. 2, *Fusion Fiasco*)

Tempers escalated during Packham's defense and the following day. An embittered Soriaga resigned from the graduate committee, according to the *Dallas Morning News*, leaving a letter in which he called Packham's Ph.D. final oral examination "a grotesque curtailment of academic freedom and an unbridled travesty."

Michael Hall, the head of the Chemistry Department, refused to sign off on the dissertation and would not allow Packham's Ph.D. to be awarded. After two or three days of turmoil and confusion, a compromise was reached: Packham would be awarded his doctorate if he eliminated the troublesome tritium-producing experiments from his thesis and mentioned them only in the appendix. (Bockris, 2000)

First Inquiry Committee Exonerates Bockris

The university inquiry into possible research misconduct by Bockris and his colleagues concluded three months later, in November. On Nov. 19, 1990, Jennifer Nagorka reported the conclusions: "Hasty work, overzealous claims, but no evidence of fraud."

Nagorka wrote that "panel members chided researchers for not verifying their results before disseminating them." This is puzzling because the researchers had their tritium samples confirmed by four laboratories: General Motors Corp., Argonne and Los Alamos National Laboratories, and Battelle Memorial Institute, a company that administers some of the national labs.

Bockris' troubles were just beginning.

.

CHAPTER 3

Fear and Loathing in Texas

On a warm July day in 2004, "cold fusion" scientist Dennis Letts picked me up from the Austin, Texas, airport and brought me to the modest, ranch-style home of John and Lillian Bockris, in College Station. Letts had admired Bockris for years and wanted to join us for the interview.

Bockris greeted us cordially and ushered us into his home office. He took his seat behind a massive wooden desk, covered with stacks of papers and unfinished projects. Bockris' longtime personal assistant, Trish Schulz, sat off to his side. Ever efficient, Schulz seemed to know immediately where to find whatever scientific paper came to Bockris' mind.

We first discussed Bockris' 1989-1992 electrochemistry research that showed production of excess heat and the production of tritium and helium. Then he began to explain the experiments he had done in collaboration with a man of questionable character named Joseph Champion.

The Bockris-Champion collaboration was unusual. The former was an accomplished scholar and scientist. The latter, whom Bockris accepted as his guide, was not a scientist. Most people perceived their experiments as attempts to perform medieval-style alchemy and transmutation of base metals to gold. Bockris' involvement with Champion was depicted by news media as a serious error of judgment.

Bockris had an illustrious 50-year career in electrochemistry and he seemed defensive about his activities with Champion. I told him that I was more interested in his other findings — excess heat, helium, and tritium — which were well-corroborated with other laboratories.

In 2004, I had already formed a judgment about the purported gold-transmutation experiments: They were complete nonsense, I thought. In retrospect, I missed the point of his story in 2004 and didn't truly grasp it until 2015, when I began examining this history in detail.

I did not understand Bockris and his sincere interest in scientific exploration. Nor did I realize that he had been assailed at Texas A&M University and investigated three times for doing experimental research that other scientists were afraid to do. Despite media scrutiny and the academic inquiries he faced — the last one spanning an entire year — not one ethical or legal violation was ever substantiated. Most important, I did not yet understand the strength of his unshakable commitment to scientific integrity.

Later, I recognized that stories I had heard of Bockris from his colleagues in the field — with few exceptions — were colored by professional jealousy and, to some extent, fear. Their fear was based on the same misunderstanding that I had experienced: Bockris was practicing pseudoscience, and his experiments beyond the boundary of accepted science would hurt the credibility of the field. In fact, for many years, people said that Bockris' claims of transmutation were evidence that the entire field was pseudoscience. Nevertheless, Bockris and his colleagues were among the first scientists to report observations of significant nuclear transmutation products created in low-energy experiments.

A Legend in Electrochemistry

Bernhardt Patrick John O'Mara Bockris was recognized as one of the world's top electrochemists. He was the author of more than a dozen textbooks and another dozen reference books, some of which were used to train an entire generation of electrochemists. His most well-known set of textbooks was the important *Modern Electrochemistry* series.

During his professorships at four universities, his reputation attracted graduate and post-doctoral students from around the world. In the 1960s, while at the University of Pennsylvania, he led the largest electrochemistry research group in the Western Hemisphere. He was one of the founders of the International Society of Electrochemistry, started in 1949.

Back Row, from left: John W. Bowler-Reed, John W. Tomlinson, Hanna Rosenberg, Edward C. Potter, Allan Wetterholm, H. Martin Fleischmann Front Row, from left: Ahmed M. Azzam, Roger Parsons, John O'Mara Bockris, John F. Herringshaw, Brian Evans Conway, Harold Egan

Bockris, born in Johannesburg, South Africa, was the son of Alfred Bockris and Emmeline Mary MacNally. His parents separated soon after his birth, and Bockris and his mother, a tailor, moved to England. His father, as Bockris told Schulz, was a wealthy man until World War II. He lived in a huge home, with a staff of servants, in Hamburg, Germany.

At 6, Bockris entered a Catholic school and, later, Withdean Hall, a preparatory school. In secondary school, he attended Xaverian College, Brighton, another Catholic school. He learned discipline early, as Schulz wrote on his Wikipedia page.

"Students underperforming in mathematics were beaten by the headmaster, Mr. Hamilton," Schulz wrote, "while his wife would sometimes compel pupils to go outside for runs at night when the weather was bad."

Bockris finished his academic training at the Royal College of Science, a higher-education institution that was a constituent college of Imperial College London. As soon as he completed his Ph.D. degree in 1945, Bockris was appointed to the faculty of Imperial College. By 1946, 10 graduate students had asked to study under his supervision.

Bockris was one of many accomplished scientists in the 1947-48 Royal College of Science Electrochemistry Group (pictured above). His colleague, Martin Fleischmann, was an accomplished chemist and electrochemist who discovered Surface Enhanced Raman Scattering (SERS) and developed the ultramicroelectrode, in addition to playing a seminal role in "cold fusion" research.

Brian Conway was a world-renowned electrochemist who wrote 400 refereed scholarly publications and four books. Roger Parsons had an illustrious electrochemistry career and 200 publications to his credit. Ahmed Azzam worked with the International Atomic Energy Commission in Vienna.

Rex Watson (not pictured in the photo), the electrochemist on the Bacon fuel cell project in 1955-56, was the first person to build a workable fuel cell. This led to the fuel cells later used on NASA spacecraft. Watson also served as director of the secret Research Station in Porton Down, England, which was used for sensitive defense activities.

Another group member was Hanna Rosenberg. She originated important concepts in the area of proton mobility/conductivity in non-aqueous solutions, and her work proved valuable in subsequent research. However, she aspired to a more glamorous career than chemistry.

She adopted the stage name Hedda Linton, taking her name from both the character Hedda Gabler in the play of the same name and from the character Cathy Linton in *Wuthering Heights*. Linton secured acting roles in 11 movies in Italy between 1953 and 1957 and appeared in cinema pinup magazines. Meanwhile, a delay in publication of the proton conduction research led to H. Linton's credit as a co-author in a widely cited scientific paper.

In 1979, Bockris was appointed to his final professorship, at Texas A&M University. In 1982, the university named him a Distinguished

Professor, and in 1983, he was appointed the director of the National Science Foundation's Center for Electrochemical Systems and Hydrogen Research.

To Go Where No One Else Dared

Bockris and his colleagues were among the first researchers in the field to achieve several major scientific accomplishments. As mentioned in Chapter 2, they include the LENR production of 1) tritium, 2) helium-4, and 3) heavy elements.

John O'Mara Bockris (Photo: Jean Wulfson)

In 1992, Bockris, using Champion's method, which he called "thermal experiments," produced what he thought were trace amounts of gold in Bockris' chemistry lab. Such ideas went back hundreds of years and were common among practitioners of alchemy. (Vol. 3, *Lost History*)

They sent the pre- and post-experiment materials to two independent laboratories for analysis. However, the experiment was not independently reproducible, even in Bockris' laboratory.

Bockris did not get credit — nor should he have — for credible experiments demonstrating nuclear transmutations at low energies.

However, he, more than anyone, advanced this crucial aspect of the research and encouraged open discussions about the results. Other researchers who had detected transmutations or isotopic shifts at low energies, such as Farmer and his colleagues at Lawrence Livermore National Laboratory, and Rolison and O'Grady, at the Naval Research Laboratory, had either ignored their own findings or found dubious reasons to dismiss them.

Bockris gave courage to other researchers who, in the mid-1990s, did indeed confirm that the experiments could produce trace amounts of new noble metals as well as a broad spectrum of new elements and isotopic changes.

In 1994, Bockris reported low-energy nuclear transmutations using a different method, known as a carbon-arc experiment, and this was independently confirmed by researchers in India and China. (Bockris, 1994; Singh, 1994; Jiang, 1998)

However, a Russian group preceded Bockris by a few months and, in 1993, presented the first paper that reported heavy-element transmutations in the field. The Russian researchers were circumspect in their claims, identifying the newly produced material not as "transmutation" products but as "impurities." (Savvatimova, 1993)

Bockris ventured into uncharted scientific terrain where no other scientist of his stature, during his time, was willing to go. He led the expansion of scientific knowledge but was not recognized for doing so in his lifetime.

He never complained that the scientific community didn't give him the recognition he thought he was owed. Instead, in the years before he died, he did his best to help me understand his accomplishments. My clear sense was that he was at peace knowing that recognition of his achievements likely would not occur in his lifetime.

Alchemy? Really?

In the spring of 1992, Bockris and his group began performing experiments that appeared to create not just light elements, such as tritium and helium, but also heavier elements, specifically noble metals.

It sounded like alchemy, but Bockris had reason to consider it possible. If the nucleosynthesis of light elements could occur in room-temperature experiments, he thought, why not heavy elements, too?

Bockris was neither timid nor apologetic about his willingness to engage in research that seemed to conflict with accepted science — or "good sense." He explained his thoughts to me in a May 5, 2004, letter:

> The attitude of the scientific community [toward our research] was due to the fact that the results which we were presenting were totally against all that had been learned by physicists up to that time ... and were, frankly, thought to be impossible. Therefore, when they heard that papers were reporting that "cold fusion" did exist and it was a nuclear process (producing tritium no less), they simply closed their eyes. Their attitude is, in my view, understandable, though not wise. If you were told by an earnest colleague who had a tremendous record of publications in refereed journals that by using special shoes he could jump over his house and land safely on the other side, you would not ask for publications. You would smile and conclude that he had become insane.

You've Got to Be Kidding!

Perhaps nothing provoked Bockris' peers more than his apparent interest in turning lead and mercury into gold. Champion's claimed thermal method of transmutation was not as simplistic as many of the newspapers at the time made it sound. It did not involve turning a lump of mercury or lead into gold; instead, it involved a complex mixture of materials that were supposed to increase the trace levels of gold only in the parts-per-per-million range.

Nevertheless, to outsiders, this type of research, which only superficially resembled alchemy, was unwelcome at Texas A&M University, as Sharon Begley, writing for *Newsweek* in 1994, wrote: "In the revered name of academic freedom, universities tolerate faculty members who are avowed communists and lifelong fascists, outspoken racists and anti-Semites, radical lesbians and rabid homophobes. But alchemists?"

The story began in 1992, when Joseph E. Champion (born approximately 1950) approached Bockris and discussed the idea of conducting some unusual experiments at Texas A&M University. A few weeks later, Champion brought his associate William Leon Telander (b. 1949) to meet Bockris.

Champion, an electronics technician by trade, said he had a recipe that could turn inexpensive materials into gold. For someone with such abilities, he seemed to be remarkably cash-poor. Bockris described Champion as "a big chap who looked more like a football player than a scientist." Champion, according to Bockris, had performed successful experiments at the University of Guanajuato, in Mexico, but the professor Bockris spoke with at the university was skeptical because he didn't know the details of how Champion had handled the material. (Bockris, 2004)

Champion explained his wild ideas and piqued Bockris' curiosity. Telander, 42, was the money guy, the wealthy entrepreneur-turned-philanthropist — he claimed — who was interested in supporting basic science research. Telander explained that he wanted Bockris to independently verify Champion's recipe. Bockris' impression was that Telander, a congenial man, was independently wealthy, thanks to the sale of a chain of restaurants he had inherited from his mother. "My wife was skeptical of Mr. Telander's story," Bockris wrote, "because his shoes and watch were of a quality less than that expected for a wealthy man. We drove to College Station Airport one evening and found that a private jet described by Telander was indeed parked there."

After Bockris explained how gifts worked at Texas A&M, Telander offered $100,000. Telander seemed so relaxed and genial, Bockris wrote, that he immediately asked for $200,000, to which Telander promptly agreed.

Bockris had met Champion once before. In September 1989, Champion had heard about Bockris' tritium results and contacted him to suggest an improved method that would more reliably lead to successful experiments. Bockris believed that Champion's 1989 idea had been effective.

Even though Champion had no formal training in science, he had learned enough about science on his own to engage with Bockris, as letters between the two men revealed. Bockris even wrote that Champion understood nuclear chemistry better than he did. When it came to esoteric topics in science, the two of them shared a passion for exploring science beyond the accepted scientific paradigm.

Telander was a different breed. Bockris welcomed the research funds for his group with open arms but made it clear to Telander that he should not expect the experiment to demonstrate a cost-effective way to produce gold. Telander assured Bockris that, of course, his interest was only for the noble pursuit of scientific exploration. It isn't clear whether Bockris believed Telander or whether Bockris simply didn't care. Here is how Bockris explained his reasoning to me on July 10, 2004:

> You see, nearly every scientist who heard this idea from Joe Champion would have said, "Nonsense! That's alchemy!" I didn't have that reaction at all because we had got tritium. So I thought, "If you can take deuterium, with a neutron of one, and make tritium, which has two neutrons, you are making a transmutational reaction, and we know we can do that. We can prove it." So I was extremely interested in this statement by Champion. I didn't think it was crazy at all.
>
> Gold has 79 protons, mercury has 80, and lead has 82. So going to gold (79) from mercury (80) is just one extra [proton]. So it suddenly became not quite so crazy. I pricked up my ears about it. I knew no one else would touch it, and I knew there would be a lot of fuss about it.
>
> Nevertheless, I was very honest about it. I went to see my boss, Michael Hall, and I told him what the situation was. I said, "It's very, very unlikely that we're going to get anything out of this, but this guy Telander wants to give $200,000 to the university." Hall's reply was, "So long as there's no publicity, and if you do succeed, you must get it verified by another lab, then it's alright."

The experiments began, and there was no publicity.

CHAPTER 4

Exploring the Atom

It may be useful at this point to offer a brief tutorial on atomic structure and basic concepts in radioactivity. For nearly a century, scientists have known that all matter is composed of elements, elements are composed of atoms, and atoms are composed of three primary subatomic particles: the proton, the neutron and the electron.

Protons have a positive electrical charge, electrons have a negative charge, and neutrons have no charge. Protons and neutrons sit close together in the center of the atom and compose the nucleus. (See diagram of Carbon-12) Electrons, which have a negative electrical charge, remain outside of and orbit the nucleus.

Different Elements

All matter exists in the form of specific elements — for example, hydrogen, oxygen, and carbon. Each element is distinguished by the number of protons in its nucleus. At present, 98 elements are known to exist in nature, and a few others have been synthesized in laboratories.

When the number of protons inside the nucleus increases or decreases, the atom changes from one kind of element to another. For example, a proton added to a nitrogen atom changes it to an oxygen atom. This is called a nuclear transmutation.

The first diagram below shows, on the left, one of the simplest atoms, a form of hydrogen called deuterium. Normal hydrogen has only one proton in its nucleus, but this variety of hydrogen, called deuterium, has a neutron, as well. On the right, the diagram shows a nucleus with two protons; this is the element helium. In this case, it's a variety of helium called helium-3, with has three particles in its nucleus: two protons and one neutron.

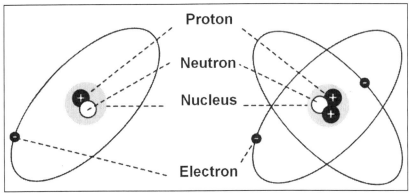

Deuterium atom (left): one proton and one neutron in the nucleus. Helium-3 atom (right): two protons and one neutron in the nucleus.

Different Isotopes

Most elements exist in a variety of forms. Just as chocolate comes in different varieties, so do elements. However, different varieties of chocolate are still chocolate. A variation of an element is called an isotope. Each isotope is slightly different from other varieties of the same element. The difference between isotopes is that they have different numbers of neutrons in each nucleus, but the number of protons in each nucleus stays the same. An isotope of helium is still helium; an isotope of hydrogen is still hydrogen.

The pair of diagrams below provides an example. The one on the left shows a variety of helium called helium-3. It has three particles inside its nucleus: two protons and one neutron. The one on the right shows helium-4, which has two protons and two neutrons in its nucleus. Both are different isotopes of the same element, helium.

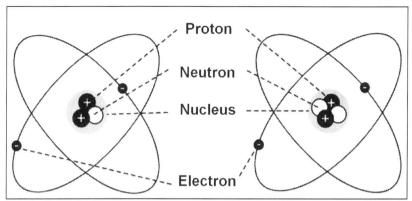

Helium-3 isotope (left): two protons and one neutron in the nucleus. Helium-4 isotope (right): two protons and two neutrons in the nucleus.

Nearly all elements have a variety of isotopes. Some elements have many isotopes; some have only a few. Some isotopes of a given element are more abundant than the other isotopes. For example, a lump of coal is mostly carbon. Isotopic analysis of the carbon reveals that most of it exists as the carbon-12 isotope.

Carbon-12 has six protons and six neutrons in its nucleus. Carbon has a total of 15 isotopes. Stable carbon-12 usually makes up 98.93% of the total amount of any carbon sample. Another stable isotope, carbon-13, for example, makes up only 1.07% of naturally occurring carbon.

The ratio between carbon-12 and carbon-13 is normally very nearly the same, whether it is measured in Colorado or Kiev. This phenomenon applies to all elements, not just carbon. The percentage of each isotopic abundance does not usually vary from its natural state. Because of this, these ratios act like scientific fingerprints. If scientists find a sample of an element that contains abnormal isotopic ratios, they know that an unusual event has occurred.

Generally, there are three types of events. First, environmental and biological factors can segregate some of the isotopes and cause minor shifts in the ratios between isotopes at certain locations over long periods, and forensic scientists can use this data to correlate biological samples to specific geographical locations. Second, a wide variety of man-made processes can be used to separate and concentrate isotopes. Methods include diffusion mechanisms, centrifuges, electromagnets, or lasers. Isotopic separation is the way low-grade uranium (containing very little U-235) is enriched to make high-grade uranium (containing much more U-235). Third, nuclear reactions can cause isotopic shifts that add or remove neutrons from isotopes. Although the first two type of events cause isotopic fractionation, only nuclear reactions change the number of protons or neutrons in the nuclei.

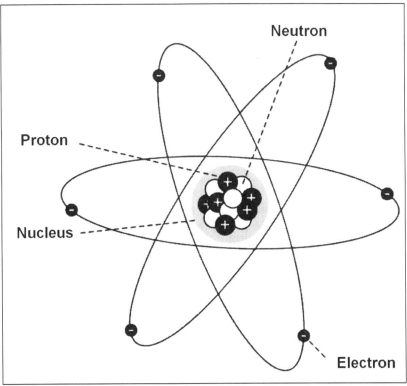

Basic diagram of a carbon-12 atom. The diagram isn't to scale; the actual distance between the orbiting electrons and the nucleus is much greater than shown here. Also, the specific arrangement of electrons in their valence shells is not depicted here.

A Matter of Power

Elements change, or transmute, into other elements by the addition or subtraction of protons. Isotopes change into other isotopes by the addition or subtraction of neutrons. Both kinds of changes are nuclear. Chemical changes involve either the addition or subtraction of electrons of a single atom, or the regrouping of atoms among themselves; they do not involve changes inside the nucleus. Not all reactions that occur in nature or science are equal. There is a big difference between reactions that can cause a chemical change and reactions that can cause a nuclear reaction. One of the biggest differences is the amount of energy required to initiate a reaction. A nuclear reaction typically requires one thousand to one million times more energy than a chemical change.

Primarily, two fundamental physics forces affect protons: the electromagnetic force and the strong force. Neither relinquishes its power and control over protons without a fight.

These forces, the electromagnetic and the strong, prevent protons from jumping from one atom to another and help provide stability for matter.

Under most circumstances, the electromagnetic force repels protons from each other like the north poles of magnets repel each other. It is no easier to squeeze two protons together than it is to try to press the north poles of two magnets together with bare hands and expect them to stick. With protons, the exception is that, if they are squeezed incredibly close to each other, the strong force overcomes the electromagnetic force, which then slams the protons together. The strong force works only at very short distances. Conversely, to separate protons from each other, an immense force is required to free protons from the stranglehold of the strong force.

But no chemical process ordinarily has enough energy to bring protons together or to separate them. For a century, scientists have known that the sheer muscle required to make these kinds of changes can be triggered only by high-energy physics. This typically takes place through the use of particles that are emitted with high levels of energy. A moving projectile such as a bullet may be small, but the rate at which it is traveling gives it enough power to shatter dense material.

At the turn of the last century, chemists and physicists began to understand the nature and types of radioactivity.

Types of Radiation

The first radioactivity that scientists observed was two particles; which were initially labeled with the Greek characters "alpha" and "beta." Researchers later figured out that alpha particles are helium nuclei and beta particles are energetic electrons. The third type of radiation was the gamma ray. Gamma rays are highly penetrating forms of electromagnetic radiation emitted from nuclear reactions. On Earth, they are emitted by devices or radioactive materials and a few rare terrestrial events. Gamma rays are a class of photons (a larger group of massless entities) that, according to quantum mechanics, behave both as waves and as particles. A range of various-energy gamma rays can be depicted in a gamma spectrum.

Researchers also understood the concept of radioactive decay, which describes how radioactive elements spontaneously disintegrate and emit alpha particles or beta particles. As radioactive elements decay, the alphas and betas fly from the parent elements into the surrounding space. In their place, they leave smaller and slightly different child elements or isotopes. Gamma radiation is different in nature and is not a mechanism for radioactive decay, as alpha and beta particles are. For more information on the types of radiation and their characteristics, see Appendix C, Basic Types of Radioactive Emissions.

CHAPTER 5

Risky Business

John Bockris was eager to begin the transmutation experiments at Texas A&M University. Two administrators at the university, Kenneth W. Durham, the vice president and director of development for the Texas A&M Development Foundation, and W. Michael Kemp, the dean of the College of Science, decided to accept William Telander's offer. The money arrived in April 1992.

Telander named the experiments "The Philadelphia Project" after a mythical tale that, in 1944, scientists at the Philadelphia Navy Yard conducted esoteric magnetic degaussing experiments on a U.S. Navy ship that somehow resulted in invisibility and teleportation.

Durham sent Telander a letter on May 14, 1992, confirming receipt of the first of two $100,000 payments. "It is only through the support of enlightened and interested people such as you," Durham wrote, "that much of the research on the cutting edge of science can be accomplished."

Alchemy research was welcome, apparently, so long as it brought money into the university and created no publicity. Three of Bockris' post-doctoral researchers — Ramesh Bhardwaj, Guang Lin, and Zoran Minevski — worked with Champion in Bockris' electrochemistry lab while Bockris managed the group.

In a May 25, 1992, letter to Ron Mallett, the director of the Mintek analytical science division in South Africa. Bockris explained that he did not expect that the process was some kind of magic formula to suddenly produce large quantities of gold. Instead, he told Mallet, to whom he later sent samples for analysis, that he was expecting amounts as little as 0.1 parts per million (ppm) and up to 100 ppm of gold and platinum. He also emphasized the importance of accurately characterizing the impurity concentrations of the starting materials.

Champion's Golden Recipe

The two key ingredients in what Champion called the "thermal experiments" were mercurous chloride (Hg_2Cl_2) and lead oxide (PbO). The two composed 150 of the 1,700-gram formula, which included specific proportions of C, KNO_3, S, SiO_2, $FeSO_4$, Cd, Ag, and CaO.

The experimental procedure began with weighing each of the chemicals, followed by the tedious task of mixing them for three days. After pouring the mixture into a coffee can and placing it under a fume hood, they ignited the mildly explosive mixture with a propane-oxygen torch. The burn, accompanied by a "whoomp" sound, lasted about 60 seconds and glowed red hot. The mixture needed to be cooled for three days, after which the researchers ground the mixture with a mortar and pestle. During the time the mixture was cooling, the researchers took radiation measurements.

After each experiment, they sent virgin as well as ignited samples to three analytical laboratories for independent analysis: one in Australia, one in Canada and one in South Africa. They also gave samples to the nuclear reactor staff at Texas A&M. A couple of the experiments also produced unexplained beta radiation that dissipated within a day or two after the experiments were ignited.

In their 1996 paper, Lin and Bockris thanked Champion for the recipe and for numerous discussions but said nothing of his hands-on involvement. However, a March 11, 1993, summary of the experiments written by Bockris says that Champion was always present when the experiments produced anomalous gold, and this cast doubt on the experiments. (Bockris, 2004)

Champion's Golden Presence

Three of the five thermal experiments performed between April 21, 1992, and June 7, 1992, showed an increase of several hundred ppm of gold. Bockris was concerned because Champion and another alchemy enthusiast, Roberto Monti, were present during the experiments in which excess gold was found.

As Bockris noted in the summary, Champion was not present when the results were null. The control variable seemed to be Champion! On June 10, 1992, Bockris, frustrated with the situation, sent Champion a terse letter in which he admonished Champion that the university's duty was to test Champion's claim independently. Bockris was having difficulty keeping Champion away from the tests, which the post-docs were supposed to conduct independently of Champion. Bockris made it clear to Champion that he did not want Champion or Monti present during the tests.

Lost Luster

Between June 8, 1992, and Aug. 21, 1992, Bockris' post-docs performed six more experiments. Champion and Monti stayed out of the lab. No excess gold was found. Bockris' team did, however, sometimes observed a specific signature of beta radiation, indicative of a platinum isotope, that they could not explain.

Midway through that series of experiments, on July 8, 1992, Bockris notified Robert J. Nowak (b. 1947), at the Office of Naval Research, of the results. Nowak was the program manager who had funded University of Utah electrochemist Stanley Pons before the University of Utah "fusion" announcement on March 23, 1989. Nowak had also funded Bockris' research in the corrosion of alloys.

Bockris told Nowak that four of the first five experiments showed excess gold, in the 250-450 ppm range. Bockris mentioned that, on two of the runs, the group also observed some radioactivity, the start of which coincided with the time of the experiment, then dropped off significantly after two days. Bockris said that his group performed another five runs but that these all ended with null results.

With Champion in the lab, Bockris' team had produced results that he thought were genuine. On Aug. 5, 1992, Bockris wrote a letter "to whom it may concern" about the project. He explained the situation and discussed the results:

> It appears that, when Mr. Champion carries out his transmutational procedure, there is indeed radiation emitted from the mixture that has been treated. This radiation appears to correspond to a radioactive platinum isotope. My colleagues tell me that they have found that this radiation has been observed in all ten experiments which we have done. ...
>
> In most, but not all, of the experiments we have done, we have found a considerable amount of gold in the mixture which has resulted. ... Because the word transmutation is often associated with alchemy, there is bound to be a question of fraud. We have, of course, been particularly careful with respect to this possibility. Scientists who study this document should ask themselves how a person who wished to perpetrate a deception would be able to introduce the unstable platinum isotope into the mixture.

In his letter, Bockris also took liberties in how he reported the results. Although two of the runs up to this date had shown radioactivity, now he wrote that 10 runs had shown radioactivity. Although four of ten runs had shown new gold, now he wrote that "most" revealed new gold. Bockris then issued a plea to keep the news quiet:

> Although there is no question of results obtained in the university environment being secret, we do not wish to have these results disseminated widely because of the furor which occurred at the beginning of the observation of cold fusion phenomena. We believe that the best thing that can be done is to let others come to our laboratory and carry out tests for themselves and, of course, carry out similar work in their own laboratories. After parallel confirmations, we shall be willing to make our work public.

From Oct. 27, 1992, to Jan. 6, 1993, Bockris' researchers ran six more experiments, then a final experiment on March 11, 1993. No extra gold was produced in any of the experiments.

About That Radiation ...

The chances of Champion having pulled the wool over Bockris' eyes seem virtually impossible. Platinum-197 (Pt-197) is an extremely unstable isotope, losing half of its radioactivity in 20 hours. This would make it very difficult just to acquire the isotope, let alone quickly and surreptitiously place it into the experiment while it was still radioactive.

As a result, it would have been extremely difficult for Champion to have obtained Pt-197, given its rapid decay rate. Pt-197 cannot be created by exposure to nuclear sources available to civilians. It would require either a neutron source from an accelerator to bombard natural platinum or a fission reactor to break apart a heavier isotope.

If Champion had spiked the experiment with Pt-197, he would have had to persuade someone at a nearby reactor to take naturally occurring platinum and bombard it with neutrons to make Pt-197, or he would have had to purchase freshly enriched Pt-197. The cost would have been about $1 million per milligram. And he would have had to repeat this for the two experiments that showed the beta radiation.

As it turns out, Pt-197 is an intermediate product between stable isotopes of platinum and the stable isotope of gold-197. Theorist Lewis Larsen depicted such nucleosynthetic reactions in a May 19, 2012, slide presentation.

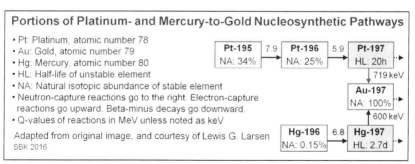

Nucleosynthetic pathways from platinum and mercury to gold

Larsen produced the slide for an analysis he did of LENR experiments in Italy that had synthesized gold from tungsten. Domenico Cirillo and Vincenzo Iorio reported these experiments in 2004. (Cirillo, 2004)

If You Lie Down With Dogs ...

Telander appeared to be flush with cash. According to the *Dallas Morning News*, he arrived at the College Station airport in a Learjet with two full-time pilots, two lawyers and a secretary. He booked a fancy suite at the local hotel. It was easy for Telander to spend money on such luxuries and gamble on the alchemy: It wasn't his money.

On Sept. 28, 1992, midway through the series of the thermal experiments at Texas A&M, the Federal Bureau of Investigation served a subpoena on the Texas A&M University Development Foundation, requesting all documents related to the Philadelphia Project. The Securities and Exchange Commission had learned about the project and suspected that it had been funded from $8 million that Telander had pilfered from investors in a fraudulent 1990 foreign-currency exchange racket.

Liska Lusk, general counsel of the Texas A&M Development Foundation, said the FBI subpoena "asked for all records or documents related to gifts by William Telander," according to a Nov. 24, 1993, article in the *Dallas Morning News*. But the FBI did not tell the university its reasons for the subpoena.

The SEC had been investigating Telander for conducting an international finance swindle, forging bank documents and defrauding private investors. He had sold the investors limited partnerships in his company Southwest International Exchange and promised returns of 25 percent to 41 percent, risk-free, according to the *Dallas Morning News* on Nov. 15, 1993. The FBI inquiry did not initially have any impact on Bockris or his group's experiments, nor did they learn about the FBI's activity in September 1992.

In mid-October 1992, Champion or Telander (accounts vary) set up a scientific meeting in Mexico City that was followed by a press conference, or only a press conference (accounts vary). Bockris told Jeffrey Weiss, of the *Dallas Morning News* (Nov. 17, 1993), that Bockris believed he had been invited to a scientific meeting.

Bockris and Telander took the opportunity to promote their project: Bockris on the science side, Telander on the business side. On Nov. 24, 1993, the *Independent Review* published excerpts of the Mexican news stories:

> In a Nov. 8, 1992, interview with *The News*, an English newspaper in Mexico, Bockris praised the Philadelphia Project. "This is the greatest advance in modern science," Bockris said. "The problem is that most people are so ingrained with traditional scientific theory that they have no room in their philosophies for such a totally revolutionary concept."
>
> The Mexican newspaper also said that Bockris had obtained 170 milligrams of gold [from] previously unmarketable ore and implied that research at Texas A&M had confirmed the legitimacy of the project.
>
> In a Nov. 17 article in *Novedades*, Telander said that $3.5 billion had been spent to develop a commercial alchemy process and that private investors and the Mexican government were being sought to invest in the project, which, in turn, would create thousands of jobs in Mexico.

Bockris had allowed his faith in Champion — and his enthusiasm — to take him far beyond his data. Bockris' only saving grace in November 1993 was that a) he had stopped the Philadelphia Project experiments six months earlier and b) he made no comments to the press about the research for 12 months.

Years later, when Bockris wrote about the situation, he recalled his behavior during the Mexican press conference as less damaging than it was. "I said that, if metal-to-metal transmutation were confirmed," Bockris said, "it would be a major innovative step in nuclear science and have consequences for the theory of the nucleus." (Bockris, 2004)

Telander's Scheme Unveiled

It is unclear whether Telander convinced any of the reporters at the Mexican press conference that this story was real. The only strong scientific evidence Bockris could offer was the 1989 electrolytic work in which he had shown the synthesis of tritium from deuterium. But synthesizing heavy elements appeared less believable.

Fortunately, the dubious claims of Champion and Telander and the ill-advised remarks by Bockris at the Mexican press conference didn't then make their way to the U.S.

Telander did not have altruistic scientific motives, as Bockris explained to me when I interviewed him in 2004:

> Telander's great aim was to make cheap gold, but we never got more than the tiniest bits of gold; on perhaps two or three occasions, we saw tiny specks. But Telander, who wasn't, of course, a scientist, thought once we had detected tiny amounts of gold [Bockris chuckled], we could make huge amounts. I did explain to him very gently that, if he was able to sell it very cheap, then the prices of gold would fall, and he wouldn't be able to make much money, anyway.

Although Telander was focused on producing gold at low cost, Bockris understood that demonstrating the ability to create one element from another was the most significant aspect because it was considered scientifically impossible.

Telander's Plan Falls to Pieces

When Bockris broke the news to Telander that the results were positive — but only in the parts-per-million range — Telander became enraged, according to a letter quoted by reporter Brian Wallstin of the *Houston Press* in 1994.

"I am not a philanthropist wishing to fund research projects," Telander said. "I am a businessman wishing to fund industrial science projects." Champion, as it turned out, was no less a criminal than Telander, as Wallstin wrote:

> In 1985, Champion talked two Arizona men into giving him $150,000 to go to South America to liberate a load of gold from the estate of a jailed drug dealer. Instead, he split the money with a partner. In 1987, he was charged with three counts of indecency with a child. Later that year, the sex and kidnapping charges were dismissed when he pleaded guilty to felony theft for writing a $2,000 bad check. He was sentenced to four years in Huntsville, [Alabama].

After Telander realized that he had been scammed by Champion and was not going to get the scientific endorsement from Bockris that he had hoped for, he turned on Champion and reported him to the FBI for his South American scam.

Sharon Begley reported in *Newsweek* that Champion and Telander were again convicted of felony-theft and financial fraud, respectively. "These guys," Begley wrote, "are enough to give alchemy a bad name."

Bockris never seemed to lose faith in Champion. Nor was he sufficiently bothered by Champion's activities. Bockris continued to exchange letters filled about intriguing science with Champion while he served his time in the Maricopa County Jail in Phoenix, Arizona.

Telander later pleaded guilty to four counts of securities fraud and admitted that he scammed $7.8 million from 380 investors, not to trade currency but to fund an extravagant lifestyle and to finance "other personal investments," Wallstin reported.

In May 1993, the SEC notified administrators at Texas A&M that Telander was under investigation. The Chemistry Department froze the remaining money in the account. The Philadelphia Project was over.

"Up to this time," Bockris wrote, "we had thought that the financial support from Telander came from his own pocket." (Bockris, 2004)

The FBI and the SEC were interested only in Telander and Champion. The federal government had no problem with Bockris or administrators at Texas A&M, and none of this — for the moment — surfaced in the local news.

CHAPTER 6

Light Water; Hot Tempers

Controversy arose from something as innocuous as ordinary water.

Toward the end of 1992, as John Bockris and his group at Texas A&M University finished their experiments using Joe Champion's "thermal reaction" recipe, a new research direction emerged: the search for and measurement of excess heat from the use of light (normal) water and normal hydrogen gas. Most researchers in 1992 had been searching for excess heat or nuclear products from heavy-water or deuterium-gas experiments.

Heavy water, D_2O, is made of deuterium and oxygen. Light water, H_2O, is composed of ordinary hydrogen (also called protium) and oxygen. Deuterium and hydrogen atoms each contain one proton, but deuterium also has a neutron, making it twice as heavy, and the resulting heavy-water molecule is slightly heavier than normal water.

Experiments with light water had been a sensitive topic since March 1989, when Martin Fleischmann and Stanley Pons submitted their manuscripts to the *Journal of Electroanalytical Chemistry* and *Nature*. They had claimed that their heavy-water electrolytic cells demonstrated deuterium-deuterium (D+D) fusion, but they didn't report any control experiments using light water. Some critics wrongly assumed that Fleischmann and Pons had committed a sophomoric blunder.

Instead of running control experiments with light water, Fleischmann and Pons had used other controls and measurements to confirm that their excess heat was real. (Vol. 2, *Fusion Fiasco*)

Fleischmann and Pons assumed that the excess heat in their cells originated from D+D fusion; they didn't imagine that excess heat could possibly occur with light water. Therefore, with one exception, they didn't run light-water tests before announcing their D+D fusion claim.

The calibration of their experiments and their other controls proved to their satisfaction that their excess heat was measured correctly and that it was real. It did not, however, prove that the heat was specifically from D+D fusion, and this is what the light-water control tests would have helped to demonstrate.

Sometime after March 23, 1989, Pons ran more light-water tests. On April 12, 1989, speaking at the American Chemical Society meeting in Dallas, Texas, Pons confirmed that he had seen some excess heat with light water.

Four days later, at a talk at Los Alamos National Laboratory, Pons confirmed again that experiments performed with light water did "not give total thermal balance" but instead gave "slightly higher amounts of heat out"; in other words, they produced excess heat.

Fleischmann, on the other hand, contradicted Pons about the light-water excess heat more than once. When asked directly by Representative Ronald C. Packard (R-CA) about light-water excess heat during an April 26, 1989, congressional hearing, Fleischmann refused to answer. In a follow-up letter he sent to the congressional committee, Fleischmann directly contradicted Pons' statements and denied any observations of light-water excess heat.

Heinz Gerischer, a prominent electrochemist from Germany who was not involved in the research, attended the Second Annual Conference on Cold Fusion, in Como, Italy, in June and July of 1991. After considering all the research presented at that conference, he summarized the observations of the research. "Anomalous phenomena have been observed in several laboratories (excess neutrons, excess tritium, excess heat)," Gerischer wrote. "[However], no comparable effects have been found with H_2O or H_2." (Gerischer, 1991)

Gerischer did not mention in his summary any reports of anomalous isotopic shifts or heavy-element transmutations, which, like light-water excess heat, would also argue against a fusion mechanism.

The consensus among the researchers in mid-1992 was that light-water experiments did not produce any excess heat or nuclear effects.

Out of Hiding

Fleischmann's opinion about light-water experiments was challenged in October 1992 at the next international "cold fusion" conference, which took place at the Nagoya Congress Center, in Nagoya, Japan. The conference was the third in the series and marked the first use of the new name, International Conference on Cold Fusion (ICCF). The first conference had taken place in Salt Lake City, Utah, and the second in Como, Italy. Hideo Ikegami, a professor of plasma physics at the National Institute for Fusion Science, was the conference chairman.

Most of the 346 registered attendees were from Japan, with scientists coming from 17 other countries. The conference, co-sponsored by seven scientific societies, received a remarkable level of support for a subject considered so controversial.

At ICCF-3, a prepared speech was delivered on behalf of Minoru Toyoda, a founding member of the Toyota automobile conglomerate, who explained why these scientists were so intensely passionate about the research. "Cold fusion is not just something to be studied by one single enterprise or a single nation," Toyoda concluded. "I am confident that it will become a precious asset to all mankind, as the ultimate, ideal form of energy, so that must be shared among all of the nations of the Earth." (White, 1992)

As was typical of the ICCF conferences, there were many reports of replication of Fleischmann and Pons' heavy-water excess-heat results. Heavy-water research has been discussed in detail in Charles Beaudette's 2000 book *Excess Heat & Why Cold Fusion Research Prevailed.*

The possibility of light-water excess heat was unwelcome among most of the researchers because it contradicted their D+D fusion idea. Robert Bush, a professor of physics at California State Polytechnic University at Pomona, one of the early light-water excess-heat researchers, told me why in 2010:

> The hostile attitude from the heavy-water cold fusion researchers toward the light-water researchers was similar to the hostile attitude of outsiders to the field toward the

heavy-water researchers. In fact, I can remember when we first heard of the light-water excess heat claims by Randy Mills (b. 1957). The heavy-water people said, "Oh, my God, this is insane."

Look, if you're one of the heavy-water people, and you're saying to people that light water is your control — just think of the logical challenge that presents. Not only was there hostility, but there was good reason for it.

Light-Water Scientist in Hot Water

Controversy involving light water had started a year earlier, thanks to Randy Mills, a physician in Lancaster, Pennsylvania, with an undergraduate degree in chemistry. On April 24, 1991, he announced that he had performed a light-water electrolysis experiment, using nickel for the cathode rather than palladium. Nickel and normal water were far less expensive than palladium and heavy water.

For the electrolyte, where Fleischmann and Pons had used lithium deuteroxide, Mills used potassium carbonate. Mills said publicly that the excess heat in his experiments, as well as that in heavy-water electrolytic experiments, was the result of some form of powerful chemistry rather than nuclear reactions. Mills didn't want to associate his work with that of the "cold fusion" community because of the stigma against the research. Additionally, most of the heavy-water researchers didn't want to associate their work with that of Mills because his results cast doubt on the D+D "cold fusion" idea.

Mills established a high public profile and, in 1991, founded a company named HydroCatalysis Power Corp. In 1996, he changed the name to Blacklight Power Inc. In 2015, he changed the name again, this time to Brilliant Light Power. He continued to attract significant funding with the aim of rapidly commercializing the idea. Year after year, his enthusiasts and backers believed that Mills would soon be selling a commercially available light-water heater device.

Mills also announced that he had developed a "grand unified theory" that united chemistry and physics. The theory, in which he suggested major revisions to existing quantum physics theory, was met with intense hostility and rejection. Journalist Erik Baard, who wrote many news stories about Mills, captured the essence of the dispute. "Telling physicists that they've got that wrong is like telling mothers across America that they've misunderstood apple pie," Baard wrote. "It's that fundamental." (Baard, 1999)

Reactions to Mills' theory from the physics community were vicious. Phillip Anderson, a condensed-matter theorist at Princeton University, told journalist Erik Baard that he was sure it was a fraud. (Baard, 1999) Robert Park, the former spokesman of the American Physical Society, called it "bullshit." (Park, 2008)

Mills' theory aside, at the ICCF-3 conference, several scientists reported successful Mills-type experiments, using light water, nickel cathodes and potassium-carbonate-based electrolyte. One of them, Reiko Notoya, was even willing to publicly demonstrate her claims.

The Chemist in the Hallway

The most significant presentation during the ICCF-3 conference in Japan did not take place in the meeting room but in the hallway. There, Notoya, a chemist with the Catalysis Research Center of Hokkaido University, in Sapporo, one of the few women researchers in the field, did not just talk about her experiment that produced excess heat — she showed a working demonstration.

Notoya had two cells running side by side, with thermometers in each one. The experimental cell was running 15° C hotter than the control cell. That amounted to three times more power, in heat, than the electrical power going into the cell.

Input power was about two Watts; output heat was about six Watts. Notoya was using the basic Mills concept: light water, nickel, and potassium carbonate. There were two differences, however. Whereas Mills had used plain nickel for the cathode, Notoya used a porous form of nickel. Mills had provided power to the cell with an intermittent current, but Notoya applied a constant current.

Most excess-heat-producing experiments were limited by a general inability to reproduce results on demand. For this reason, Notoya's feat demonstrated her remarkable confidence. Researchers who were open-minded about light-water excess heat, like theorist Peter Hagelstein (at the time), were impressed. He wrote in his summary of the conference that Notoya's cell produced so much excess power that improperly accounting for recombination would not substantially change the result. "You can put your finger on the tubes of Notoya's demo to convince yourself that a very significant temperature difference occurs," Hagelstein wrote. "The power excess [she] demonstrated at Nagoya would defeat a chemical explanation in tens of minutes, and the cell ran for many hours."

Reiko Notoya and Douglas Morrison (1996) Photo: Unknown

Hagelstein thought it was ironic that senior researchers who worked with heavy water were convinced that the light-water claims were the result of sloppy experimental technique. "It is almost humorous," Hagelstein wrote, "to find senior members of the cold fusion community sounding very much like their critics and tormentors of 1989." (Hagelstein, 1993)

For many years, outspoken people in the "cold fusion" community had complained that people in the broader scientific community had treated them unfairly. These researchers said that the outsiders had dismissed their heavy-water claims because they didn't believe the extraordinary claims.

Now, the more-prominent members of the "cold fusion" community, with Fleischmann as its figurehead, were doing the same thing to the subset of researchers who were proposing light-water excess heat and nuclear effects.

This behavior reflects the observation of sociologists and historians of science that scientists who form a new community of specialty soon develop the same conservatism that existed in the larger parent community from which they came, and that the pattern repeats over time. Hagelstein explained the light-water problem:

> The experimentalists have grown used to the idea that deuterium gives anomalies and hydrogen does not; the theorists who believe in fusion mechanisms are comfortable with positive effects in deuterium and negative effects in hydrogen. A light-water heat effect causes consternation in both camps; it would be exceedingly difficult to reconcile with a fusion mechanism.
>
> The neutron transfer model which I have been looking at needs a neutron donor (usually deuterium) and an acceptor nucleus and, therefore, has somewhat fewer constraints; nevertheless, I do not relish the prospect of attempting to explain an apparently general light-water heat effect where the nuclei present are widely different from one cell to another.
>
> An experimental determination (and confirmation) of the ashes in any of these experiments would of course greatly improve the situation. (Hagelstein, 1993)

Notoya also measured an apparent heavy-element transmutation: the increase in the concentration of calcium in the electrolyte after the experiment. The amount of calcium she found was enough, she thought, to correlate with the amount of excess heat, within an order of magnitude. She based her calculation on a speculative reaction mechanism suggested by Robert Bush. (Notoya, 1992)

NASA Confirms Mills

A few years later, 1996, researchers at NASA's Lewis Research Center in Cleveland, Ohio, observed excess-heat in a light-water cell using a nickel cathode and potassium carbonate electrolyte, confirming Mills' experiment. (Niedra et al., 1996)

The team consisted of three physicists: Janis M. Niedra (b. 1938), Gustave Clarence Fralick (b. 1942), Ira Thomas Myers (b. 1925) and an electrical engineer, Richard S. Baldwin (b. ~1934)

The NASA researchers reported excess heat of 11 Watts maximum for a 60 Watt electrical power input. The team published the results in a technical memorandum that was not widely accessible and received scant public notice, and these findings had little effect on the development of the field at the time. NASA's minimal dissemination of this result also shielded it from controversy.

The Light-Water Brigade

Notoya and her colleagues were not the first scientists to report confirmation of Randall Mills' idea of using a nickel cathode in light water.

Half a year earlier, in March 1992, Vesselin C. Noninski, a Bulgarian electrochemist and visiting scholar at Franklin and Marshall College, in Lancaster, Pennsylvania, reported confirmatory results for a nickel cathode in light water in *Fusion Technology*. (Noninski, 1992) Noninski's paper was followed by another report in *Fusion Technology*, by Robert Bush. (Bush, Robert, 1992)

At ICCF-4, researchers from the Bhabha Atomic Research Centre (BARC), in India, and a Russian team led by Yuri Bazhutov reported that they had observed excess heat in nickel-light-water experiments. (Bazhutov, 1993)

Despite the results of Bush, Mills, Notoya, Noninski, the BARC researchers and Bazhutov's team, the mistrust of light-water reactions continued because they didn't fit the "cold fusion" narrative.

CHAPTER 7

The Light-Water Problem

It was becoming apparent that nuclear reactions did not require heavy water; ordinary (light) water produced these effects as well. In October 1992, at the Third International Conference on Cold Fusion (ICCF-3), in Japan, in addition to the Japanese group that reported light-water excess-heat results, two independent groups from Bhabha Atomic Research Centre (BARC) also reported light-water excess-heat results. Mahadeva Srinivasan delivered the news.

In addition, in some of the cells, the BARC researchers detected direct products of nuclear reactions: Samples from 18 of 29 light-water experiments indicated anomalous levels of tritium. Srinivasan reported that the nickel-light-water system had many advantages, including higher rates of reproducibility and detectable excess heat, if present, within the first day of electrolysis. (Srinivasan, 1992)

Many experimentalists as well as theorists didn't believe the light-water excess heat — they had claimed that excess heat happens only with heavy water. They disputed the results, arguing that the one in every 6,000 molecules of D_2O that exists naturally in H_2O could be responsible for the light-water excess heat. In an attempt to resolve that argument, Bush and his colleague Robert Eagleton ran tests with specially prepared de-deuterated water (where only one of every 500,000 molecules is D_2O), but they still observed excess heat. (Bush and Eagleton, 1993)

In a paper, Bush pointed out the immense economic ramifications of this development: Light water costs pennies per gallon; heavy water, at the time, cost $1,000 per gallon. Bush projected that, if the research continued to be confirmed, it would lead to a new branch of science, the physics of low-energy nuclear transmutations. (Bush, Robert, 1992) A decade later, when I met theorist Scott Chubb (1953-2011) and asked him why he believed that only heavy-water experiments could produce excess heat, he still argued that the 1:6,000 D_2O molecules in H_2O could be causing room-temperature fusion.

Bush, as well as Notoya, found an increase in calcium. Both of their experiments used potassium carbonate in the electrolyte. Calcium is one atomic number higher than potassium. Bush also found the presence of strontium after an electrolytic experiment using a rubidium carbonate solution. Strontium is one atomic number higher than rubidium. (Bush, Robert, 1992) If a source of neutrons existed in the experiments, a neutron-capture process, in which an atomic nucleus captures one or more neutrons to form a heavier nucleus, could explain the observed data.

As mentioned in Chapter 6, in 1992, Notoya reported that she had observed an amount of produced calcium that might have quantitatively corresponded to the energy production. Bush had observed the same phenomenon in two cells a year earlier. (Bush, Robert, 1992)

In December 1993, at ICCF-4 in Maui, Hawaii, the BARC researchers elaborated on the excess-heat and tritium results from their light-water cells. Srinivasan reported that BARC researchers measured excess heat in 14 of 28 Ni-H_2O cells, ranging from 0.2 to 0.8 Watts. (Ramamurthy, 1993)

Srinivasan delivered another slide presentation on behalf of his colleague Thevarmadhom Krishna Sankaranarayanan. The presentation seemed to address all of the obvious questions about the origin of the tritium, including potential leaching from the glass apparatus or from the nickel cathode, electrolytic enrichment, and even the odd speculation that "Trombay air is full of tritium." His paper included a poke at critics who offered vague suggestions that the tritium measurements were somehow mistaken:

> While a sudden, sharp increase in tritium level can be understood as a production [of tritium] in the first phase [of the experiment during the first] 5 to 10 days, the decreasing phase, lasting again for about 5 to 10 days, is difficult to understand. Electrolytic enrichment cannot account for this

behavior. Critics, please make up your mind; you can't invoke contamination and electrolytic enrichment to explain this decrement [while invoking the same to explain the increase]! (Sankaranarayanan, 1993)

Pouring Cold Water on Light Water

Challenges to the claims of light-water results were inevitable. A large fraction of the limited funding was going to researchers doing work with palladium and heavy water. But solidarity and camaraderie — or at least the appearance of them — was deeply rooted in the community. Heavy-water researchers were very careful about openly criticizing light-water colleagues.

Since 1989, Michael McKubre, at SRI International, had been leading a large research group in testing the Fleischmann-Pons palladium-heavy-water system, funded with several million dollars from the Electric Power Research Institute. Whenever McKubre took to a podium and mentioned Fleischmann, it was always in the most honorific terms. McKubre dedicated many years of his life to defending and confirming the excess-heat claims of Fleischmann and Pons.

The electrolytic experiments performed by McKubre's group in the 1990s were the best-documented examples of closed-cell calorimetry in the field. A three-volume set of research reports, nearly 700 pages, serves as an exemplary model of electrochemistry calorimetry. (McKubre, 1994, 1998; Jevtic, 1999)

Sometime before September 1993, McKubre invited Mahadeva Srinivasan to come to SRI International as a visiting researcher for a six-month sabbatical. Srinivasan and his colleague Padmanabha Krishnagopala Iyengar, the director of BARC between March 10, 1984, and Jan. 31, 1990, had led a massive effort — 50 researchers working in a dozen independent groups — that conducted "cold fusion" research in 1989. In December 1989, Iyengar and Srinivasan published the collected work of all the groups in the landmark BARC-1500 report.

When Iyengar finished his term as the director of BARC and became the chairman of the Indian Atomic Energy Commission, Rajagopala

Chidambaram (b. 1936) took his place as the director of BARC. Unlike Iyengar, Chidambaram was opposed to the research, and he stopped all official funding for it the day he became director. As a result, Srinivasan accepted McKubre's invitation and worked at SRI International from September 1993 to March 1994 as a visiting researcher.

In late 1993, Bose Corp. increased its participation in the research, according to Srinivasan. Bose Corp. had already been involved; Robert Bush's September 1992 paper in *Fusion Technology* acknowledged Bose Corp. funding and thanked Joe Veranth, Bose vice president of engineering, for his help.

In an e-mail to me, Srinivasan wrote that he had been invited to visit the Bose laboratory and that he had a private audience with the founder, Amar Bose. Srinivasan told me that Bose Corp. researchers were enthusiastic about "cold fusion."

Unconfirmed Confirmation at SRI International

On Nov. 1, 1993, after several weeks of planning and cell fabrication, Srinivasan started running his first nickel-light-water cell at SRI. That first phase of experiments ended on Jan. 31, 1994. The ICCF-4 conference was held in December 1993, while the experiments were running. In the report of Srinivasan's experiments at SRI, the authors wrote, "Of the 22 cells in which calorimetry was carried out, 10 cells appeared to indicate some apparent 'excess power'."

The cautious language made clear that whoever wrote the report was not convinced that the measured excess heat in the 10 cells was real. Srinivasan told me that he thought he might have given a draft version of his results and conclusions to McKubre but that the final document probably was written by McKubre and SRI chemist Francis Tanzella.

The authors explained their doubts that the experiments truly showed excess heat: "The excess-heat margin was such that it always fell short of 1.48 x I," referring to input voltage, current, and energy involved in recombination of the electrolyzed gases. Their concern was that recombination of gases inside electrolytic cells, especially in the presence of catalytic electrode materials, could cause unaccounted-for side reactions that could affect accuracy of excess-heat measurements. This is described by the measure called "Faraday efficiency," the details of which are beyond the scope of this book.

At the same time that Srinivasan was performing his experiments at SRI, researchers at Bose Corp. were also working on heavy-water and light-water experiments. The Bose researchers failed to see any excess heat and instead "found evidence" (according to the SRI authors) for recombination effects in their Ni-H_2O experiments.

The SRI authors wrote in their report that the Bose researchers concluded that recombination effects were giving false-positive results in the Bose Ni-H_2O cells. The SRI authors suggested that the same problem could explain the apparent excess heat Srinivasan measured in his Ni-H_2O experiments at SRI. The Bose researchers' paper had not published yet, but McKubre had read a copy because, as the paper says, he and Peter Hagelstein helped the Bose authors to review the draft manuscript.

Sometime before Feb. 22, 1994, McKubre shared with Srinivasan the discouraging news from the Bose researchers about the recombination concern. At that point, SRI researchers began testing the Bose researchers' recombination hypothesis with their own set of light-water experiments. The details of Srinivasan's Phase 1 Ni-H_2O experiments at SRI and the SRI researchers' Phase 2 experiments are described in an appendix to the 379-page report on the EPRI-sponsored deuterium-based experiments: "Development of Energy Production Systems From Heat Produced in Deuterated Metals." (McKubre, June 1998)

The SRI researchers offered no plain-language conclusion for their Phase 2 experiments. However, given the way the authors of the report depicted the results of Srinivasan's Ni-H_2O experiments at SRI, the authors clearly felt justified in dismissing Srinivasan's excess-heat results as error. To my knowledge, McKubre and his colleagues published no additional studies on this light-water issue, let alone a peer-reviewed journal article.

In the SRI report, the authors made no attempt to discuss any of the data reported by Bush, Notoya, Mills or the BARC researchers or to evaluate whether the data reported by those researchers was affected by a possible recombination-based calorimetry error. In the absence of such analysis, the speculation by the Bose group and by the SRI authors was unwarranted and probably was irrelevant. Nevertheless, the speculation by the SRI authors provided them cover, had they been asked by their EPRI sponsors about light-water excess heat.

Bose Corporation Speaks

Srinivasan returned to India in March 1994. Within three months, the Bose Corp. researchers submitted their paper to *Fusion Technology*. The lead author was physicist Zvi Shkedi. The other authors were Robert C. McDonald, a chemist, John J. Breen, an engineer, Stephen J. Maguire, an electrical engineer, and Joe Veranth, the vice president of engineering at Bose who had helped Robert Bush with his light-water research. The Bose Corp. authors thanked McKubre, Hagelstein, and Amar Bose — whom Srinivasan had met in late 1993 — for supporting their research.

The paper was published in November 1995. In it, the authors boldly claimed that the "excess heat" reported by all researchers working with $Ni-H_2O$ cells might be explained by conventional chemistry. The Bose authors also wrote that they tried heavy-water experiments but failed:

> In the heavy-water arena, we have performed many experiments using the original $Pd-D_2O$ Fleischmann and Pons configuration. None of these experiments revealed the presence of excess heat. ... To find out whether excess heat in heavy-water cells can also be explained by simple chemistry, all reports claiming the observation of excess heat should be accompanied by simultaneous measurements of the actual Faraday efficiency. (Shkedi, 1995)

But Fleischmann and Pons were well aware of Faraday efficiency and recombination factors and had specifically addressed this issue five years earlier in their own *Fusion Technology* paper. Although Shkedi and his co-authors cited two papers by Fleischmann and Pons, they failed to cite the Fleischmann-Pons paper titled "Calorimetric Measurements of the

Palladium/Deuterium System: Fact and Fiction." Fleischmann and Pons discussed the recombination issue:

> Recombination does not occur because there are no exposed metal surfaces in the gas head spaces in our cells and because special care has always been taken to ensure that the palladium cathodes and platinum anodes remain totally immersed throughout all of the measurement cycles. Furthermore, the volumes of gas evolved from the cells correspond to that predicted from Faraday's law to better than 99%. (Pons and Fleischmann, 1990)

Fear, Uncertainty and Doubt

I met Srinivasan in Cambridge, Massachusetts, in August 2003, and we had many interactions during the next 13 years. During that time, whenever the topic of the BARC light-water experiments performed came up, he made it clear to me that he had abandoned those results.

In 2008, Srinivasan wrote to me about the Bose Corp. light-water electrolysis experiments. "I have published papers earlier on such light-water systems," Srinivasan wrote, "but during my sabbatical at SRI international, I found that the apparent excess heat was due to recombination." Bose also came to a similar conclusion and dropped the electrolytic work from their research program in 1994.

Going back to 1994, in the same month that Bose Corp. researchers submitted their paper to *Fusion Technology,* McKubre and Srinivasan published an article in *Cold Fusion* magazine in which they mentioned the many reports of excess heat in nickel-light-water cells. As a physicist, Srinivasan was not in a strong position to question or argue electrochemistry and calorimetry with McKubre. Without providing any specific facts or critique, McKubre, with Srinivasan in tow, cast a cloud of doubt on the entire lot, based solely on the as-yet-unpublished Shkedi paper, suggesting that recombination effects could explain away the excess heat measured in $Ni-H_2O$ experiments. (Srinivasan, 1994)

While writing this book, I re-examined the ICCF-4 proceedings and read the papers presented in Hawaii by BARC researchers. I was surprised to find Halasyam Ramamurthy's slides discussing the BARC group's light-water heat results. Srinivasan had presented the paper in Hawaii for him. There is no ambiguity; they were already fully aware of the recombination/Faraday efficiency issue, and they still could support their claim of light-water excess heat.

> · Faraday efficiency measured in some cells $> 95\%$. Hence recombination not cause of excess power; especially when excess power margins > 50 to 60%
> · Remember, apparent excess power fraction
>
> $$= \frac{f}{\left[\frac{V}{1.482} - 1\right]} \quad \text{where } f = \text{recombinat-ion fract}^n$$
>
> In our cells V often > 1.482 & hence denominator > 1.0. Hence to show 50% EH need $> 50\%$ recombination!

Slide discussing BARC light-water research. (Ramamurthy, 1993)

I showed the slide to Srinivasan in January 2016 and discussed it with him by e-mail. I asked him whether anyone had ever found a specific error of protocol, a mistake in the data analysis, or an unstated assumption for the BARC light-water experiments that showed excess power margins greater than 50% to 60%. He said no.

Srinivasan had a deep loyalty to and friendship with many researchers in the field. He was well aware of the conflict posed by light-water excess heat, particularly when factionalism among the researchers began to grow in the mid-1990s. Additionally, my communications with Srinivasan showed that he was philosophically attached to the idea of nuclear fusion as the explanation for their anomalous results.

These factors may have contributed to Srinivasan's willingness to abandon his and his colleagues' light-water experimental work.

CHAPTER 8

Clues From Russia

At the outset of the "cold fusion" conflict, neither the experimenters nor the theorists could have foreseen an entirely new nuclear reaction mechanism. But growing experimental evidence started accumulating, indicating that nuclear fusion was not a tenable explanation for the results.

In 1989, Peter Hagelstein was one of the first theorists to propose a fusion-based idea to explain the excess heat reported by Fleischmann and Pons in their experiments. Hagelstein used the equation $d + d \longrightarrow Helium\text{-}4 + 23.8\ MeV\ (heat)$ to express the mechanism.

No such mechanism is known to exist in nature, but something similar takes place in deuterium-deuterium (D+D) nuclear fusion. In D+D fusion, the reaction paths occur through one of three possible branches.

The first branch produces helium-3 and a neutron. The second branch produces tritium and a proton. The third branch produces helium-4 and a gamma ray. In D+D fusion reactions, the first branch takes place, on average, almost 50% of the time, as does the second branch. The third branch, which produces helium-4, occurs far less than 1% of the time.

Deuterium+Deuterium Fusion

D+D $\rightarrow$ Helium-3 (0.82 MeV) + $n_{neutron}$ (2.45 MeV) *[50%]*
D+D $\rightarrow$ Tritium (1.01 MeV) + p_{proton} (3.02 MeV) *[50%]*
D+D $\rightarrow$ Helium-4 (0.08 MeV) + g_{amma} (23.77 MeV) *[10^{-6}]*

As described in Vol. 2, *Fusion Fiasco*, experimentalists in 1989, with the exception of Fleischmann and Pons, avoided claiming or asserting that their results were due to D+D fusion; rather, they simply reported that they had observed excess heat or nuclear products.

In October 1989, researchers from the Naval Research Laboratory (NRL) and Lawrence Livermore National Laboratory (LLNL) reported persuasive experimental data that was inconsistent with D+D fusion. By March 1990, Hagelstein had abandoned his fusion idea in favor of a non-fusion neutron-based idea. At the end of 1992, Hagelstein remained in the non-fusion camp.

In his 1993 report about the Third International Conference on Cold Fusion (ICCF-3), Hagelstein described how other researchers began to adopt the "cold fusion" belief. (Hagelstein, 1993) They hypothesized that a "cold" fusion process released nearly all of its energy as heat, during the production of helium-4. But there was a problem, as Hagelstein wrote: There was almost no evidence of experimentally measured helium-4 that could account for the amount of heat produced.

Searching for Clues

Hagelstein now thought that the clues to the mechanism likely would be revealed in isotopic shifts and elemental transmutations from the experimental data.

Researchers from Johnson Matthey, the British supplier that had loaned palladium to Fleischmann and Pons, analyzed the cathodes used by Fleischmann and Pons that had produced excess heat. At ICCF-3, Johnson Matthey researchers reported that they had measured significant isotopic shifts between lithium-6 and lithium-7 on those cathodes. (Coupland, 1992)

Hagelstein knew that, in 1989, researchers at LLNL as well as researchers at NRL had also seen lithium-6 and 7 isotopic shifts.

When the Johnson Matthey researchers examined the cathodes, they also noticed that one of the electrolyzed rods showed damage to the microstructure, which indicated temperatures far greater than 200° C. Nevertheless, the temperature of the electrolytic bath could not have been much greater than 100° C, thus revealing an unexplained localized heating process in the experiments. Portions of another rod showed recrystallization of the palladium, which indicated an even higher temperature, more than 300° C.

The isotopic shifts gave Hagelstein valuable clues about what was happening. "Unfortunately," Hagelstein wrote, "very few groups are currently pursuing the lithium isotope shift problem; I consider it to be an important question."

As mentioned in Chapter 1, two diverging philosophical schools emerged among people attempting to explain the underlying mechanism. Hagelstein explained this bifurcation in his ICCF-3 review:

> The theories may initially be divided into two general categories: those involving (modified) fusion mechanisms, and those not involving fusion mechanisms. Papers considering fusion mechanisms face the two basic problems of (1) arranging to get nuclei close enough together to fuse, and (2) possibly modifying the fusion reaction profiles. ...
>
> A number of theorists, including myself, have gone away from fusion reaction mechanisms. The motivation for this is to avoid the Coulomb barrier (if possible) and to find reactions with signatures that hopefully more closely match the experimental observations. (Hagelstein, 1993)

In the early 1990s, Hagelstein sought a reaction mechanism that involved virtual neutrons and weak-interaction processes. He didn't see how real neutrons might be created in electrochemical experiments so he instead hypothesized a process based on virtual neutrons. (Hagelstein, 1993)

Hagelstein's attempt to develop a theory based on virtual neutrons was misdirected because virtual particles do not leave permanent evidence of their presence. Transmutation data from low-energy nuclear reaction (LENR) experiments show evidence of more-neutron-rich isotopes, which means real neutrons were created in the system and they were captured by nearby nuclei. Normally, temperatures found only in the core of stars provide the required energy to get a free electron and a free proton to react and make a neutron.

Years later, theorist Lewis Larsen (b. 1946) conceived a plausible mechanism for the creation of real, ultra-low-momentum neutrons. Larsen, together with condensed-matter physicist Allan Widom (b. 1942) published a detailed, mathematically supported, neutron- and weak-interaction-based LENR theory in 2006. (Widom et al., 2006) According to Larsen, weak interactions, despite the name, are not always energetically weak, and significant reaction rates and energy output can take place in LENRs under room-temperature environmental conditions. Widom and Larsen released another paper in 2007 that gave a more detailed first-principles calculation of the reaction rates. (Widom et al., 2007)

Endless Frustration

Like so many of his colleagues in the field, Hagelstein experienced tremendous frustration with colleagues outside the field. He could not understand why uninvolved scientists had so much difficulty with the topic. But he could also see that identifying the research as "cold fusion" was wrong:

> For such a significant conference, it has been largely ignored by the scientific community. Wrongly so, I think. The majority of scientists are currently ill-informed of the experiments, the implications, the arguments, or the goals of ongoing research in the field. At some point, this needs to change, but I confess that I do not see how it might happen in the foreseeable future.
>
> The name "cold fusion" has been adopted by the field to some degree by default. This name implies a generic physical reaction mechanism (fusion), and because the experiments involve deuterium, the name further presupposes specific reactions (D+D fusion reactions).
>
> But D+D fusion is expected to produce [high rates of energetic] neutrons and tritons, neither of which is

quantitatively present with the excess heat. Scientists who are not in the field are discouraged because the expected fusion products are not present in quantities commensurate with the observed energy production, and scientists working in the field have not come up with an explanation in three and a half years that is acceptable to the [broader] scientific community as to why deuterons should fuse [at room temperature].

There have been proposals to change the name of the field: "solid state nuclear physics" has been suggested; "nuclear effects in metals" has also been put forth. I would strongly endorse a name change.

The experimental evidence was strongly against a fusion process, and it gave the impression to people outside the field either that people in the field were poor scientists or that their wisdom was clouded by their belief in "cold fusion." Hagelstein was perplexed about why, after four years, the controversy was still unresolved:

> This simply must change. ... I do not know how this controversy is to be ended, but I know that it does need to be ended in a satisfactory manner. The basic experiments have been done, they have been repeated in many different ways by numerous groups, and the effect is observed with considerably better signal-to-noise ratio than in 1989.
>
> Scientists in the field have gone to extremes in attempts to satisfy [critics]. Cells were stirred, [controls] were done, extremely elaborate closed-cell calorimeters have been developed in which the [excess heat] effect has been demonstrated, the signal-to-noise ratio has been improved so that positive [excess heat] results can now be claimed at the 50 sigma level, the reproducibility issue has been laid to rest; but still it is not enough. I have heard some [critics] saying that a commercial product is the next hurdle to be jumped through before any significant funding can be justified. This is simply not right.

It was not right. The scientific method does not require a commercial product to establish the validity of an experimentally observed phenomenon. Hagelstein didn't anticipate the level of incredulity of most scientists toward the idea that nuclear reactions could take place in such small devices in such moderate environmental conditions. Equally important, he underestimated the bitter hostility that was engendered by the continued and mistaken use of the term "fusion." He suspected that the lack of a viable theory was a major hindrance to progress in the area, and he may have been right. To most outsiders, everyone in the field appeared to be a "true believer."

Hagelstein's comments shed light on the confusion and dissatisfaction associated with the often-messy early stages of a paradigm shift in science.

Not a Believer

By the end of 1993, Hagelstein wrote in a December letter to John Bockris that he was finally convinced that there was no experimental evidence for "cold fusion":

> There are numerous anomalies that have been reportedly observed in this field that many [people] refer to as "cold fusion." I do not believe that fusion (especially D+D fusion) can occur in electrochemical cells. The idea that a lattice could somehow squeeze deuterons together sufficiently hard to get them to fuse seemed to me to be absurd when I first heard about it. ...
>
> I first considered a scenario in which fusion could occur as a coherent process. Due to the local politics at MIT, even the mere consideration of such a scenario attracted media attention and came very close to costing me my job.
>
> Since the early days of claims [in the field] of heat, tritium and neutron production, we have [also] seen reports of alpha, beta, and gamma emission and [isotopic changes to heavy elements]. Kucherov claims to have seen energetic

fission products emerging from his cathodes in glow discharge experiments. The heat and tritium results have proven generally difficult to understand theoretically, since so very little positive information is available. Heat is not accompanied by energetic products, by quantitative neutrons or tritium or, so far, by quantitative helium-4. ...

It is immediately clear that fusion is not the source of the anomalies. Fusion would not lead to significant [isotopic changes] of Pd cathodes ... and to the production of what appears to be fission products of Pd.

As you recall, at ICCF-4, I presented a theory that seems to have the prospect for accounting for these various effects. Neutron transfers from Pd to Li and from K to Ni are, at the moment, leading candidates for explaining heat production.

The Russian Trio

The experiments of Yan Kucherov (1951-2011) that Hagelstein mentioned were in fact a milestone in the field. In 1992 and 1993, Kucherov, Irina Borisovna Savvatimova (b. 1942), and Alexander B. Karabut (1945-2015) made science history in Russia at the Federal State Unitarian Enterprise Scientific Research Institute, also known as "LUCH." The laboratory is located in Podolsk, just outside Moscow.

Yan Kucherov, Alexander B. Karabut and Irina B. Savvatimova (circa 1990)

Kucherov, born in Kharkov, in the Ukraine, had a background in solid-state physics, nuclear physics and heat physics. His Ph.D. dissertation was "High Current Cathodes for Plasma Generators," a classified document. He was credentialed as a patent attorney, an engineer and an inventor. In time, Kucherov's work in Russia, and later in the U.S., grew more interdisciplinary, reflected in his 58 patents and more than 100 publications in various fields of technology. His work encompassed a variety of nuclear materials problems, and in 1989 he focused on research in plasma physics and "cold fusion" phenomena at LUCH. He was honored by LUCH as one of its most outstanding scientists and inventors.

Karabut was born in Shostka, in the Sumi Region of the Ukraine. He studied nuclear engines and heat physics, and his Ph.D. dissertation concerned high-power plasmotrons. From 1974, he worked at the LUCH Nuclear Rocket Department on plasma sciences, receiving a distinguished national scientific prize in 1982. When funding for nuclear rocketry dried up as the political regime changed in 1991, he moved to the Nuclear Materials Science Department and worked on low-temperature transmutation research.

Savvatimova was born in Gorki, Russia. She studied welding technology and nuclear physics at Moscow Engineering and Physics Institute and received her Ph.D. in materials science and heat treatment of metals at LUCH. She remained there throughout her career.

From 1964 until 1974, she worked on research involving nuclear reactor fuel elements. In 1974, she joined the Nuclear Materials Science Department, where she worked on materials science, damage by low-energy ions, plasma technology and eventually "cold fusion." In 1982, she was recognized as the Best Scientist at LUCH. In 1991, she was awarded a silver medal at the Exhibition of Achievements of the National Economy. She has nine patents.

I met Savvatimova in 2003 at ICCF-10. Of the several hundred scientists who have worked in this area of research, Savvatimova is one of only a dozen women. In a later e-mail conversation, I asked what that was like for her. "That's the way it's always been in my work," Savvatimova wrote. "I did not have a choice. It was normal for me. I know it is necessary to be very strong."

Savvatimova and I have not been able to properly converse in person because I do not speak Russian and her English is limited. Savvatimova, about 5 feet tall, is a demure and unassuming woman. At podiums giving science presentations, her voice is barely audible. Throughout her presentations, she struggles with English but eventually finds the right words. Her determination is evident.

I came to appreciate this when I arrived in Sochi, Russia, in 2007 for a science conference, and she met me at the airport. Before we could get out of the terminal, local police decided that I was a good candidate for an on-the-spot "fine." After they detained me for a half-hour for no apparent legitimate reason, Savvatimova, while sitting with me in the dingy police barracks, started to call the U.S. embassy. Instantly, the police handed my passport back to me and told us we could go.

Irina Savvatimova (1998) Photo: Unknown

Teamwork

The three scientists formed a productive team and published several joint papers in peer-reviewed journals. They submitted their first manuscript on Aug. 2, 1989. (Karabut et al., 1990)

They performed their experiments in what is known as a "glow discharge" tube, which uses the same concept as a fluorescent lamp. A glass tube with electrodes at each end, filled with a gas, is subjected to an electric current, which causes the gas to form a plasma and glow. In the experiments, the researchers used palladium cathodes and deuterium gas. Their November 1992 paper in *Physics Letters A* provides a detailed explanation of the experimental configuration. (Karabut, 1992)

Using a variety of instruments and analytical modalities, they measured heat output five times greater than the electrical input, as well as evidence for charged particles, tritium, helium-4, heavy-element transmutations, and gamma-rays with energies of about 200 keV.

Using a transmission electron microscope, they also observed unusual damage to the cathodes: small bubble-like objects about 0.1 micron in size at a depth ranging from 0.1 micron to 1.0 micron, which suggested that the reactions took place on the surface of the cathodes rather than within them.

My first conversation with Karabut took place a decade later, in 2003. I had asked him what inspired his interest in fusion. In the 1980s, he labored for two years on the challenging problem of cooling the first wall of Tokamak (thermonuclear fusion) reactors.

"In 1989," Karabut wrote, "the Cold War was finishing, my big plasma installation was destroyed, and I was looking for a new scientific area in which to work. It's a new and very difficult area of physics, and I hope to receive a Nobel Prize."

By 2003, he told me, he and his colleagues had run between 3,000 and 4,000 "cold fusion" experiments. He was remarkably confident. "The problems of the critics are their own problems. We can reproduce the excess heat and elemental transmutations always with 100% reproducibility," Karabut wrote.

I asked whether his group's experiment had been replicated in other laboratories. "My experiments have been replicated in a big international company," Karabut wrote, "but these results are secret."

Scientists for Hire

In late 1991, the former Soviet Union dissolved into its component states and the Russian Federation. The Cold War had just ended, the

Berlin wall had come down, and communist regimes in Europe were replaced with democracies. The economy of the former Soviet Union had been devastated by the financial burden of its role in the arms race with the U.S.

Russian weapons and nuclear institutes were virtually bankrupt and could barely pay their employees a fraction of what they had been paid before. During that financial crisis, the U.S. Department of Energy (DOE), which is responsible for U.S. nuclear weapons, seized a unique opportunity to prevent wholesale migration of specialized expertise to potential adversaries. DOE began buying the services of key Russian scientists at a relatively inexpensive cost. A source who hired some of these researchers for private research and does not wish to be named, told me that DOE had two objectives: 1) to acquire all kinds of specialized information and consulting services at low cost and 2) to employ the Russian scientists so their nuclear knowledge didn't transfer to Middle Eastern countries. Although financial support from the U.S. helped the Russian scientists put food on their families' tables, many of them felt resentful toward the U.S.

Most of the Russian scientists I got to know seemed, by U.S. standards, to live at or below the poverty level. It was a strange feeling to walk through the University of Moscow, be introduced to scientists with distinguished histories and discoveries and see the dilapidated, dimly lit offices in which they worked, surrounded by worn instruments and crusty furnishings many decades old.

Scientists Needed

ENECO, a Salt Lake City, Utah, company, knew about the opportunities to hire Russian nuclear talent. The company made financial arrangements to have Kucherov, Karabut and Savvatimova conduct research in Russia on its behalf.

ENECO was founded in 1991, originally as Future Energy Applied Technology, with the aim of acquiring intellectual property in the "cold fusion" field and developing its research into commercial technology.

The founding members were scientists and inventors Robert Bass,

John O'Mara Bockris, Robert Bush, Dennis Cravens, Robert Eagleton, Samuel Faile, Avard Fairbanks, Steve Gregory, Robert Huggins, and Edmund Storms. The group was organized by Hal Fox (1923-2012), a prolific publisher of news and scientific research in the subject, who also ran what he called the Emerging Energy Mutual Fund, to promote research and facilitate investment in new energy technologies. His news archives, available at the *New Energy Times* Web site, are a useful window into this history, but the independence of his journalism was compromised by his financial conflict of interest.

On Dec. 1, 1993, ENECO acquired the rights to the University of Utah's patent applications based on Fleischmann and Pons' research. Also in 1993, ENECO sponsored 22 scientists to attend the ICCF-4 conference, eight of them Russians. ENECO hired Kucherov full-time in 1993 and was able to bring him (and his wife and daughter) to the U.S. on a specialized visa for nuclear scientists from the former Soviet Union.

In 2006, according to Graham K. Hubler (b. 1944), a nuclear physicist and the former head of the Materials and Sensors Branch at NRL, Kucherov left ENECO and became a senior staff scientist at Nova Research Inc. in Alexandria, Virginia, a company that outsources skilled workers to federal laboratories. In 2010, Kucherov was hired by NRL as a research physicist in the Materials Science and Technology Division. In addition to his research at NRL on "cold fusion," he conducted research in the areas of ballistics, personal protective equipment and vehicle armor for the military and electromagnetic launchers. He worked there until he succumbed to pancreatic cancer when he was 60.

Hubler shared the news of Kucherov's death in a December 2011 e-mail: "Yan was a true friend, outstanding human being and a talented colleague whose honor and integrity knew no bounds." Savvatimova described him not only as a brilliant interdisciplinary physicist but also as a man with the courage to work on controversial research.

ENECO never produced any commercial technology, and it filed for bankruptcy in 2008, leaving unpaid debts to, among other people, Kucherov for $9,000 and Hagelstein, a consultant, for $90,000. I met Kucherov that year at a science conference in Washington, D.C. He was soft-spoken and easy-going. He had a sense of humor, too, as was evident from an e-mail he sent me about people who were trying to own the hoped-for technology.

"There are two groups trying to get this technology in their hands," Kucherov wrote. "Peter Hagelstein and I define them as crooks and lesser crooks. [ENECO name redacted] sides with crooks, and we are not talking to [ENECO name redacted] and will not work with him or his group. Hopefully, the other group will get it."

Stunning Data

At the December 1993 ICCF-4 conference in Hawaii, the three Russian scientists reported the first credible evidence in the field for heavy-element transmutations: a wide range of anomalous isotopic shifts and newly found elements that were not present before the experiment. The researchers did not use the word "transmutation." Instead, they identified their observations as "impurities." (Savvatimova, 1993)

They observed significant increases in the concentrations of lithium, boron, aluminum, titanium, rubidium, zirconium, molybdenum, niobium, silver, and indium.

They also observed significant isotopic changes to lithium-6 and 7, boron-11, vanadium-51, chromium-53, iron-54, -56 and -57, nickel-60 and -61, copper-63, strontium-87 and -88, and zirconium-90 and -91.

For some elements, they saw an astonishing increase in concentration of up to 10,000 times. For some isotopic ratios, they saw an increase of up to 20 times. They also measured an increase in helium-4 in the cathodes by up to 100 times the starting value.

The benefit of the glow-discharge system was that it provided a far cleaner experimental environment than that associated with electrolytic experiments. Therefore, the likelihood of unrecognized contamination was significantly reduced.

The trio knew that the results could not be explained by conventional fusion and made a guess that perhaps they were explained by some sort of "fusion-fission" reaction. In 2016, at my request, Savvatimova revealed the previously secret company that Karabut said had confirmed their 1993 results: Honda.

CHAPTER 9

The Johnny Appleseed of LENR Transmutation

In December 1993, at the Fourth International Conference on Cold Fusion (ICCF-4), the Russian team of Yan Kucherov, Alexander B. Karabut, and Irina B. Savvatimova reported some of the first credible evidence in the field for heavy-element transmutation. So did an American scientist who, for two decades, shared his knowledge with the next generation of researchers.

John Dash (1933-2016) earned his Ph.D. in metallurgy from Pennsylvania State University in 1966 and began teaching in the Physics Department at Portland State University in Oregon. From 1998 until 2015, he was professor emeritus and continued to mentor students.

The head of the Physics Department, Dash told me, asked him to look into the Fleischmann-Pons claims. They performed their first experiment on April 24, 1989, and the results were positive; Dash was hooked, as he told me. "I've been doing research for 50 years now, and I've not seen anything like this," Dash said. "You can make a new discovery in this work every day. There's no competition; you have this big frontier all to yourself."

Dash told Eugene Mallove (1947-2004), the editor of *Infinite Energy* magazine, more details about his first experiments:

> [Before 1989, we had] studied the electrolysis of water with an acidic electrolyte, so we used the same composition, except we substituted heavy water for light water in the electrolyte.

Using a small (about 1 sq. cm), cold-rolled palladium foil cathode about 25 μm thick, we observed macroscopic plastic deformation of the cathode soon after the start of electrolysis. I had never seen such behavior in my 30 years of research on electrolysis, so I was immediately intrigued.

We believe that an acidic electrolyte inhibits the deposition of impurities and enhances the deposition of hydrogen isotopes on the cathode, in comparison with the basic electrolyte used by many investigators. The use of thin foil cathodes reduces the time required to saturate the cathode with hydrogen and deuterium, so it is possible to demonstrate an excess-heat effect in hours rather than days.

John Dash (2008) Photo: D. Tran

We also commonly observe localized concentrations of unexpected elements, in particular silver, on the surface of the palladium after electrolysis. Using mass spectrometry, we found inversions of palladium isotope concentrations on the surfaces of palladium cathodes after electrolysis, compared with isotope concentrations on unelectrolyzed palladium from the same batch.

In recent years, we have been studying the use of titanium cathodes in an acidic electrolyte. We find excess heat and localized concentrations of unexpected elements (usually vanadium, chromium, and iron) on the titanium cathodes after electrolysis. Through mass spectrometry, we also find changes in isotopic abundance, compared with the natural abundance. For both palladium and titanium, it appears that excess heat is associated with hot spots, which are marked by localized concentrations of unexpected elements. We believe that the amount of excess heat is directly related to the number of hot spots.

The challenge we have is to determine the characteristics of the metal (thickness, grain size, purity, etc.) to maximize the number of hot spots. (Mallove, 1999)

Doubly Doubted

Dash was regarded with great skepticism within the field, twice over. At the ICCF meetings, he told his peers that he could get excess heat with light-water electrolysis, despite the conviction of many scientists, including Martin Fleischmann, that it was not possible.

Dash also told his peers that he could perform the equivalent of modern alchemy. In 1993, most researchers, including Fleischmann, thought that heavy-element transmutations in the experiments were impossible. Most scientists in the field with whom I spoke had the impression that Dash was inept or suffering from self-delusion. Nevertheless, I'm not aware that any of them identified any specific errors of protocol, mistakes in his data analysis, or unstated assumptions in his work. And no one did as much as Dash to teach the new science to younger generations.

Spreading the Seeds

From 1989 to 2014, Dash was a sort of Johnny Appleseed, spreading seeds of knowledge about his transmutation work to students around the world. As of 2005, he had advised seven students who completed master's theses and two who completed Ph.D. dissertations on the research. Some of them traveled internationally to study with him.

During the summers, through a program called Apprenticeships in Science and Engineering, Dash taught science to high-school students from all over Oregon and southern Washington, using "cold fusion" as the area of investigation. By 2005, 40 high school students had apprenticed with Dash and learned, among other topics, chemistry, physics, metallurgy, and spectroscopy.

In 2003, when I attended the ICCF-10 conference in Cambridge, Massachusetts, I met three of his apprentices: Corissa Lee (Gresham High School), Shelsea Pedersen (Clackamas High School) and Ben Zimmerman (Wilson High School), and a graduate student, Abhay Ambadkar. I had never met more enthusiastic high school students. Zimmerman summarized for me their typical electrolytic experiment and their results: 0.5 Watts, on average, of excess heat and up to 20% unaccounted-for silver after electrolysis from cathodes that produced excess heat.

Corissa Lee, 2003 Photo: S.B. Krivit

Corissa Lee had complete confidence in her group's work. Her spunk and excitement infused the room during my filming of their demonstration:

> Most of the time when people ask me what I did with my summer vacation, I tell them "cold fusion." And they're like, "What is it?" And then the second question I get is, "Do you believe it?"
>
> I tell them, "Well, I DID it. And it WORKS. So yeah, I believe it!"

Dash also visited Leonardo da Vinci Science Center, in Milan, Italy, and ChangChun University, in ChangChun City, China, and taught students and professors how to perform the experiments.

Family Secrets

When I wrote *The Rebirth of Cold Fusion,* in 2004, scientists in the field told me that they could reproduce excess heat with 100% repeatability. I later found that this was not true. Excess heat has, however, been observed in carefully measured experiments hundreds of times and by many dozens of researchers around the world.

Electrochemist Michael McKubre, at SRI International, told the *BBC Horizon* film crew in 1994 that his team had cracked the reproducibility problem. The narrator said, "McKubre found that they could always reproduce the excess heat if they managed to achieve extreme conditions within the palladium electrode by forcing it full of deuterium and maintaining this for a long time."

McKubre explained:

> The issue of reproducibility, or its lack, has been frequently misunderstood in this field. The difficulty has been reproducing the *conditions* which produce the results, not reproducing the results.

> So if we meet the conditions that we have established, then we, in every case, observe excess heat. (*BBC Horizon*, 1994)

McKubre's explanation was problematic. John Bockris wrote to me on May 5, 2004, about this idea, although he didn't want to name McKubre. McKubre, and only McKubre, was well-known in the field for this idea. Bockris was the only person to challenge McKubre:

> Of course, you get a lot of people stretching truth when they claim they have reproducibility. For example, I have a colleague whose statements I respect. He will tell you that he can get reproducible results, and then he will add quietly "when I get sufficiently high concentrations of deuterium." Of course, this becomes the major point. Everybody knows that you can get [excess heat] if the concentration of deuterium in the metal is sufficiently high.

Many electrochemists attempted to accurately reproduce Fleischmann and Pons' experiment. In doing so, they used thicker cathodes (rods) than those used by Dash (foils). Dash could therefore test, obtain results and, if necessary, reconfigure much faster than people working with thicker cathodes.

Dash seemed to be able to make cathodes that produced excess heat more reliably than most scientists in the field. He used an old-fashioned hand-operated "cold roller" tool to make his foil cathodes. Not all of them would make excess heat, he told me. However, if he found that a newly created cathode worked once, it usually worked until the cathode broke or otherwise suffered damage.

Dash was so confident that he made arrangements to demonstrate excess heat at science conferences in Cambridge, Massachusetts (2003), Marseille, France (2005), Salt Lake City, Utah (2009), and perhaps elsewhere. Dash's Cambridge demonstration succeeded, the Marseille demonstration failed, and he succeeded again in Salt Lake City.

CHAPTER 10

Battery Acid on a Saturday Night

In March 2007, in Chicago, the American Chemical Society hosted a New Energy Technology Symposium for low-energy nuclear reaction (LENR) research at its spring meeting. The symposium was organized by Jan Marwan, a German electrochemist. Marwan was new to the subject, but he had fresh ideas on how to bring the new research back to mainstream science; the symposium was his idea.

He didn't know many people working in the field, but through my news reporting on the subject, I knew nearly everyone in the field so he asked me to help organize the next two symposia: 2008 in Philadelphia, Pennsylvania, and 2009 in Salt Lake City, Utah.

A month before the Salt Lake City symposium, I asked John Dash to show his excess-heat demonstration during the symposium. I thought that, if Dash had the courage to do another demonstration, I was willing to include it in the program.

The Dash demonstration would be fitting, I thought, because the symposium was taking place in Salt Lake City, where, 20 years earlier to the day, Martin Fleischmann and Stanley Pons publicly introduced their research.

I knew that the experiment could be set up just about anywhere so I made arrangements at a hotel in downtown Salt Lake City. I told the hotel that we needed a room for a business meeting. I didn't think we would be well received if I said that we planned to demonstrate a nuclear reaction experiment. I planned the demonstration for the end of the conference sessions on Monday evening, March 23, 2009. Initially, I had an agreement with people at the University of Utah Chemistry Department to host the demonstration on campus, but they changed their minds. To minimize further unhelpful surprises, I told as few people as possible.

The topic came up in a phone call with Pamela Mosier-Boss, an analytical chemist at the U.S Navy's Space and Naval Warfare Systems Center (SPAWAR). She was enthusiastic about the idea. She was planning to drive from her San Diego home to Salt Lake City with her husband, Roger Boss, and she asked whether we could provide another table for her to display her group's recent results. I said sure, the more the merrier. She and Roger were now part of the stealthy operation.

Mosier-Boss also came up with the idea of purchasing inexpensive poster boards at a local office supply store to allow other scientists to put up posters about their research. She volunteered to discreetly coordinate the other scientists who wanted to participate in our little Science Fair, as we eventually called it.

I checked with Dash to verify my understanding that the safety risks were minimal. As long as the recombiners in the cells worked, he told me, the hydrogen and oxygen separated by electrolysis would reunite into water and never leave the cells. The only danger, he advised, was proper handling of the sulfuric acid. I was confident that he could handle the chemical safely, although the hotel room was not equipped with the safety features normally found in a legitimate laboratory.

The only problem was getting the chemicals from Portland, Oregon, to Salt Lake City. Dash needed to get 250 ml of a sulfuric acid and H_2O electrolyte mixture and 250 ml of a sulfuric acid and D_2O electrolyte mixture to Salt Lake City. Dash wasn't going to put sulfuric acid in his carry-on or checked baggage when he flew down to Utah. Electrochemist Melvin Miles, who lived near Los Angeles, was planning to go to the meeting. He had grown up in Utah and had family there. Miles said that Dash could send the chemicals by Federal Express to his sister Mary's house 20 miles outside of Salt Lake City. The chemicals were scheduled to arrive there on Friday, and I was supposed to go to Mary's house and pick them up on Saturday morning. Dash would need at least 24 hours to set up the experiment and get it running correctly.

On Friday evening, I received bad news from Linda, Mel Miles' wife.

She had heard from Mary that the chemicals had not arrived in Utah. I asked Dash, who was already in Salt Lake City, if he had a tracking number. He didn't; he had simply given the box of chemicals to a clerk at Portland State University with shipping instructions. On Saturday morning, Dash put together an alternate plan, and I went into action. I called Mosier-Boss to see whether she could stop by her lab and pick up some heavy water, but she and Roger were already on the road. However, she called her colleague Larry Forsley, who had not yet left San Diego, and he was able go to the SPAWAR lab and pick it up. Dash was thrilled. Getting the light water was easy: We simply purchased distilled water from a supermarket.

Then I began calling people in the Chemistry Department at the University of Utah to try to obtain sulfuric acid. Because it was a Saturday, I reached nobody and left messages. By Saturday evening, nobody had called me back. Dash was still calm and optimistic but my disappointment was growing. Then I saw a glint in his eye.

"You know, Steven," Dash said, "battery acid is essentially sulfuric acid. It's sold in auto parts stores for car batteries. I've never used it before for experiments, and it's a long shot, but it might work."

Moments later, we were in my rental car cruising the streets of Salt Lake City on Saturday night to find an auto parts store that was still open. Dash seemed to be enjoying the adventure. The clerk behind the counter was understandably puzzled when I, in a business suit, and Dash, an old greybeard dressed in clothes much too nice for an auto mechanic, asked for battery acid. Yes, the clerk said hesitantly, they had it.

A smile came over Dash's face. He contained his excitement until the clerk brought us the container of sulfuric acid from the back and completed the transaction. Dash couldn't resist telling the clerk that we were going to make a nuclear reaction with it. All I could think of was how to get out of the store quickly before the clerk decided to phone the police.

By 9 p.m., Forsley arrived, heavy water in hand. Dash had everything he needed, and he began setting up the experiment. Sunday morning, Julie Yurkovic, one of his students from Oregon, arrived and worked with him the rest of the day on the setup.

My hope was that, at the end of the Sunday session at the ACS, I could announce to the attendees that a) an excess-heat demonstration was set up in a hotel a few blocks away, b) it was producing excess heat, and c) it would be open for public viewing on Monday night. By 6 p.m. Sunday, there was still no word, and all I could tell people was that the experiment was being set up and that, if we were lucky, people might see the experiment producing excess heat on Monday evening.

Late Monday afternoon, I got the phone call from Yurkovic: It was working. After the scientific sessions, most of the people in the room walked over together to see the demonstration. One of them was Peter Hagelstein, and despite the temperature readouts on the electronic displays, he wanted to feel the temperature with his own hands. Dash warned him that it was hot. As Hagelstein put his finger on the cell, he learned that Dash was not joking. When Hagelstein abruptly pulled his finger away, he jostled the cell, and a thermocouple came loose. Other than that, the demonstration went without a hitch. Dash was pleased that the battery acid worked, and a few years later, one of his master's degree candidates replicated the experiment in the lab.

Youthful Enthusiasm

In 2008, in Cadarache, France, Mathieu Valat (b. 1982) was working on seismic analysis and engineering of an experimental neutron detection device for a public company (National Radioprotection and Nuclear Safety Institute). He had just completed his master's degree in engineering. He later explained his aspirations to me:

> I wanted to work on sustainable energy. At the time, the International Thermonuclear Experimental Reactor (ITER) was the big thing in Cadarache; teams were forming, and engineers were coming in from all over the world.
>
> So I looked at the literature of thermonuclear fusion (Tokamak reactors, in particular) to see if it would fulfill my eagerness and motivation. I visited the ITER design offices, met teams and visited the TORE-SUPRA reactor site (the

direct predecessor of ITER) multiple times. I read a couple of books, got information from scientific papers and publications, and discussed the project with engineers working on ITER and TORE-SUPRA. As I learned more, I became discouraged and failed to see how it would ever become anything more than a huge laboratory toy.

Valat soon found out about LENRs and learned that Jean-Paul Biberian, a professor who had been involved in the research for many years, was only a two-hour drive away. Biberian invited Valat to study under him and obtain a degree, but limited financial resources made this impossible. Instead, on Valat's request, Biberian directed him to Dash's well-funded U.S. lab to pursue his graduate studies. Dash responded to Valat's inquiry immediately and enthusiastically. Most of Dash's funding came from a private philanthropist who had given more than a million dollars to Portland State University so Dash could train students in the new science. A year later, in 2009, Valat and his wife moved to Portland, Oregon.

During his time at Portland, Valat learned a lot from Dash and repeated the basic excess-heat experiment that was Dash's specialty. In his lab, Dash had a scanning electron microscope with an attached energy-dispersive X-ray spectrometer (SEM/EDX). This was his primary instrument to analyze palladium cathodes.

Valat also collaborated with others in the Physics Department. Dash held emeritus status by this time; therefore, he was not officially permitted to mentor graduate students. Instead, Valat's official advisor was Erik Sánchez, an associate professor of physics. Sánchez, a specialist in microscopy and nanotechnology, had an array of sophisticated microscopes that he made available to Valat for his research.

Valat understood that evidence of isotopic shifts would provide direct evidence of nuclear reactions. He also knew that isotopic information would be useful in revealing the underlying mechanisms, so he developed his skill in microscopy. He repeated the battery-acid electrolytic experiment and, as happened in Salt Lake City, was able to measure the production of excess heat.

More important, he measured significant shifts in isotopic ratios found on the palladium cathode after the experiment, as compared with the isotopic ratios for virgin palladium from before the experiment.

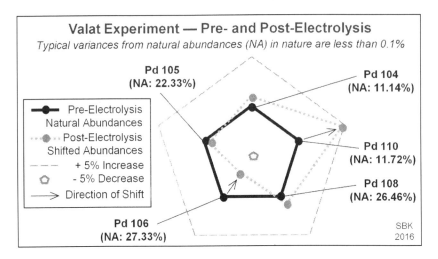

Valat learned in his research that geologists observe variations in isotopic abundances in nature, but they typically vary no more than a few tenths of a percent, thus making his observed changes of a few percent scientifically significant.

In the diagram above, the dashed-line pentagon depicts the natural abundances of five palladium isotopes, before electrolysis. The deformed pentagon depicts the shifted abundances of those same isotopes after the electrolysis experiment. The data show a significant decrease in Pd-106 and a significant increase in Pd-110, a slight decrease in Pd-105 and increases in Pd-104 and Pd-108. These isotopic shifts cannot be explained by chemistry or contamination; they can result only from nuclear transmutation of palladium isotopes.

Extraordinary Claims

The story of John Dash, metallurgist, now returns to December 1993. At the Fourth International Conference on Cold Fusion (ICCF-4), in Hawaii, Dash presented his results and called them transmutations.

As noted in Chapter 8, at the same meeting, the Russian team of Yan

Kucherov, Alexander B. Karabut, and Irina B. Savvatimova, two nuclear physicists and a rocket scientist, respectively, presented their evidence of heavy-element transmutations and isotopic shifts but, instead of calling them "transmutations," called them "impurities."

Most meeting attendees were unwilling to accept that heavy-element transmutations could occur in room-temperature experiments. Regardless, the two sets of research complemented each other.

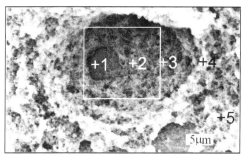

Silver was found in the newly formed craters. Concentration was five times greater inside the craters than in spots outside of craters. (Zhang et al., 2007)

At ICCF-4, Dash reported increased concentrations of silver and gold on his cathodes after electrolysis. His cathodes were palladium, and the anodes were platinum. Silver is one atomic number higher than palladium; gold is one higher than platinum. (Dash, 1993)

In later experiments for the U.S. Army Research Office, Dash used titanium cathodes as a lower-cost option to palladium. He saw the unexpected elements sulfur (atomic number 16), potassium (19), calcium (20), vanadium (23), chromium (24), iron (25), and nickel (28). All are within six atomic numbers of titanium (22). (Dash, 2001)

According to Dash's U.S. patent application US 2005/0276366, "the chemical analysis provided by the supplier of the titanium used to make the cathode showed 0.910 ppm vanadium and 1.150 ppm chromium. The EDX spectra of the cathode following electrolysis showed concentrations of these metals that were more than four orders of magnitude higher."

When values differ by one order of magnitude, they differ by an amount of about 10. Two orders of magnitude represent a difference of about 100. Four orders of magnitude represent a difference of about 10,000.

Dash was tipped off, and guided, as he was peering through his scanning electron microscope, by the odd topographical features on the surface of the cathodes after electrolysis. He saw things that he described as "rimmed craters" or "volcanoes." Valat later recorded excellent images of such topographical features on his palladium cathodes.

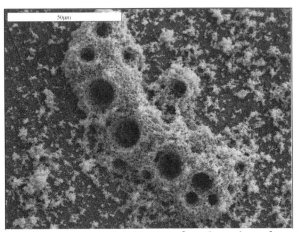

Volcano-like structures in a SEM image of an electrode surface after an experiment. Scale: 50 micrometers, magnification: 1,000x. (Valat, 2011)

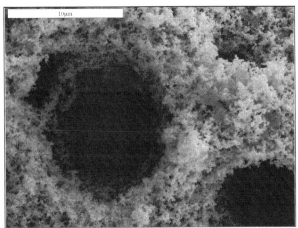

Volcano-like structures in a SEM image of an electrode surface after an experiment. Scale: 10 micrometers; magnification: 5,000x. (Valat, 2011)

In them, he saw evidence of highly localized melting, even though only a few Watts of electrical power had been applied to the system. Palladium melts at 1,555° C. On the side of the cathode that faced the anode, he observed "spherical globules" that had formed on the surface of the cathode.

Theorist Lewis Larsen told me that such spherical globules result from metal that has been flash-boiled and quickly cooled and solidified. Larsen considered flash boiling likely because the reactions are known to occur on small "hot-spot" sites, and they can exceed 4,000° C and cool down very rapidly. Palladium boils at 2,963° C.

Dash observed a correspondence between the presence of new elements and the measurements of excess-heat production. Although he saw results in both light-water experiments as well as heavy-water experiments, the effects were stronger with heavy water.

"Our system behaves as though low-energy neutrons are produced," Dash wrote. "Although we have tried, we have never detected radiation outside our cells. Our results also indicate that substitution of an electrolyte containing H_2O and H_2SO_4 [sulfuric acid] gives the same type of results but to a lesser degree." (Dash, 1995)

Dash likely realized that most other scientists would not have had the courage to report that they had observed the possible synthesis of gold or silver. In his final report to the Army, Dash reported isotopic versions of Pd-108 and Pd-106, and Pd-110 with Pd-104, from a palladium cathode. For a titanium cathode, Dash reported the reduction of Ti-50 by 4% to 13%.

"We realize that our claim of changes in isotopic ratios is extraordinary," Dash wrote. "Therefore, we have preserved their samples so that our results can be checked by any independent laboratory." Perhaps his cathodes are still stowed away somewhere in a drawer at Portland State University. According to Dash's calculations, the transmutation reactions released a lot of energy:

> Our preliminary experiments indicated that an electrolytic cell with a titanium cathode ran at 4° C higher than a control cell of the same size containing a platinum cathode. Calculations showed that the cell with the titanium

cathode was producing about one Watt excess thermal power compared with the control cell.

This means that the 10 mg titanium cathode produced about 83,000 joules more energy than was consumed by electrolysis. This is [100 times] more energy than could be produced by any known chemical reaction.

Like Dash, Valat also observed the presence of anomalous silver in his post-electrolysis cathode when he analyzed it with SEM/EDX. Valat also chose to look for silver using secondary ion mass spectroscopy (SIMS). But he didn't see any silver when he looked for it on the same cathode using SIMS. The two analytical methods vary dramatically and thus provide qualitatively different data.

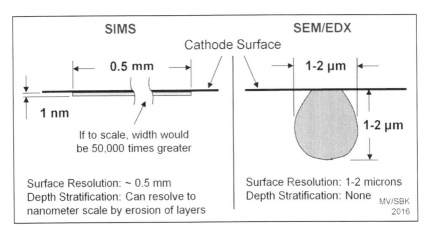

In July 2011, Valat had two things to celebrate: the successful defense of his master's thesis for his LENR work under Dash; and three days later, the birth of his first child. He and his wife soon moved back to France with their daughter.

Dash's results were intermittent, and he didn't identify key experimental parameters. Nevertheless, he succeeded in encouraging and training others and in moving the science forward.

CHAPTER 11

Disturbing Data

Professor John Dash and his graduate students, including Mathieu Valat, reported scientific results that contradicted what most scientists thought was possible.

In a room-temperature electrolysis experiment, they measured isotopic shifts (changes in the isotope ratios) of palladium, and the presence of elements that weren't there before. For other researchers who lacked the academic freedom Dash had, it wasn't so easy to challenge the status quo.

In October 1989, Debra R. Rolison (b. 1954), an electrochemist at the Naval Research Laboratory (NRL) in Washington, D.C., saw similar transmutation results and reported her data years before Dash did. However, by 1991, she had dismissed her anomalous results without offering a compelling scientific reason for doing so.

Pamela Mosier-Boss, at SPAWAR, told me that Rolison was NRL's best electrochemist at the time. "Rolison saw changes in the palladium isotope ratios," Mosier-Boss said. "She ended up saying that it was just a natural re-distribution of isotopes which didn't make any sense at all, but that's what she claimed."

At the Oct. 16-18, 1989, Workshop on Anomalous Effects in Deuterided Metals, in Washington, D.C., co-sponsored by the National Science Foundation and the Electric Power Research Institute, Rolison presented results from her experiments. As Dash had done, she used foil cathodes and got fast loading times and positive results. She observed an increase in the palladium isotopes 103, 106, 107, 109, 111 and a decrease in isotope 105. She reported that Pd-106 had increased by nearly 50%.

She performed depth tests, and all evidence confirmed that the isotopic shifts in palladium were taking place only within the 0.1-micron surface layer of the 127-micron thick foil. The interior (bulk) of the cathodes showed normal isotopic abundances for the palladium. She mentioned in passing that "other processes are occurring, including enrichment of palladium bulk impurities such as rhodium and silver at the [cathode] surface." Rolison knew that many unanswered questions about the rhodium and silver remained, and she wrote that she would be addressing them in a manuscript she was preparing. The palladium isotopic shifts obliged her to conclude that her data indicated evidence of an as-yet-unidentified nuclear reaction. (Rolison, 1989)

The workshop proceedings include a transcription of the discussions that followed each presentation. Edward Teller (1908-2003), a preeminent physicist who made many contributions to science but is best-known for his key role developing the first U.S. hydrogen bomb, participated in the workshop.

Teller thought he could explain the data. After hearing about the NRL isotopic shifts and other similar results at the Lawrence Livermore National Laboratory, he proposed a conceptual explanation. "Perhaps a neutral particle of small mass and marginal stability is catalyzing the reaction," Teller said. "You will have not modified any strong nuclear reactions, but you may have opened up an interesting new field." (Teller, 1989)

Caltech chemist Nathan Lewis (b. 1955) asked Rolison, "What other elements did you find in the SIMS?" "Chromium and iron were always present," Rolison replied, "even in the light-water blanks." One spectroscopy method (XPS) in her 1991 paper showed that iron was not present before the experiments, and another method (AES) showed it only at miniscule parts-per-billion levels. Neither method showed any presence of chromium before the experiments.

Half a year later, in March 1990, at the First Annual Conference on Cold Fusion, in Salt Lake City, Rolison presented her research progress. She explored the rhodium and silver issues in detail. She wrote that she had not been able initially to detect the rhodium or silver as pre-existing contaminants in the palladium foil using X-ray photoelectron spectroscopy (XPS). But with an atomic emission spectrometer (AES), she found that traces of rhodium and silver, and four other elements, at parts per million levels, existed in the sample material: Ni (200-300 ppm), Pt (200 ppm), Ag (100 ppm), Rh (50 ppm), Cu (50 ppm), Si (20-40 ppm). (Rolison, 1990)

After electrolysis, using XPS analysis on the surface of the cathode, she was able to detect silver and rhodium atoms at 1% and 3%, respectively, in relation to the total number of palladium atoms. This compares to her measurements in the bulk of the cathode before electrolysis for silver at 100 ppm (0.01%) and rhodium at 50 ppm (0.005%) — a difference of 100 and 600 times, respectively.

To reach this level of rhodium in the surface layer, Rolison said, every rhodium atom from inside the cathode would have had to migrate to the surface. Her paper shows that she was also clearly aware that neutron reactions were a plausible explanation for her data:

> Reaching this rhodium level in a 0.1-micron surface layer from a bulk value of 50 ppm in a 127-micron foil implies a great deal of segregation. In fact, it indicates that all of the rhodium is in the 0.1 micron surface layer sampled by XPS. Because our palladium was not 99.999 percent pure, and we did have a bulk presence of these elements [at trace levels before electrolysis], we either must say that they are completely segregated near the surface, making the metallurgists unhappy, or that they were produced there by nuclear transformations. Rhodium and silver are near neighbors to palladium in the periodic table, and there are certainly a number of neutron processes that can generate their nuclei. (Rolison, 1989)

Rolison tested and ruled out the possibility that rhodium and silver had been electrodeposited from the platinum anode. She also wrote that she found rhodium and silver after light-water electrolysis as well as heavy-water electrolysis. She did not, however, provide details about how the amounts differed between H_2O and D_2O electrolysis. On this information alone, and absent direct tests, she concluded that every atom of silver and rhodium within the cathode had migrated to the surface of the cathode:

As both Ag and Rh appear at the surface of Pd charged in either D_2O or H_2O, a mechanism based on surface segregation and forced diffusion as the Pd lattice is filled with hydrogen or deuterium atoms seems more probable than one relying on known neutron activation reactions of Pd isotopes, some of which yield stable Ag and Rh isotopes. ... Experiments with 99.999% Pd would be required before a physical transport mechanism could start to be discounted. (Rolison, 1990)

In 1990, there were no known radiation-free mechanisms capable of creating real neutrons that could occur in room-temperature experiments. The lack of a theoretical explanation added to Rolison's conviction that a hypothetical "physical transport mechanism" was a more likely explanation than a nuclear process.

She arrived at this conclusion despite the fact that she performed no explicit tests to see whether low-power electrolysis had the extraordinary ability to cause all rhodium and silver atoms to permeate bulk palladium in an experiment that, macroscopically, operated at less than 100° C. Nor did she suggest why only rhodium and silver might have that unique ability. In contrast, atoms of nickel, copper and silicon that were present at ppm levels before the experiments didn't migrate to the cathode surface.

Although Rolison wrote that the increased presence of rhodium and silver was from migration, rather than an as-yet-unexplained nuclear process, she concluded that the anomalous peak at mass 106 (most likely palladium-106) could not be "attributed to heretofore-identified plausible chemical interferents." In other words, she could not explain the data by known chemistry. (Rolison, 1990)

In 1991, Rolison published a final paper on her "cold fusion" experiments. She acknowledged that rhodium and silver were "tantalizingly close in atomic number to palladium." She also identified scientifically credible nuclear concepts by which the rhodium and silver could have been created by neutron reactions with palladium isotopes. "The decay products of the Pd-103 and Pd-109 radioisotopes are the stable isotopes Rh-103 and Ag-109, respectively," Rolison wrote.

Rolison saw trace levels of nine other elements after electrolysis, some of which were detected before electrolysis. Yet none of these showed the massive increase in local abundance on the surface as did rhodium and silver. She even remarked that "silver forms a continuous series of solid solutions with palladium through the entire temperature range, while rhodium forms solid solutions with palladium above 850° C, so it is somewhat unexpected that either element would readily move through palladium to segregate in a near-surface layer."

She was aware that segregation of solutes and migration toward the surface of solids was known to occur, but only after forceful treatments such as annealing, fracturing, corrosive oxidation or energetic ion bombardment. Rolison could readily have tested her hypothesis of segregation and migration. The AES detected 50 ppm of rhodium and 100 ppm of silver in the foil before the experiment. The AES had a sensitivity of at least 10 ppm. Thus, she proposed — without testing — that, in her in electrolytic cells, segregation and migration occurred because rhodium and silver are dissolved within the cathode by the action of electrolysis, at not much more than 100° C outside the cathode.

Despite all the indicators of nuclear processes and the lack of any test evidence, she concluded that the "most plausible mechanism" for the rhodium and silver was "most likely" segregation of existing impurities from within the bulk of the cathode and migration to the surface of the cathode, "rather than one based on nuclear chemistry." (Rolison, 1991)

Rolison saw that something new, different and extremely important was occurring in her experiments, and although she had all the requisite data, she did not step forward and make such an assertion.

Hatred and Hostility

To understand Rolison's possible reasons for offering conclusions that seemed antithetical to her data, it may be useful to acknowledge the unfriendly political climate at the time. By 1991, "cold fusion" was an "untouchable" science. It had been dismissed by the Department of Energy's (DOE) "cold fusion" panel members in November 1989. It was not dismissed by DOE staff members. (Vol. 2, *Fusion Fiasco*)

John Maddox, the editor of the prominent journal *Nature,* said on BBC television on March 26, 1990, "My own belief is that Pons and Fleischmann, for a very particular reason, had come to nurture a delusion." Caltech physicist Steven Koonin had declared Fleischmann and Pons incompetent in front of the world's news media.

Three highly visible science authorities advocated both openly and behind the scenes against "cold fusion." They were convinced that all such research had no scientific merit and that it threatened to tarnish the good name of science and waste the public's money. One of these men was Robert L. Park (b. 1931), a professor of physics at the University of Maryland who also held the influential post of spokesman for the American Physical Society. Park enjoyed expressing his caustic remarks in his weekly "What's New" newsletter. In his book *Voodoo Science: The Road from Foolishness to Fraud,* Park denigrated the field and its researchers, described their work inaccurately, and used character attacks in portraying his subjects. (Park, 2000; Krivit, 2008)

Another vocal critic was Richard L. Garwin (b. 1928), a prominent U.S. physicist, a research fellow at IBM, and a consultant for many high-level science and nuclear projects for the U.S. government. In 1989, as a key member of DOE's "cold fusion" panel, Garwin played a significant role in dismissing the new science.

The third scientist was physicist Peter D. Zimmerman (b. 1941), who for many years was the science advisor for the Department of State's Arms Control and International Security Division. According to a transcript of Zimmerman's talk at the "Pseudoscience Session" at the American Physical Society meeting in Atlanta, Georgia, on March 22, 1999, he, Park and Garwin shared their opposition to the controversial research. In his 1999 talk, Zimmerman spoke with pride about his role in killing a scheduled science conference at the State Department that included "cold fusion" research. (Zimmerman, 2009)

Zimmerman wrote an opinion piece published in the *Los Angeles Times* in 1989 professing the virtues of scientific skepticism. "Science works by institutionalizing skepticism," Zimmerman wrote. "Rare but understood chemical reactions can easily account for what Pons and Fleischmann actually saw." (Zimmerman, 1989)

Actually, Zimmerman's comment was neither skeptical nor scientific.

There was no known chemical reaction that could account for the levels of excess heat Fleischmann and Pons had measured. Nevertheless, by condemning this new science, these three men misused the authority and respect accorded them by the scientific community and the public.

Rolison was in a precarious situation. She had measured isotopic shifts in palladium that had shown irrefutable evidence of nuclear reactions by chemistry. She had obtained evidence that some people might have interpreted as the transmutation of metals, bringing with it the stigma of alchemy. After Rolison's 1991 paper appeared in *Analytical Chemistry,* no NRL researchers published any of their own experimental "cold fusion" results in the peer-reviewed literature for two decades, until 2012. (Knies, 2012)

"Cold fusion" was feared equally by upper management and by staff scientists. One former Navy researcher told me flatly, "It was not a career-enhancing area in which to do research." Staff scientists, like Rolison, avoided being associated with the research to preserve their scientific reputations. Navy researchers told me that, when David Nagel (b. 1938) and Graham Hubler, mid-level managers at NRL, asked more-experienced scientists to assist less-experienced scientists in their "cold fusion" experiments, the more-experienced researchers refused or did their best to avoid getting involved.

The only NRL researcher who published "cold fusion" papers in peer-reviewed journals in the early 1990s was Scott Chubb, but, as he told me, he was forbidden to list his affiliation as NRL.

When SPAWAR researchers published a compendium of such research in 2002, it included a paper by Chubb listing his affiliation as NRL. After Zimmerman contacted Chubb's supervisor, Chubb almost lost his job.

Larsen's Nucleosynthetic Pathways

In 2004, Lewis Larsen, while developing the ideas that led to his and Allan Widom's theory (Chapter 25), created one of his LENR nucleosynthetic reaction networks that shows textbook neutron capture and decay processes in which palladium can transmute to silver. It reveals why the appearance of Pd-103 and Pd-109 isotopes in Rolison's experiment are significant: They are unstable relatively short-lived intermediate isotopic products on the way to stable rhodium and silver by way of neutron capture, k-shell electron capture, and beta-minus decay.

Portions of Larsen's reaction network below depict what can happen to nuclides, starting with palladium-102, if they capture neutrons. Stable isotope Pd-102 (white background) captures a neutron and changes to unstable isotope Pd-103 (shaded background). Pd-103 can go in one of two directions: upward to stable rhodium-103 by way of electron capture or across, eventually arriving at unstable Pd-107, which has a half-life of 6 million years.

Subsequent neutron-capture processes advance the reaction network toward the right, up the periodic table to progressively higher values of atomic mass. When these hard-radiation-free LENR processes take place, elements can transform into other elements without the need of a heavily shielded nuclear reactor or high-energy accelerator.

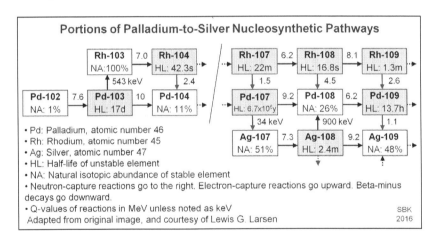

NRL had the tools, the expertise and the opportunity to seize a leading role in studying and understanding the transmutation activity associated with this new science. But it was not to be. Institutional caution prevailed.

CHAPTER 12

Transmutation on Trial

Our story now brings us back to 1993. Electrochemist John Bockris, at Texas A&M University, found himself on trial yet again, this time for what the press called alchemy.

Bockris knew that he wasn't the only researcher observing evidence of heavy-element transmutations. Both the Russian group from the Kurchatov Institute and John Dash at Portland State University had reported such evidence in December 1993, at the Fourth International Conference on Cold Fusion (ICCF-4) conference, in Hawaii.

But Bockris' claims were bolder: He reported the production of trace amounts of gold in his experiments. It was a surefire trigger for backlash because the mention of gold conjured images of alchemy. This made him an easy target for a disgruntled former employee of Texas A&M University.

Back in September 1992, the Federal Bureau of Investigation (FBI) subpoenaed records from Texas A&M about William Telander, the benefactor for Bockris' early transmutation research, which Telander named the Philadelphia Project. In March 1993, the Securities and Exchange Commission (SEC) filed suit against Telander for defrauding investors in a currency-trading scheme.

When Michael Hall, the head of the Texas A&M Chemistry Department, was notified of the lawsuit by the SEC in May 1993, he froze the remaining funds and halted the research. Throughout 1992 and most of 1993, Bockris and Texas A&M had managed to remain out of the media spotlight. But six months later, in November 1993, all hell broke loose.

Dawn Lee Wakefield (b. 1957), one of Bockris' former students, was the point of origin. Wakefield earned her master's degree and Ph.D. in physical chemistry at the university, then took a job as the director of development (fund-raising) for Texas A&M's College of Science.

On July 15, 1993, Wakefield met with Durwood Lewis, the director of College Programs for the Texas A&M Foundation, at 9 a.m. On her arrival, Lewis gave her the option to resign by 4 p.m. that afternoon or be terminated effective July 31, 1993. She argued with Lewis about the cause of her impending termination and chose to wait and be fired.

When Wakefield was later contacted for a story in the *Chronicle of Higher Education*, she said she was fired because she warned her superiors not to accept Telander's $200,000 gift. She sued the university in Brazos County, Texas, District Court on July 29, 1994, but her complaint was dismissed in June 1997. (Wakefield, 1994)

Wakefield's lawsuit was rooted in her complaints about how the foundation leadership was doing its job and her perception that she wasn't receiving equitable compensation compared with people in similar positions at other institutions. The court record indicates that Lewis actually fired Wakefield because of continuing failure to do her job.

But she also claimed that Bockris and his graduate student Nigel Packham, when they reported finding tritium in 1989-1990, had performed fraudulent research. The story made national headlines, thanks to Gary Taubes' article in *Science* magazine. Bockris, she wrote, "had a solid history of scientific misrepresentation." The ensuing negative publicity from their tritium research was entirely their fault, Wakefield wrote, and she said this damaged her ability to attract donations for the college. In her lawsuit, Wakefield explained her attempts to thwart the Philadelphia Project.

"I developed a plan," Wakefield wrote, "to prevent further damage to the university and the foundation." She contacted Robert E. Wiatt, the director of the Texas A&M University police, and told him that she "was extremely concerned that the science of the Philadelphia Project was fraudulent."

Wiatt had heard about "a recent planeload of people from the People's Republic of China" who had arrived at the College Station, Texas, airport. Wiatt thought that the presence of Chinese visitors to an internationally recognized university was suspicious. He had one of his police units tail them from the airport. The suspicions of Wakefield and Wiatt were rewarded: The Chinese visitors went to Bockris' lab. Wiatt and Wakefield compared notes, and he told her that "the FBI would be interested in looking at Telander and these visitors from China to know if, and how, they may have become involved."

According to the complaint, Wiatt also told Wakefield that the FBI would be interested in investigating the Bockris research, and Wiatt asked Wakefield to quietly deliver to his office any and all records on the Philadelphia Project. Wakefield, who had refused to handle the Telander gift, did as Wiatt had asked and gave him copies of letters and documents about the project dating between June 1, 1992, and July 1993.

On July 15, 1993, the day Wakefield was advised of her impending termination, all foundation development officers were asked to come to the foundation offices at 4 p.m. for a meeting. With everyone gathered, Kenneth W. Durham, Wakefield's immediate supervisor and one of the two men who approved the Telander gift, announced that in one hour he was no longer going to be employed there.

According to Wakefield's complaint, Durham offered no explanation; he said that he wasn't in bad health, didn't have another job, and didn't have specific plans for the future. He said that maybe he and his wife would do more traveling. He thanked everyone who worked with him for the past 10 years.

Wakefield's Vendetta

According to the legal complaint, on August 18, 1993, William S. Thornton Sr., one of Wakefield's fund-raising volunteers, called Wakefield and accused her of being "consumed with the Bockris situation, with a vendetta against Bockris and an emotionally driven vengeance against him."

In her defense, Wakefield told Thornton that she was "committed to protecting her university against incompetence of administrators [who were] too cowardly to take corrective action to problems caused by actions of one single professor."

On Aug. 20, 1993, former employee Wakefield met again with officer Wiatt. He told her that "the FBI was no longer interested in the situation, they had gone home and not found anything, and that the FBI was interested only in Telander, not Bockris." Wiatt told Wakefield that she was "stupid to think that Bockris had anything to do with her termination, [that she had] not done her job well and warned that she would only cause trouble by telling people about Bockris, ... and that she was paranoid and needed to get on with her life."

Wakefield rebuked Wiatt and told him that, instead, she intended to write a book about the drama and that she was planning to "go to the newspapers to expose the situation for correction." She told Wiatt that she would also go to the FBI. Wiatt reminded her that the FBI had been there and left and that the FBI would think Wakefield was "some kind of a nut" and would "throw her out of their offices." Wiatt asked her what she was going to do next. "No longer be a victim," Wakefield said.

With a Vengeance

Three months later, on Sunday, Nov. 14, 1993, the local newspaper, the *Bryan-College Station Eagle*, ran two stories on the front page written by Joe Toland: "Investigators Probe Gift to A&M" and "Mercury Into Gold? Did an Investor's Donation Fund a Fairy Story at A&M?"

The stories reported that the SEC was investigating Telander's $200,000 gift to Texas A&M. The main source for the news story was an anonymous "high-level administrator," which incorrectly intimated that the source was a current Texas A&M employee. (Toland identified Wakefield as the source five days later in a follow-up story.)

But the March 1993 SEC inquiry was old news. The real news was Bockris' transmutation work, which, for more than a year and a half, had remained out of the news. As expected, the mixture of a federal investigation with the odor of medieval alchemy caused a stir. University administrators and professors began damage control.

Hall deflected blame and told a reporter that Telander's gift had been accepted in April 1992 by Durham. On Monday, twenty-four hours after the story published in the *Eagle,* Texas A&M interim president E. Dean

Gage asked Robert A. Kennedy, the vice president for research and graduate studies, to begin an inquiry into Bockris' research. Bockris faxed a letter to the newspaper defending himself, but the editors waited until Thursday to publish it.

On Tuesday, the *Eagle* ran a second story on the dramatic news: "Bockris' Research Involved Turning Mercury Into Gold," and within hours "Mercury to Gold" stories appeared in newspapers throughout Texas and California.

On Wednesday, Jeffrey Weiss, reporting for the *Dallas Morning News,* spoke with W. Michael Kemp, the dean of the College of Science, who was the other man who had approved the Philadelphia Project. "I'm at a loss, frankly, to understand how this came about," Kemp said. "This is not the sort of thing that should be happening at a university." Kemp told Weiss that he would not have approved the project "had Telander been honest about what the money was for."

Weiss was quick to highlight the irony: "Mr. Champion is in jail in Phoenix on unrelated criminal fraud charges. The man who contends that he can turn mercury into gold is represented by a court-appointed lawyer because he can't afford to pay one."

Bockris maintained his scientific perspective. "I don't believe in any way that they can produce commercial, large amounts of material," Bockris said. "But I think they have stumbled upon reactions that are not consistent with normal physics."

Michael Hall, the chairman of the Chemistry Department, praised Bockris in the *Dallas Morning News* story. "To Bockris' credit," Hall said, "my impression is the reason Champion and Telander left here was they could not get Bockris to support them at the level that they wanted. My impression, in hindsight, is that Telander and Champion came here to use Bockris and Texas A&M in some kind of scheme."

On Thursday, Nov. 18, the *Eagle* published the letter from Bockris. It was typical of Bockris: reasoned, scientific thinking. He wrote that the headline of their Sunday story "implied ridicule" for his research project.

"The legend of alchemy," Bockris wrote, "was replaced by reality more than 50 years ago. For much of the century, high-energy physics has made it possible to convert one element into another."

Bockris explained that a new avenue of nuclear research had been going on all around the world for the last four years, showing a variety of unexpected results and nuclear reactions. He wrote that the work potentially had widespread societal benefits and applications, including the remediation of nuclear waste.

Bockris made it clear that Telander had exaggerated the weak support that his experiments had provided and that "there was no basis whatsoever from Texas A&M's work to support synthetic gold at commercial levels."

With administrators and university professors scrambling to recover from the public relations disaster, and Bockris' reputation once again under attack, Wakefield had succeeded. As she had told officer Wiatt three months earlier, she was no longer going to be a victim.

Bockris Faces Second Inquiry Panel

If Wakefield had wanted revenge on Bockris, she certainly got it.

On Saturday, Nov. 20, 1993, the *Dallas Morning News* headline read "A&M Professor Faces Audit, Is Urged to Quit."

Bockris was immediately confronted with a two-pronged assault. Although the university had initiated an inquiry that Monday, Dean W. Michael Kemp, who, along with Durham, had accepted the $200,000 from Telander, turned on Bockris. Kemp made formal allegations to the administration that Bockris was guilty of misconduct:

> Bockris knowingly and intentionally conspired with Telander to obscure the true intention of the research related to the Philadelphia Project and later, by the omission of action, allowed his collaborators Telander and Champion to exaggerate the results of his work on this project. This exaggeration resulted in the reputation of Texas A&M University being used to legitimize and to lend validity to the project as a means of encouraging investment in the Philadelphia Project. (Texas A&M Committee of Inquiry Report, 1994)

Please Leave!

The second assault facing Bockris, according to the *Eagle* on Nov. 19 and the *Dallas Morning News* on Nov. 20, 1993, was a demand he received from 11 of the 38 full professors in the Chemistry Department to leave the university: "We strongly feel that you should sever your relationship with the university immediately." The *Eagle* said the letter was dated Nov. 14, the same day that paper broke the news. The person who drafted the letter hadn't wasted any time.

John Fackler, a former dean of the School of Science, was one of the chemistry professors who signed the letter asking Bockris to resign from the university, Weiss reported.

"Bockris accepted money to explore a research project that doesn't make sense," Fackler said Friday. "As scientists, we presumably have to police ourselves and only conduct research that makes sense."

It was a wholly regressive attitude, one guaranteed to end the expansion of scientific knowledge. Fackler's statement did, however, reflect his unease with the disturbing implication of a paradigm shift revealed by nuclear reactions in room-temperature experiments.

But the reporter understood the implication. "The science did make sense to Bockris," Weiss wrote, "and it would mean that much of what is now accepted in atomic theory must be fundamentally flawed."

Two days later, on Nov. 22, the administration appointed an inquiry committee. One of the members was Joseph Natowitz, the former head of the Chemistry Department, who later became the director of the Texas A&M Cyclotron Institute.

Violent Outrage

On Tuesday Nov. 23, the *Eagle* editorial board published a thoughtful article about Bockris titled "Scientists Always Push the Envelope." It is reproduced here it in its entirety:

> In 1633, Galileo Galilei, the Italian astronomer and physicist, was called before the Inquisition and forced

publicly to recant his belief that the earth revolves around the sun. As he was leaving the Inquisition chambers, though, under his breath, Galileo whispered, "But it does move!"

Today, it seems hard to believe that at one time scientists such as Galileo and, before him, Copernicus, would be ridiculed and reviled for stating that the sun and not the earth is the center of the universe. But at one time, scientists who held that the world was round were scorned, as were those who proved that gravity exists, that electricity can light the night, that man, indeed, could fly to the moon.

Scientists are forever pushing the line that separates the possible from the impossible, the probable from the improbable. That is what they do, and every advance we take for granted today is because some scientist was willing to bear the scorn of his colleagues and the public to try the untried.

John Bockris, the distinguished professor of chemistry at Texas A&M, has been held to public ridicule in recent days because of his experiments described by some people as trying to turn mercury into gold, that old alchemists' dream. Bockris describes his work somewhat differently, less flamboyantly, as research into the electrochemical transmutation of elements.

Everyone knows you can't turn mercury into gold, right? And yet — at one time everyone knew that voices couldn't travel by wire, let alone on beams of light, that talking pictures couldn't be broadcast through the air, that pneumonia and other dread diseases couldn't be cured by medicines.

This isn't the first time that Bockris has been involved in a public controversy over his work at A&M. Remember the cold fusion stir of five years ago? Or the hydrogen fuel from water a decade ago? Perhaps Bockris is so controversial because he dares to dream, because he dares to push the envelope of known science.

We aren't qualified to judge the results of Bockris'

experimentation; few people are. But we do know that, because of John Bockris and dreamers like him, the world is a far more advanced place today than it was yesterday and all the yesterdays before.

The newspaper received and published a vitriolic response from Texas A&M "distinguished professor" of chemistry Frank Albert Cotton (1930-2007). It is reproduced here in its entirety:

> The *Eagle* editorial heroizing J. O'M Bockris reveals that the editorial staff of the *Eagle* is not only abysmally ignorant but intellectually and morally bankrupt.
>
> To liken a man who has perpetrated three of the most egregious examples of sick science ever seen with Galileo is an act of such stunning inanity as to deserve the top position on a list of the all-time most misguided editorials.
>
> The facts are that Bockris has repeatedly violated every known canon of responsible scientific research. He has published unreproducible trash in the newspapers instead of the scientific journals, made claims that are completely irrational, and has now capped a disgraceful career by consorting with known stock swindlers to support his sleazy so-called research.
>
> Prior to the present fiasco, he has spent thousands of the taxpayers' money on his irresponsible pseudo-research. I wonder if the taxpayers who read your newspaper will be pleased to hear your high opinion of this "benefactor" of theirs? The one bright spot for him is, no doubt, that there are ignoramuses like your Editorial Board who are so pitifully stupid as to liken him to Galileo. Shame on you all. Your editorial is a disservice to your readers and a new low in journalism.

A Lump of "Coal"

Cotton did not stop there. He realized that Bockris was not going to leave the university voluntarily, as requested by the Nov. 14 letter from his colleagues in the Chemistry Department. Cotton was not content to leave the Bockris matter in the hands of the deliberating university inquiry committee.

On Dec. 21, 1993, Bockris received another attack, led by Cotton, from professors at the university. Cotton had collected signatures from 23 of 28 distinguished professors at Texas A&M on a petition demanding the revocation of Bockris' title of Distinguished Professor:

> A REQUEST: Professor J. O'M. Bockris' activities since 1989, (the inception of the "cold fusion" imbroglio) and particularly recent allegations that he lent his name and that of our university to a fraudulent scheme to promote a bogus engineering enterprise, has brought this university into disrepute. Note that on page 6 of the Policies and Procedures Regarding Distinguished Professor Appointments "it is stated that "the Distinguished Professors ... bring honor and recognition to the university."
>
> Instead, we believe that Bockris' recent activities have made the terms "Texas A&M" and "Aggie" objects of derisive laughter throughout the world among scientists and engineers, not to mention a large segment of the lay public.
>
> The "alchemy" caper is, everywhere, a sure trigger for sniggering at our university. And so it should be. For a trained scientist to claim, or support anyone else's claim, to have transmuted elements is difficult for us to believe and is no more acceptable than to claim to have invented a gravity shield, revived the dead, or to be mining green cheese on the moon. We believe it is sheer nonsense, and, in our opinion, could not have been done innocently by one with a lifetime of experience in one of the physical sciences.
>
> In view of the above considerations, we the undersigned

Distinguished Professors of Texas A&M University, hereby request the provost to take steps to revoke the title of Distinguished Professor now carried by John O'M. Bockris. We do this because of our belief that Bockris' alleged disregard of the accepted standards of scholarly and professional behavior has brought great embarrassment upon this university and his colleagues. In our opinion, he no longer merits the title of Distinguished Professor.

Beneath Cotton's vitriol, the letter shows that the basis of his outrage was not the fact that Bockris had claimed to have transmuted elements — for nuclear transmutations in large, high-energy physics devices had been commonplace for decades. They even took place at the university's own cyclotron. Rather, the root cause of Cotton's vitriol, though he was too blinded by his own rage to see it, came from the fact that Bockris was challenging the scientific paradigm, the paradigm that held that nuclear reactions could take place only through high-energy physics.

Meanwhile, the committee began its inquiry. It included an oral hearing, somewhat like a trial, in which Bockris explained and defended his research. Concurrently, Bockris sent out letters to his colleagues worldwide, let them know that he was under attack, and asked for their letters of support. One such letter, addressed to the inquiry committee chairman, came from Eugene Mallove, the editor of *Infinite Energy* magazine. Here is an excerpt:

If Texas A&M University decides to violate the most basic tenets of academic freedom by granting the petitioners any satisfaction, it will be committing a serious injustice. Action against the legitimate scientific investigations by professor Bockris would have horrible repercussions for Texas A&M, particularly because these low-energy transmutations are now being increasingly verified in various cold fusion experiments around the world. For the administration of a great university to have acted in haste against the expansion of scientific knowledge would be an indelible black mark on that university.

Bockris also received a sympathetic letter from Peter Hagelstein, who wrote that the "last thing in the world that the field needed was to be associated with alchemy." Even the tiniest association, he thought, would be "seized by critics of cold fusion to hammer yet another final nail into the coffin they have constructed for it." Hagelstein then offered sardonic but sympathetic words for Bockris:

> I assume that by the time you receive this letter your colleagues will have stoked the kindling at the base of a very large stake with your name on it, and will ultimately succeed, in this way, in cleansing your soul of the scientific errors that, in their view, you have made. And after you, others.

Hagelstein conceded, however, that much of the work that had been reported in "cold fusion" could indeed be classified under one of the definitions of alchemy. Hagelstein thought that the radiation that immediately followed the "thermal" experiments was important:

> Whereas I would be worried that someone might be motivated to spike a sample with gold for one reason or another, it is a very different matter to spike a sample with a radioactive impurity that has an 18-hour half-life, especially if you have tested for its presence beforehand.
>
> Should you succeed in demonstrating chemically induced radioactivity with mercury, I have to say that I am interested, even if it happens to come from replications of "experiments" that were first carried out 500 years ago.
>
> Good luck in the trials ahead that are facing you. Judging by historical precedent, I suspect that you will, in fact, get burned at the stake, in spite of my input or input from anybody else.

Before the inquiry committee came to its conclusion, Bockris received an unexpected package in the mail. It contained feces.

Exonerated, for the Second Time

On Jan. 31, 1994, Bockris received better news. For the second time at Texas A&M, the university inquiry committee exonerated Bockris from all allegations of research misconduct. (Bockris faced his first inquiry committee in 1990 as a result of the insinuations, published by Gary Taubes in *Science* magazine, that Bockris and his graduate student had faked their tritium results.)

The second inquiry committee sent its conclusions to Provost Robert A. Kennedy. According to the committee's report, it examined more than 1,000 pages of material, audio recordings, and financial documents. The committee interviewed Kemp and Durham, the two men who had accepted the money from Telander. It interviewed Bockris and his direct supervisor, Michael Hall. It also interviewed Lane Stephenson, the director of public information for the university.

The committee wanted to interview Dawn Wakefield, the woman who had made the accusations in the newspapers and started the whole mess. She declined to be interviewed.

The committee found that the administration had been well-informed of the nature and intent of Bockris' research, including the potential production of precious metals. The committee learned that Bockris had told Telander in September 1992 that he should not use the research results at the university to make exaggerated announcements that implied any sort of commercial endorsement.

By a unanimous decision, the committee exonerated Bockris of all of Kemp's allegations. Instead, the committee "recommended that members of the Texas A&M University community allow the process of experimentation and peer review of published data to resolve any scientific issues."

Financial Audit

Although Bockris was exonerated of the charges of misconduct, the university was conducting two other investigations. According to Jeffrey Weiss, reporting for the *Dallas Morning News* on Feb. 4, 1994, the university was reviewing whether any rules were violated in how Bockris spent the research money and how the administrators accepted the money.

A month later, Weiss reported that Bockris, on the advice of his lawyer, Gaines West, did not fully cooperate with the audit. According to Weiss, West questioned the legality of the audit.

"If the university was not proceeding with a legal audit," West said, "then nobody should cooperate or supply information."

Two Days; Two Pleas

On Feb. 8, 1994, Thomas M. Brown, an assistant United States attorney in the major fraud section, announced that Telander pleaded guilty to four counts of the fraudulent sale of securities and two counts of failing to file federal income tax returns. Telander admitted that, in 1992, he received more than $500,000 in reportable adjusted gross income.

On Feb. 9, 1994, the Texas A&M provost, Benton Cocanougher, received a letter from "The Silent Majority of the Chemistry Department." Its authors voiced their support for Bockris but felt too intimidated to identify themselves. They wrote that they had been subjected to "extreme coercion and intimidation," tactics for which Frank Cotton was well-known.

"We do hope that there is enough backbone in the university administration to stand up to the attacks that inevitably will continue from this one individual, if not at Bockris, then at anyone else that gets in his way," the chemistry professors wrote.

Request to the FBI

On Feb. 16, 1994, Gaines West, Bockris' attorney, wrote to the FBI. Bockris had been participating in the FBI's Cooperating Witness Program and assisting with its investigation of Telander, his associate Roger Briggs, and Champion. The FBI had asked Bockris to stop communicating with Telander. West told the FBI that Bockris agreed to continue providing information to the FBI about Telander, Champion, Briggs and the Philadelphia Project but that Bockris was concerned.

"Dr. Bockris," West wrote, "feels at this time that his failure to respond to the repeated letters from Mr. Telander will in some way alert Telander that Bockris has distanced himself from Telander, thus inciting him and/or Champion to violence."

As if that were not enough, the university went after Bockris again.

On Trial, for the Third Time

The university administration did not stand up to Cotton. In the first week of July 1994, Texas newspapers reported that Provost Benton Cocanougher was forming a 10-person task force to begin yet a third investigation of Bockris. According to the news reports, Texas A&M auditors had found alleged financial violations within the Philadelphia Project, and the new committee was to consider possible "personnel changes."

Bockris had a hunch that a group of professors in the inorganic division of the Chemistry Department, who were unsatisfied with his exoneration, had something to do with it.

"They felt that it had not been broad enough," Bockris wrote "and my true sin was that I had not exposed my work to peer review in an established journal. To attempt this would, of course, have been useless because no one would have agreed that the results were possible." (Bockris, 2004)

When Bockris asked about the nature of the new inquiry, the Texas A&M attorneys told him nothing. Nine months later, he had heard that the committee had had 11 meetings; he knew nothing else. The only hint, passed on to him from one of the administrators, was, "Tell Bockris he will not be the only one." Bockris explained what he did next:

> In exasperation, I returned to the lawyer who had helped me in the first investigation, and we composed a detailed letter to the American Association of University Professors (AAUP), the essence of which was to present the

evidence that the university was treating me in a capricious and unfair way. I asked for the university's treatment of me to be the subject of an AAUP investigation.

Universities fear the AAUP, which can blackball them. Texas A&M had been the subject of such a blackball for its ill-treatment of a professor at an earlier time. I do not know if this factor influenced the decisions of the *ad hoc* committee.

At any rate, two months later, on May 25, 1995, I received a letter signed by the provost informing me that the eleven-month re-investigation had concluded that no action of mine had been contrary to the rules of the Policy and Procedures Manual of the University.

My statement given here — of what was done by a university to a faculty member who published research results inconsistent with the existing paradigm of the time — does not tell of the anguish involved; of the rejection by one's peers; of the effect upon family life; of the isolation and rejection.

My wife was a victim of the Nazi occupation of Austria and a refugee who reached America in a British liner convoyed by warships. She has told me that, during the years she lived in Vienna under the Nazis, she never felt so rejected and threatened as in College Station, Texas, 1992-1995. (Bockris, 2000)

While Texas A&M University was conducting its third investigation of Bockris for what some of his peers thought was a fraud, scientists around the world were observing positive results in their experiments.

CHAPTER 13

He Moved the Science Forward

In 1994, while John Bockris defended himself against a third academic inquiry into his controversial research, scientists in Italy reported that they had observed very large amounts of excess heat from experiments that used ordinary hydrogen gas rather than heavy water.

In a seminar on Feb. 14, 1994, professors in the Physics Department at the University of Siena, Italy, announced the results of their 1992 and 1993 experiments. The Siena scientists inserted a specially prepared nickel rod, 5 mm diameter by 90 mm long, into a cylindrical stainless steel chamber, 50 mm diameter by 100 mm long, and filled the chamber with hydrogen. The nickel rod was surrounded by a ceramic spindle, on which a platinum wire was wound to provide resistance heating. The heater brought the cell up to a few hundred degrees Celsius, which triggered the reaction. Then the scientists reduced power to the heater.

One valve connected the chamber to a vacuum pump, and another valve went to a bottle of either normal hydrogen gas or deuterium gas. The scientists prepared for the experiment by pumping air out of the cylinder, then gradually introducing hydrogen. Initially, some of the hydrogen was absorbed, or loaded, into the nickel, and this occurred over the course of several cycles until the hydrogen was fully loaded into the nickel.

The scientists measured an average of 44 Watts of excess heat for 24 days, corresponding to about 90 megajoules of excess energy — a record level of excess heat in a room-temperature benchtop experiment.

The scientists calculated that the amount of energy produced was at least three orders of magnitude larger than any known chemical reaction involving hydrogen and nickel. They failed to detect any penetrating radiation, such as neutrons or highly energetic gamma rays, above the background level during the experiment.

Biophysicist Francesco Piantelli, the inventor and lead scientist of the group, submitted a paper to *Il Nuovo Cimento*, the journal of the Italian Physics Society, for peer review. It published on Jan. 12, 1994. Eugene Mallove, the editor of *Cold Fusion* magazine, the predecessor of *Infinite Energy* magazine, reported the news in May 1994, in an article titled "An Italian Cold Fusion Hot Potato":

> [This research], if substantiated by others, may soon revolutionize all of cold fusion. It could make excess energy much easier to generate than heretofore possible. In December 1989, serendipity struck at the University of Siena, which is in an ancient and beautiful city. Above the stone streets of Siena in a laboratory that specializes in biomedical applications of physics, Professor Piantelli was trying to measure the charge on an organic molecule called a "ganglioside." He was working at an ultra-cold, cryogenic temperature more than 70° C below zero (near 200° K), and magnetic fields were involved in the apparatus employed.
>
> The sample of biological material had been tagged with deuterium. ... Unexpectedly, the cooling apparatus was having difficulty maintaining the low temperature necessary to carry out the measurement. It seemed that there was a mysterious source of heat production coming from the sample, heat that Piantelli could not account for in any way.
>
> The organic sample was resting on a piece of nickel, an element whose crystal structure bears some resemblance to that of palladium and which has figured prominently in cold fusion experiments in ordinary water. (Nickel is element 28, and palladium is element 46, but they both are in the same column in the periodic table of elements.)
>
> A non-observant scientist might have dismissed the apparent heat generation and a possible link to claims of excess heat associated with palladium-heavy-water cells. But

Piantelli and his colleagues Sergio Focardi and Roberto Habel, who soon joined him in the scientific detective work, were up to the task. ...

Their goal was to design an experiment that would demonstrate, on a larger scale and in another way, the heat anomaly that Piantelli had seen in late 1989.

By the end of 1992, they had their equipment ready. Their first major success was achieved in the spring of 1993, a few tens of Watts of excess heat. (Mallove, 1994)

Despite Mallove's enthusiasm, researchers were not eager to confirm the Piantelli group's excess-heat result. The conundrum was that the reaction using ordinary hydrogen contradicted the "cold fusion" idea, which presumed deuterium-deuterium (D+D) fusion.

The Piantelli group's gas experiment presented other issues. It far outperformed the excess-heat levels (a few hundred milliwatts) observed in typical palladium-deuterium electrolytic experiments.

As with the light-water experiments reported in the past two years, if such strong results could be obtained using inexpensive nickel and hydrogen, why should money be spent on costly palladium and heavy-water supplies? Also, the fundamental design of the Piantelli group's experiment — an enclosed, solid-state, gas-filled device — seemed significantly closer to any inevitable commercial device than a potentially messy liquid-based electrolytic experiment.

Last, the gas design had the potential for a much wider range of technological applications because its working operating temperature was in the hundreds of degrees C. Electrolytic designs are limited by the boiling point of the liquid, typically water-based solutions. There is no apparent upper temperature limit in the gas system other than the melting point of the metal.

Later that year, the Piantelli group completed two more experiments. The group reported the production of 18 and 72 Watts of excess heat – 600 and 900 MJ of integrated energy – over 319 and 278 days, respectively. (Campari, 2004)

For the scientists focused on proving Fleischmann and Pons' D+D fusion hypothesis correct, it was more convenient to assume that the Piantelli group's work was a mistake, and ignore it. Which they did.

LENRS-1 at Texas A&M

As soon as Bockris was exonerated by the third university inquiry committee, he used his academic freedom to organize what would be the first of two "Low-Energy Nuclear Reaction" scientific meetings on transmutations. Texas A&M professor Guang H. Lin organized the workshop with him. This meeting — the first of its kind specifically organized to discuss heavy-element LENR transmutations — was a milestone in the history of the field.

"I went to my boss," Bockris wrote, "who at that time was Professor Emile Schweikert. He readily agreed to an international meeting on the subject being held in the Chemistry Department at Texas A&M. I'm glad to say that, when he was later criticized for allowing the meeting, he stuck to the truth and agreed that he had indeed sanctioned the meeting." (Bockris, 2004)

Bockris and Lin quietly invited specific researchers they hoped would attend and avoided advertising the meeting publicly. The June 13, 1995, announcement reflected their awareness of the controversial nature of the meeting:

> Meeting is private. No press. No cameras. No summaries to be published. If approached by press, we believe it best to speak about inorganic reactions, investigations of anomalies, etc. Unwise to talk publicly of possibilities of transmutation reactions as yet.

The meeting took place on June 19, 1995, in Room 2122 of the Chemistry Building, almost without incident. The invitation called it a "Meeting on Low-Temperature Nuclear Change." In the proceedings, published by Hal Fox, it was called the "Low-Energy Nuclear Reactions Conference." Bockris appears to have been the first person to use that term publicly although, in 1992, Robert Bush used the term "low-energy nuclear transmutations."

The workshop helped to establish the legitimacy of heavy-element transmutations and helped this subset of scientists in the field learn about one another's work. It drew researchers from the U.S., Japan, Canada, Italy and the Ukraine.

The only disruption, according to a 2004 memoir Bockris wrote, was that Frank Cotton, Bockris' arch nemesis in the Chemistry Department, showed up very angry and yelled at the participants that they were "kooks." (Bockris, 2004)

Cotton went to the *Bryan-College Station Eagle* later that day and spoke with reporter Shelly Smithson." [The meeting perpetuated] the black eye this university has in the scientific community," Cotton said. "I am outraged, and so are other professors that this sort of nonsense is still going on here."

Cotton used the same descriptive language two years later. John Kirsch, reporting for the *Dallas Morning News*, on April 15, 1997, wrote that, when Bockris was involved in a planned alternative energy seminar, Cotton forced the meeting off the campus. Cotton told Kirsch that the speakers for the planned meeting were "all kooks and charlatans."

Peter Hagelstein, whom Fox described as "brilliant and persistent," was there and described his neutron-hopping theory. (Fox, 1995) Significantly, Joseph Natowitz, the head of the Texas A&M Cyclotron Institute, attended the meeting — he had been one of the members of the second inquiry committee appointed to investigate Bockris. (Fox, 1995) Hagelstein's paper was not published in the proceedings. He may have been wary of additional aggravation from his MIT colleagues. Fox, taking a broad view, introduced the proceedings by praising the adventurous scientists:

> We especially acknowledge and congratulate all of those true scientists and experimenters who have been responsible for forging ahead with investigations of new anomalies when assured by lesser lights that what they were trying to do was scientifically impossible. This is the same

professional category of pioneers to which Faraday, the Wright brothers, the Curies, Goddard, Steinmetz, Tesla, Kervran, and (more recently) Pons and Fleischmann belong. More important than the winners of Nobel prizes amidst the acclaims of their peers, are those who, against the "better judgment" of their peers, persevere in the discoveries of new truths and change both scientific knowledge and the way in which we were abundantly live, work and play. (Fox, 1996)

The Strange Story of Dr. Wolf

Many experimentalists who spoke at the workshop had seen some evidence of heavy-element transmutations, or anomalous isotopic changes and/or excess heat in light-water systems. Two of the experimental reports were not confirmed by other researchers. The first was that of Roberto Monti, who claimed heavy-element transmutations using a formula he declined to disclose. The second report was given by nuclear chemist Thomas Passell (b. 1929), who at the time was a program manager for the Electric Power Research Institute.

Passell reported results from experiments performed by Kevin Wolf, a nuclear chemist and a professor at Texas A&M. Bockris described Wolf's strange story in a 2004 paper.

In late 1992, Bockris had heard a rumor that Wolf had observed new radioactive isotopes in an electrode that he had used in his experiments. There are many unsolved mysteries about this story, the least of which is why Wolf performed any more electrolysis experiments in 1992.

In early 1990, after reporting for the previous year that he had found tritium in his experiments, Wolf did an about-face just before Gary Taubes' article was published in *Science*. Wolf depicted a scenario, without offering any evidence, that the tritium (a radioactive isotope) was the result of contamination, that his palladium cathode had come from a nuclear reactor, and that, the reactor technicians had allowed radioactive palladium to enter the recycled metals market. In late 1990, Wolf became hostile to other researchers who had reported positive results from their LENR experiments. (Vol. 2, *Fusion Fiasco*)

But in 1992, Wolf's newly transmuted elements were radioactive. This was odd because nobody else in the field had ever reported appreciable amounts of unstable isotopes. Wolf never published or reported his data, Passell told me when I interviewed him in 2004. Passell didn't understand Wolf's reluctance, and in 1995, Passell reported Wolf's data, without his consent, at ICCF-5. (Passell, 1995) Passell thought the data was important and, as Wolf's program manager at EPRI, felt entitled to report it.

In 2001 or 2002, Passell had the opportunity to resolve the mystery, at least for himself. Here's what he told me:

> Suffice it to say, we proved that he had faked the results. Well, let's put it this way: We proved that you could get the same results by using the cyclotron that was up on the first floor while Wolf was in the basement.
>
> I proved that you could get the same results by irradiating palladium with 80 MeV protons. As much as I hoped that it was true, I have to admit that it looks like the same spectrum. The money for this project came out of the secret world so I'm not at liberty to publish this fact out in the open.
>
> Some of Wolf's friends couldn't believe that he would do such a thing and said that someone on the first floor must have played a trick on him while he was running his experiment in the basement.

Wolf died in 1997 at age 55. Neither the news reports nor his obituary in the local newspaper provided the cause of death.

Artificial Distinction

In a memoir Bockris wrote in 2011, he recalled the highlights of his 1995 LENR workshop in a letter he wrote to Joe Champion:

The major thing, in my opinion, on the good side, which came out of all the struggles we went through in those years, was that a well-known nuclear engineer, [professor] George Miley (b. 1933) of the University of Illinois, became convinced of the field of transmutation, not by anything we had done in respect to gold, but because he had his other irons in the fire and it certainly gave rise to convincing research. (Bockris, 2011)

Bockris wrote more about Miley in a 2004 paper:

Miley, who was the editor of *Fusion Technology*, had [initially] been a doubting spectator on the [edge] of the work on transmutation. I had a number of discussions with him, and his attitude was that he was interested but he could not publish transmutation work because he would be soon be out of his job as editor of the journal.

Gradually, however, Miley's attitude changed, and of course, as is now well-known, he became a forefront person in research on transmutation reactions. He has published several confirmatory papers himself in which his own ability as a nuclear engineer has come to the fore. He has utilized more-advanced methods, particularly determining the isotope abundance frequency of the new material; he is now a leader of the new world of transmutation. (Bockris, 2004)

By 1995, the two philosophical camps continued to diverge. Some researchers thought that the underlying processes were caused by neutron-based transmutations. The majority of researchers, in the fusion camp, assumed that light elements, such as helium and tritium, were products of deuterium fusion. Both camps knew that the heavy-element transmutations were unlikely to be explained by fusion.

Therefore, members of the fusion camp created a false distinction. They assumed and asserted that the production of elements heavier than tritium resulted from an entirely different process. They called experiments which produced changes in heavier elements "transmutation experiments." In fact, all of the experimental results were better explained by nuclear transmutations, or, more accurately, nucleosynthetic processes.

Grievance Against Texas A&M

In September 1995, a few months after Bockris and Lin's workshop, Bockris filed an administrative grievance against Texas A&M University. He described many of the incidents reported in this book, but he also revealed some unusual aspects of the consequences of his science research:

> It is difficult to describe the extraordinary and widespread effects of the long investigations by the university, lasting from December 1993 to May 1995. The suspicions voiced by Dean Kemp, and repeated in the audit, upon which judgment was given on Jan. 2, 1994, still cling.
>
> A number of happenings are unexplained. Theft of my 1993 tax files has occurred from a building in the woods outside my house. In September 1995, two more files have been removed from this building. During 1994, there were signs that this locked building, which contains many of my work files, was repeatedly entered.
>
> My office telephone has been used to place bets on horses and for calls to sex-talk agencies. My telephone logs have been sequestered by the press and individual calls investigated. My office at Texas A&M has been frequently entered, though locked, and numerous files and correspondence stolen. ... These are sad things to relate when one remembers their origin: the reporting of facts which do not fit the present paradigm in chemistry.

His grievance continued with a list of the personal consequences he endured: damage to his reputation, reduction in the number of invitations to international meetings, significant loss of grant income, and the postponement of a scheduled award. "Anyone who has read this document," Bockris wrote, "will hardly be surprised to know that the stress and tension of these 20 months has aged Mrs. Bockris significantly."

Lily Bockris filed her own grievance against the university. Although the Bockrises felt entitled to redress from the university "for the needless irreversible damage," their persecutors were not necessarily the whole university but a few faculty members and one former employee of the university.

LENRS-2 at Texas A&M

On Sept. 13 and 14, 1996, Bockris defied his adversaries again and chaired a second LENR meeting. As before, professor Lin helped to organize the conference. It was called the "Second Conference on Low-Energy Nuclear Reactions" and established the first official use of the term. In the proceedings, it was called the Second International Low-Energy Nuclear Reactions Conference. This time, Miley co-chaired the meeting. It was a bold move for a scientist who served as the editor of a prominent nuclear fusion journal. Bockris summarized the meeting in a 2004 memoir:

> [Professor Guang] Lin and I thought that it might be a good idea to hold a second meeting on transmutation, particularly as interest in the subject seemed to be growing. I once more approached the department head, Professor Emile Schweikert, but by now the rules had changed.
>
> Our colleagues in chemistry had decided that requests for such a meeting should pass through a committee which consisted of 12 members of the department.
>
> I reproduced a comprehensive review paper of cold fusion by Ed Storms which contained more than 100 references and distributed a copy to each committee member before the meeting. I made a brief presentation saying that this was new work and that it was going on around the

world, there had been a number of confirmations of new nuclear reactions. I hoped that they would allow the new science to be heard.

The vote was 12 to 0 against. I called one of the people on the committee I had known and asked him what the discussion had been about after I had left the room. He said that everyone on the committee knew that it was impossible for nuclei to be changed except under conditions of very high energy exchange. Hence, the members had concluded that the work that I wanted to have presented was either a joke or a fraud.

Because of the assault made by Professor Cotton and his colleagues on the 1995 meeting, we thought that a more violent one might be made in this meeting and therefore hired a deputy from the Police Department to be present outside the door of the meeting in order to quell any attempt by members of the Chemistry Department to suppress the presentation of new ideas by violence. (Bockris, 2004)

Because Bockris was not granted permission to have the meeting on campus, he held it at a nearby Holiday Inn hotel. Bockris reported the results of the workshop to Schweikert, the head of the Chemistry Department. Researchers presented 23 papers, and 42 people attended. Some of the papers," Bockris wrote, "were as good as any in the field. Three or four should not have been accepted."

Miley and two researchers from Hokkaido University, applied physicist Tadahiko Mizuno (b. 1945) and electrochemist Tadayoshi Ohmori, each presented results showing abnormal isotopic abundances and synthesis of heavy elements.

John Dash presented evidence of excess heat correlated with the formation of new elements on the surface of his cathodes. Physicist Thomas Claytor, at the Los Alamos National Laboratory, reported his continuing five-year success in producing tritium. (Claytor, 1996)

In his letter to Schweikert, Bockris summarized his interpretation of the significance of the meeting:

Twelve papers gave evidence that nuclear reactions take place in solid lattices in the cold. If this contention obtains still further confirmation, I suggest it constitutes a discovery of magnitude comparable with that of atomic disintegration with high-energy particles (Rutherford, 1919) and nuclear fission by neutron bombardment (Hahn and Meitner, 1939). It opens a new area of great potential [to] radically change the ideas of nuclear stability. (Bockris, 1996)

His Place in LENR History

Criticizing Bockris in hindsight is easy. He went public with his claims of heavy-element transmutations without characterizing his results as thoroughly as other people did. He went public with far fewer successful experiments than other people would have required. To his credit, Bockris was the first to acknowledge the more-advanced methods of Miley, a nuclear engineer, and Mizuno, a physicist. "The work by these nuclear physicists," Bockris wrote, "was better than the work we had done earlier. They did isotopic abundance analysis on the new materials, which we did not do." (Bockris, 2004)

Bockris was able to see his own circumstances and how they fit with the broadest of scientific conflicts. Here are some of his thoughts on various topics:

Fear

The common image that research scientists live for "The New" is true so long as the new concerns the advancing frontier and not a revision of earlier principles.

If change in fundamentals is proposed — experimental results which demand that there is change — a sense of unease develops among scientists. To really understand what is happening takes a long time. It is easier to suppose there must have been experimental error or even fraud.

For those who blithely cross the barrier, ridicule and then odium will be applied. Hence, there is a justified fear in putting out an idea which flies in the face of the ruling theory. It is a paralyzed fear, often, and leads to an attack with emotional intensity in the facts presented, hoping they will go away. Any discussion of these strange (and very uncertain) results must be done with the understanding that scientists may not behave with objective calm in considering them. (Bockris, 1993) [He wrote this before or just as his second academic inquiry began.]

Academic Freedom

Historically, big discoveries have been made by following up experiments anomalous to the theory of the time. At a first-class research university, the aim of the professors in the sciences, as I see it, should not be primarily to perfect the knowledge of the time by the publication of papers consistent with the known theory but to carry out researches which have the aim of finding anomalies in the existing paradigm. (Bockris, 2000)

Paradigm

Scientists within their time have always thought that what they know is "the final truth." They do not understand the temporary nature of the theoretical construct. Hence, a man whose proposal does not simply add to and support the theoretical constructs of the time will not be approved but, in fact, ridiculed and rejected. His funding will rapidly sink. (Bockris, 2000)

Telescopes

During the long period (about one month) in which [an] electrode sporadically evolved tritium, I invited several colleagues in the Department of Chemistry to come see a result so unexpected, so anomalous within the chemistry of 1991. (Chien, 1992)

One said that his son had a birthday party that day so

that he could not come to see the remarkable experiment, and the other that he was going to Germany to do some experiments in an institute there.

I asked two other professors, telling them that, by staying with the cell for one hour, they could themselves use the scintillation equipment in the next room and become convinced that the tritium concentration in the solution was increasing without the means of addition of tritiated water by [according to the false rumors] stealthy secret nightly visits by graduate student Packham.

These two individuals also declined to come and see, which reminded me of Galileo and the cardinals who, in the 16th Century, would not look through his telescope, because it showed an irregular mountainous moon, although at this time, the moon was supposed to be the Ptolemaic view of the Church, to be Queen of Heaven and "perfect." (Bockris, 1999)

Bockris retired — voluntarily — in June 1997. In response to an inquiry I sent to the university in 2004, H. Joe Newton, the dean of Science, wrote that Bockris was not given emeritus status. "A vote was taken of the chemistry faculty on whether he should be named emeritus," Newton, wrote, "and in fact the vote was negative."

Researcher Dennis Letts recognized Bockris' contribution to science. Letts wrote to me after the conference: "You must interview Bockris soon!" No other researchers had encouraged me to talk with Bockris. Usually, when his name came up in conversations, some researchers had unflattering things to say.

Eventually, I decided to go to Texas, and I interviewed John and Lily Bockris in their home on July 10, 2004. In September 2004, they moved to Gainesville, Florida, to be closer to their son. Lily died on Dec. 18, 2005, at 82, and John died on July 7, 2013, at 90, after a decline in health. The last letter I received from Bockris was in March 2013, and he seemed to have enjoyed a clear mind through his final months.

CHAPTER 14

Across the Periodic Table

Electrochemist John Bockris and his wife, Lily, paid dearly to defend his academic freedom at Texas A&M University: financially, politically and socially. Even so, at the conclusion of Bockris' third university inquiry on May 25, 1995, he began planning the first of two scientific meetings focused on low-energy nuclear transmutations. The researchers Bockris invited to the meetings reported substantial evidence for heavy-element transmutations.

The first meeting took place on June 19, 1995, on the Texas A&M campus. At least 31 participants from around the world attended; 17 of them presented new research. The participants knew that the data they were reporting were as controversial as the idea propounded, centuries before, that the Earth revolved around the sun.

Bockris' greatest contributions to this science were not his own discoveries but rather the opportunities for discourse he opened for scientists, such as the 1995 meeting, that allowed them to openly discuss their own low-energy nuclear transmutation research with one another.

Excess-Heat Correlation With Transmutation

Tadayoshi Ohmori and Michio Enyo, electrochemists at Hokkaido University, in Japan, reported an example of such data at the 1995 conference. They ran two sets of electrolysis experiment in light water, one using gold cathodes and another using the more-typical palladium cathodes. The experiments produced iron, one-to-two orders of magnitude greater than the maximum amount of iron that could have existed in the experiment beforehand.

In support of their claim of iron production, they observed very large changes among the four stable isotopes of iron as compared with natural abundances. After the experiment, the amount of iron-57 was 6.6 times greater than the naturally occurring abundance of iron-57.

In the early 1990s, the remaining critics who paid any attention to the research often criticized claimants, saying that their measurements of excess heat were not necessarily of nuclear origin. The critics demanded that researchers demonstrate that nuclear products were produced at the same time and in the amounts expected for nuclear reactions. Ohmori and Enyo reported such a correlation.

"The maximum total amounts of iron present in gold and palladium electrodes yield about 17 µg and 38 µg, respectively," the authors wrote. "Although the data are rather scattered, there is a proportionality between these two terms. This supports strongly that the production of iron atoms is related to the excess-heat evolution observed." (Ohmori, 1995)

They also noticed that the gold cathodes produced much larger results than the palladium cathodes. They concluded that something was special about gold, but they didn't know what it was.

The Growing Division

By 1995, the two camps in the field began to diverge further. The nuclear-evidence group expanded its search and looked specifically for heavy-element transmutations and isotopic shifts. Because these researchers were more open-minded, they more readily accepted results of experiments performed with ordinary hydrogen rather than just deuterium-based experiments.

In an essay published on John Dash's Web site at Portland State University several years later, Enyo expressed his dismay at the growing isolation of the research from the larger scientific community as well as his dissatisfaction toward the growing philosophical division among the researchers:

It is rather difficult to understand why people in the field accept that processes with heavy water produce nuclear results but those same people reject transmutation processes involving light water. Here again, people should have more open-minded attitudes. We should always anticipate wider possibilities of the science than we know at the time. (Enyo, 2000) [Edited for clarity]

Transmutation Products in Heavy Water

Researchers presented comprehensive sets of transmutation and isotopic analyses in 1995 at the first Low-Energy Nuclear Reactions Conference and at the next conference, in 1996. The first set was reported by Tadahiko Mizuno with Ohmori and Enyo, who used heavy-water electrolysis.

Mizuno had studied deuterium absorption into titanium by electrolysis for his Ph.D. dissertation; therefore, he had a great deal of directly relevant experience and understanding of the electrochemical system. In fact, he had performed similar experiments with palladium and deuterium for 20 years before 1989 but for a different purpose, as he wrote to me.

"I had seen some anomalous effects during that time," Mizuno wrote, "but I threw away the data, thinking it was noise. After the cold fusion announcement, I spent eight months preparing for my first cold fusion experiment."

In his earlier research, he had used an accelerator to study the irradiation of deuterium ions in titanium samples and the emission of neutrons. In 1989, Mizuno took the necessary time to set up his experiment carefully. Two months after he began his Fleischmann-Pons-type experiment, he saw a weak but significant neutron flux emanating from his experiment.

Mizuno published his transmutation results in the April 1996 issue of *Infinite Energy* a few months before George Miley independently published his transmutation results. (Mizuno, 1996)

Mizuno reported the results of a set of five electrolysis experiments performed in a closed stainless-steel cell coated with a 1 mm thick layer of Teflon. He obtained essentially the same results each time: synthesizing elements across the periodic table as well as an equally broad collection of isotopic shifts.

With all such claims, the greatest source of concern could be the possibility of contamination. Nevertheless, the aggregate mass of newly produced elements he detected was 10 to 100 times more than what could have pre-existed in the cathode.

Mizuno recognized the significance of isotopic shifts, and he gave an example in his paper. "Natural copper [Cu] is 70 percent Cu-63 and 30 percent Cu-65," Mizuno wrote. "But the copper found in the cathode was 100 percent Cu-63 with no detectable levels of Cu-65. Natural isotopic distribution varies less than 0.001 percent for copper." He also found evidence of gaseous xenon. He thought that it was highly unlikely to be a contaminant because metals don't absorb noble gases and because the cathode was degassed in a vacuum before the experiment.

The dotted line in the semilog graph below represents the detected elements before electrolysis. The only distinct peak on that line is palladium. The solid line represents the detected elements after electrolysis and includes newly synthesized tin, titanium, chromium, iron, copper and lead.

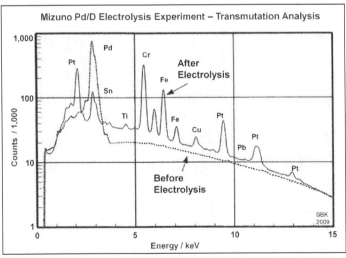

Change in abundances of elements after LENR electrolysis

In the same paper, Mizuno presented another graph that showed quantitative measurements of a wide variety of elements that were synthesized in the cell. Several of them — boron, silicon, chromium, iron, and gold — had increased three-to-four orders of magnitude from the pre-existing trace (10-50 ppm) levels.

Oddly, silver, which was at 44 ppm before the experiment, was undetectable after the experiment by X-ray spectroscopy (EDX) or Auger electron spectroscopy (AES), and it was only marginally detectable by secondary ion mass spectroscopy (SIMS).

Mizuno showed three groups of anomalous isotopic shifts across the atomic spectrum. Although Russian researchers at LUCH had preceded Mizuno with isotopic shifts, Mizuno and his colleagues were the first researchers to show such results so clearly and over such a broad range of the periodic table. After measuring and recording an array of isotopic shifts, Mizuno noticed that the abundance of newly formed elements followed a distinct distribution pattern.

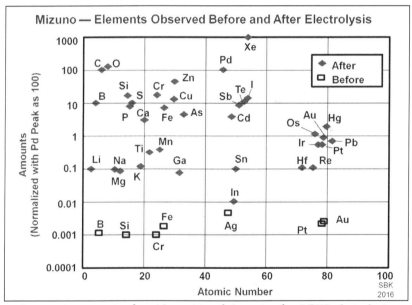

Increase in amounts of a wide variety of elements after LENR electrolysis

"The mass numbers of the evolved elements were distributed roughly into three groups: 20 to 28, 46 to 54 and 72 to 82," Mizuno wrote, "with the amounts more than 50%, 10% and less than 5%, respectively, compared to palladium."

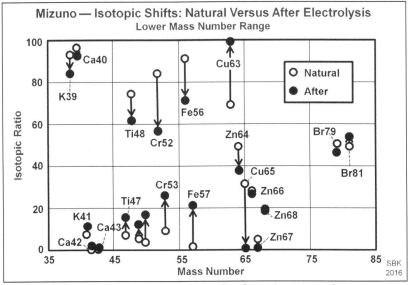

Nuclear reaction evidence: isotopic shifts of a wide variety of elements

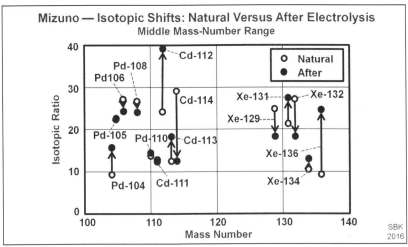

Nuclear reaction evidence: isotopic shifts of a wide variety of elements

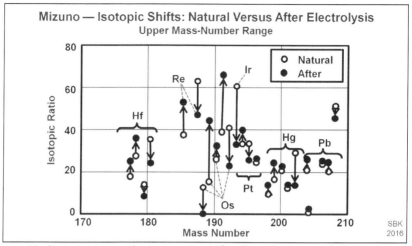

Nuclear reaction evidence: isotopic shifts of a wide variety of elements

Mizuno also took two elements, palladium and chromium, from the same heavy-water electrolysis experiments and displayed a detailed before-and-after comparison of the isotopic shifts.

As discussed in Chapter 10, variations in isotopic abundances in nature typically vary no more than a few tenths of a percent. Thus, the before and after charts below indicate massive shifts.

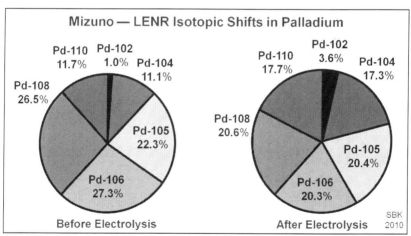

Nuclear reaction evidence: isotopic shifts on palladium cathode from an excess-heat-producing LENR experiment with heavy water

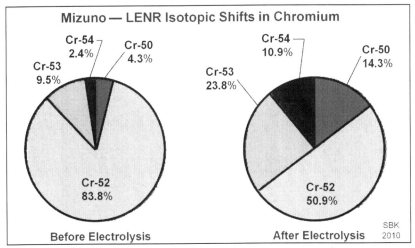

Nuclear reaction evidence: isotopic shifts on palladium cathode from an excess-heat-producing LENR experiment with heavy water

Transmutation Products in Light Water

Miley was reluctant to join the other dissident researchers. As a well-known and highly respected nuclear scientist, he seemed unlikely to become part of a fringe science group. But for many decades, he had had a fascination with alternative nuclear fusion ideas, including the use of alternative plasma confinement geometries, the use of alternative fusion fuels such as boron, and inertial electrostatic confinement fusion.

In 1996, Miley was the director of the Fusion Studies Program at the University of Illinois. He was also a consultant at various industrial and U.S. Department of Energy laboratories, the founding editor of the American Nuclear Society journal *Fusion Technology*, managing editor of the Cambridge University journal *Laser and Particle Beams* and the editor of the *Journal of Plasma Physics*. He was a past president of the University Fusion Association and served on several state technical committees for radiation protection.

As editor of *Fusion Technology*, Miley, unlike most other journal editors in the early 1990s, fostered peer review for "cold fusion" papers submitted for publication. Nearly every other journal editor at the time refused to receive the papers and send them out for peer review.

Tadahiko Mizuno (left) and George Miley (right) Photo: S.B. Krivit

Miley also joined Bockris in chairing the Second International Low-Energy Nuclear Reaction Conference, on Sept. 13-14, 1996. Here are Miley's opening remarks to the conference:

> Many of you know me as editor of *Fusion Technology* because I have taken the step of publishing papers in this subject — assuming they get through peer review. This has caused considerable criticism by my colleagues. Most of my research record is in the area of hot fusion, and hot plasmas. ... In fact, I consider myself a plasma physicist.
>
> So in the work I am doing now, ... I have been criticized by some of my colleagues in metallurgy, who say, "That guy doesn't know anything about metallurgy. What is he dabbling in this for?"
>
> This is an interdisciplinary field, which requires people to keep an open mind. I would say that, if I didn't dabble in things, I would never have gotten my other research done. In 1978, I had the first electron-beam-pumped laser, and I never built a laser before. In 1985, I had the first design for ... an inertial confinement fusion (ICF) target, and I never designed an ICF target before.

So I have no apologies for doing something that I have never done before. That's what research is all about. (Miley, 1996)

Bockris, as always, understood the relevance of the research in the context of the history of science. Here are his opening remarks to the conference:

On the one hand, my colleagues in the Department of Chemistry have declared the field to be non-existent and any pretense of a conference a HOAX! On the other hand, the work a number of you are going to present today would, in the opinion of some people who know of advances in the field, stand with the great steps in nuclear research, with Rutherford and Chadwick of 1919 and Hahn and Meitner of 1939.

But we must forever be on the lookout that we do not let our enthusiasm become overly confident and overly optimistic. Although my scientific experience has told me that more is lost through over-skepticism than too much credence, we must always try to see if it is not possible to explain our results in terms of the traditional nuclear chemistry of the textbook. (Bockris, 1996)

From Berkeley to Beads

In 1995, when Miley was in Monte Carlo attending the Fifth International Conference on Cold Fusion, he noticed an electrolysis experiment on display. The electrode configuration used millimeter-sized plastic beads coated with metals. The beads looked similar to targets he had used in his inertial-confinement fusion research. The man behind the table was James A. Patterson (1922-2008). Miley told Patterson that he knew a lot about using similar material for inertial-confinement fusion targets and that he thought he could help improve Patterson's design. It was the beginning of a good friendship.

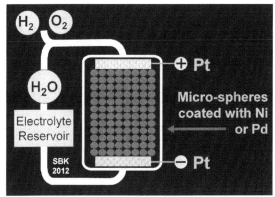

Diagram of thin-film electrolysis in packed bed method

In 1951, Patterson was studying at the University of California, Berkeley, working toward his Ph.D. to become a chemistry professor. Dow Chemical Co., however, made him a job offer that he couldn't refuse and hired him before he could graduate.

Patterson was a brilliant inventor, with dozens of patents and patent applications to his name. One of his inventions was a method for making tiny micro-spheres: small, perfectly round polystyrene beads. The beads were useful in a variety of applications, including water purifiers and cosmetics and as an artificial replacement for the talcum powder normally used inside surgical gloves.

It was natural then for Patterson to think of a way to use his beads for Miley's experiments. The first step was to take the 1 mm beads and deposit multiple thin-film layers of nickel and palladium onto them.

The second step was to construct a unique cell in which the micro-spheres, submerged in the electrolyte, would fill the space between the cathode and anode. A major advantage of this method was that it provided an extremely high total surface area on which the reactions could take place. Larry Forsley, one of Miley's colleagues, described Patterson's lab in a memoir:

> His lab in Sarasota was wonderful. It was the size of an
> oversized garage — or an undersized airplane hangar. It was
> a marvelous combination of 1950s' technology coupled with
> the best of 19th century physics and chemistry. A modern-

day Faraday would have been right at home among the variety of ovens, wires, cables, chemicals, stirrers and more. There was even an office with a recliner that Jim liked to nap on, and a dog or two to keep an eye on things. I'd say we spent equal amounts of time talking about fishing, chemistry and eating and the rest of our time futzing in his lab.

Patterson's story was dramatic and unfortunately typical of the many scientists and inventors in this troubled history who tried to commercialize their experimental discoveries without having a clear understanding of the underlying science. Patterson created a company called Clean Energy Technologies Inc. to promote and sell do-it-yourself "cold fusion" kits.

James Patterson (Photo: D. Nagel)

One reason for the failure of Patterson's device was his claim that the kits could produce enormous levels of excess heat. He gave public demonstrations and in 1996 he appeared on the ABC national television program *Good Morning America*. Even some of his team members contradicted the amounts of excess heat that he claimed. It was a flop.

By most accounts, Patterson's extravagant claims aside, his cells did produce some excess heat. But it wasn't the heat from Patterson's beads that led to a major contribution to the development of this science; it was the transmutations, and that's where George Miley came in.

CHAPTER 15

Plunging Into Light Water

T he taboo topic of light-water experiments had been unavoidable since the Third International Conference on Cold Fusion (ICCF-3), in 1992. By 1994, Michael McKubre and researchers from Bose Corp. separately cast doubt on the reported light-water excess-heat results. In 1996, researchers reported heavy-element transmutations with light-water electrolysis.

In August 1996, in *Infinite Energy* magazine, Eugene Mallove published a preprint of George Miley's paper for the Second International Low-Energy Nuclear Reaction Conference, scheduled for September. James Patterson, the inventor of the small beads that were intrinsic to the experiment, was his co-author. (Miley and Patterson, 1996)

Unlike Patterson, Miley was not primarily interested in searching for excess heat, although he did find low levels, close to the error limit of his calorimetry system. Miley, like Mizuno, was instead eager to analyze for transmutations and isotopic shifts, and he too observed similar changes. Miley presented his results at ICCF-6 (Miley et al., 1996)

Miley noticed a correspondence between his results and Mizuno's, despite the fact that the two electrolytic systems were different. Miley used light water and primarily nickel, while Mizuno used heavy water and palladium. Miley wrote that he and Mizuno had observed a similar grouping of transmutation products:

> It is interesting to compare [our] results to those reported by Mizuno et al., who ran a high-current-density Pd electrode in a cell at high pressure and temperature with

a heavy-water Li_2CO_3 electrolyte. They report a rich variety of reaction products at 1 mm depth, concentrated in groups with atomic numbers 6-8, 20-30, 46-54, and 72-82. While a one-per-one comparison is not possible since [our] study used Ni rather than Pd, this distinct grouping of products is consistent with [our] results, where major products group between atomic numbers 12-14, 20-30, and 46-56. (Miley and Patterson, 1996)

Although Mizuno, Ohmori and Enyo published their results four months before Miley and Patterson published their results, Miley incorrectly wrote in his ICCF-6 paper that he and his co-authors were the first to achieve such results:

> The authors have achieved, for the first time, a quantitative measure of the yield of transmutation products. Results from a thin-film (0.05 to 0.3 micron) nickel coating on 1 mm microspheres in a packed-bed-type cell with 1-molar $LiSO_4$-H_2O electrolyte were reported recently at the Second International Conference on Low-Energy Nuclear Reactions (Miley and Patterson, 1996). ... The transmutation products in all cases characteristically divide into four major groups, with atomic numbers 6-18, 22-35, 44-54, and 75-85. Yields of ~1 mg of key elements were obtained in a cell containing ~1,000 microspheres. In several cases, over 40 atom % of the metal film consisted of these products after two weeks' operation. (Miley et al., 1996)

Miley plotted the combined transmutation products from six of his experiments onto a single graph (shown below), then plotted a curve that revealed a distinct visual signature: an array of four roughly evenly spaced and gradually descending peaks at atomic numbers 12, 30, 48, and 82.

Another figure in Miley's ICCF-6 paper showed an overlay of his four-peaked curve with similar peaks from Mizuno's data. Miley also recognized portions of a similar four-peaked curve from yet a third data set, reported by John Bockris. Miley was aware of the significance of his, Mizuno's and Bockris' discoveries, as he wrote in 1996:

> The results presented here defied conventional views in many ways. First, chemically assisted nuclear reactions are not widely accepted by the scientific community. The present results not only confront that disbelief but add a new dimension to the issue by reporting copious light- and heavy-element reaction products that seem to imply multi-body reactions due to the formation of heavier elements, such as copper and silver from nickel. (Miley and Patterson, 1996)

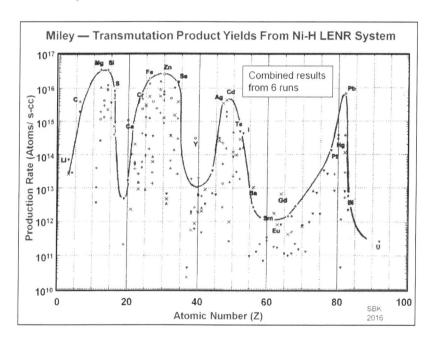

In 2016, while I was writing this chapter, I contacted Sheldon Landsberger, who in 1996 worked with Miley at the University of Illinois and had performed the nuclear activation analysis for these experiments using the university's "TRIGA" reactor. Initially, when I spoke to Landsberger about his 20-year-old data, some of which he had personally recorded, he said, "I wouldn't touch that data with a 100-foot pole." After I asked him for more detail about his concern, he wrote only that he did not agree with Miley's interpretations. However, he added, "There is no need to reaffirm or retract the data I provided."

A Formerly Astute Observer

Only two people who attended the Second International Low-Energy Nuclear Reactions Conference reported on it publicly. One was Hal Fox, the editor of *Fusion Facts*, *New Energy News* and the *Journal of New Energy*. Fox also had a personal financial interest in the research. The other person was Jed Rothwell (b. 1954) from Atlanta, Georgia.

When I came to know Rothwell, sometime after 2000, I saw him as an articulate but dogmatic defender of the D+D "cold fusion" idea.

By 2008, as I realized there was no experimental evidence that supported the idea of room-temperature fusion, I experienced a more vicious side of Rothwell. For years, he has used aggressive tactics against people who did not share his fusion belief. In 2011, for example, he wrote to randomly selected editors at John Wiley & Sons, for whom I had worked as editor-in-chief for Wiley's *Nuclear Energy Encyclopedia*. Rothwell erroneously assumed that they would be publishing my next book, and he wrote disparaging comments about me and tried to convince them not to work with me.

In conducting the research for this book, I was surprised to learn that Rothwell was not always a zealous fusion believer. In fact, in 1996, Rothwell showed no signs of alignment with the fusion faction. He recognized the validity of the transmutation results and he reported them accurately and articulately.

After Rothwell graduated from Cornell University in 1976 with a degree in Japanese language and literature, he worked as a programmer and technical writer, writing code for billing and monitoring systems for telephone exchange equipment.

He was a founder and manager of software development for Micro-Tel, Inc., a company in Norcross, Georgia. According to what Rothwell told Jerry Bishop for his 1993 article in *Popular Science*, Rothwell "lives off dividends from the software company he founded" in 1978.

Rothwell first learned about the research in 1992 after he read Mallove's book *Fire From Ice*. On Feb. 8, 1992, Rothwell launched a petition, under the (unregistered) organization name of Cold Fusion Research Advocates, to promote the research. He gathered 325 signatures, most of them from scientists, and submitted the petition to the U.S. House of Representatives Committee on Science, Space and Technology. Nothing ever came of it, Rothwell told me.

On June 2, 1992, he founded Solid State Fusion, Inc. to develop energy technology based on tabletop fusion. Nothing ever came of that, either, and the corporation dissolved in 1995. An Aug. 10, 1992, press release from Clustron Sciences Corp. says that Rothwell was the vice president of that company and that, for two years, he had been facilitating information exchange between Japanese and American researchers. Nothing ever came of Clustron Sciences Corp.

Throughout the 1990s, Rothwell was a regular contributor to *Cold Fusion* magazine and *Infinite Energy* magazine, and he attended the international conferences since ICCF-3, in 1992. In 2002, in preparation for the ICCF-10 conference in Cambridge, Massachusetts, Rothwell collaborated with Edmund Storms to establish the LENR-CANR.org online library for LENR research papers. Rothwell then worked as a volunteer to collect papers, edit and format them, and upload them to his Web site, making them freely available to the public.

The initial bibliographic data for his library came from Storms. Rothwell also used Dieter Britz's database of papers as a cross-reference. Rothwell's publication of his online library has been a benefit to the field.

When I asked Rothwell for more details about his accomplishments with Micro-Tel, he declined:

> I would prefer not to discuss my connection to the company, and I would prefer you not mention it. As I am sure you know, there are some fanatical people opposed to cold fusion. There are some troublemakers.
>
> If the company is associated with cold fusion, some of them might spread false rumors, which might be bad for business. I do not mind that they attack me personally. They

can't hurt me. There are not many troublemakers, but it only takes one to cause problems on the Internet.

In fact, Rothwell's own fusion fanaticism has caused problems for some people. In 2002, scientist Paulo N. Correa, after experiencing many public attacks by Rothwell, wrote a long article on his Web site about Rothwell called "The Serpent's Tooth and Its Egg (Or, How the Stupid Are So Often Malicious)." Correa wrote that Rothwell had engaged in "malicious and slanderous accusations."

Despite Rothwell's behavior, his first-hand reports of the science in the mid-1990s serve as useful, detailed and objective historical documentation.

At the Second International Low-Energy Nuclear Reaction Conference, Miley began by describing his earlier multilayer thin-film methods, which used alternating layers of palladium and nickel placed on a glass or quartz substrate.

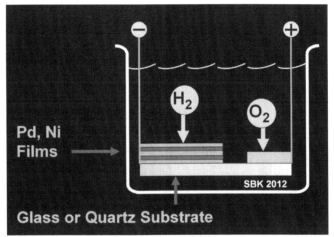

Diagram of thin-film electrolysis on substrate method

Rothwell reported on Miley's presentation:

> [Miley] is working with [Patterson-type] thin-film-coated beads packed in three-to-five layers. The beads are fabricated by the Materials Science Department at the

University of Illinois. They have devised a new type of sputtering machine that shakes the beads, levitating and rotating them to produce less than 10% deviation in coating thickness. Miley says the sputtering technique simplifies the problem because it does away with the need for a copper flashing, reducing the number of elements at the start.

In conversations, he has also told me that he thinks the coating is tougher and the beads will generate heat more readily than Patterson's electrodeposited beads. Miley has made a great effort to prevent contamination in the closed flow calorimeter.

He describes contamination as his greatest worry. He has performed 20 runs to date. He performed the most extensive analysis on run #8, done in March 1996. The analysis is very tedious, taking weeks of hard work. He presented voluminous charts comparing data from different runs. He uses four techniques to measure elements and isotopes: neutron activation analysis (NAA), X-ray spectroscopy (EDX), Auger electron spectroscopy (AES), and secondary ion mass spectroscopy (SIMS). ...

The nickel is transmuted into a wide variety of other elements including silicon, titanium, chromium, magnesium, copper, iron, silver, cadmium, lead, yttrium, zirconium, and zinc. As Professor Dash said of his own results, "you see everything but the kitchen sink." ...

Some runs produced little heat but showed massive transmutations. ... Miley has succeeded too well in finding transmutations. He has found that [cathode materials made from] palladium, nickel, titanium and gold all react to some extent. It is becoming a little difficult to select material suitable for electrolysis in a blank cell that does not transmute! ... Miley concluded that, "if fifty percent of what we are talking about is true, this revolutionizes nuclear physics." (Rothwell, 1996, ILENR2)

Rothwell summarized the issues common to many of the experiments reported at the September 1996 conference:

> The results from different experiments in different labs were remarkably similar, something we must admit we are not used to seeing in this field. Damage [to cathodes] was seen at the same depth as the transmutations. Scanning electron microscope photos of many cathode surfaces showed eruptions of material from inside, looking like the surface of the moon. (Mizuno, 1996; Kopecek, 1996; Ohmori, 1996)
>
> It was this damaged material, and the areas around it, that was transmuted. In cases where the transmutations occurred well below the surface, the material was not ejected, but damage was seen when the outer layers were removed. (Minevski, 1996)
>
> The transmuted elements were not uniformly distributed. Areas with no damage often remained highly purified metal (palladium, nickel, and gold.) This is a large effect, easily detected. In many cases, large amounts of the metal were transmuted. Miley reported that up to 40% of the total metal was transmuted, where the starting metal was better than 99% pure. Minevski said that, in the layers where transmutations occurred, up to 70% of the material was no longer palladium. Mizuno reported three peaks of new elements: light, medium, and heavy. Miley said he saw the same pattern, with the three peaks moving up or down the spectrum depending on the metal he started with (palladium or nickel).

Rothwell explained the importance of the various detection instruments:

> A variety of different detection instruments must be used to verify the transmutations. These include NAA, EDX, AES and SIMS. The speakers all emphasized the importance

of cross-checking with different methods. As Miley puts it, with SIMS "the absolute calibration is lousy." In other words, with SIMS you can detect tiny differences between very close isotopes and species (like deuterium molecules and helium-4), but when you first start out, you never know where on the spectrum you will land.

Other methods are more accurate but less precise. This problem should not be exaggerated. Any instrument will verify that 40% of the nickel cathode has turned into silver, silicon, copper, and so on. Indeed, if Miley ever runs a cell with a liter of beads for six months, he will be able to refine the material and make the new elements visible to the naked eye. He will have more silver and gold than you would find in all the contamination in the building. You do not need high precision to verify nickel has turned into iron. You need high precision to find out whether the isotopes' distributions are unnatural and to draw useful conclusions about the nuclear transformations at each step.

The NAA has the useful property of analyzing many atoms over a large mass simultaneously (four or five beads, in the case of the Patterson device). This is important because the material is inhomogeneous. SIMS looks at only one tiny spot on a sample.

Next, Rothwell explained why transmutation results were more significant than excess-heat results. One of the most important aspects of the history of LENR is the divisiveness that arose in the field. Rothwell observed it but didn't fully grasp it at the time, as he explained:

> People will argue that the excess heat might be caused by chemical reactions, even when it is a 100,000 times too big for that. But nobody will say that chemical processes convert nickel to iron. ... I believe we have turned the corner in the long search for the mechanism of the Pons-Fleischmann effect. Host-metal transmutation is the key. Years ago, everyone assumed the effect must be caused by the fusion of

light elements. Melvin Miles, [Roger Stringham], and other researchers found helium ash roughly commensurate with conventional deuterium fusion. No doubt they did see helium; we cannot dismiss these earlier results.

Yet many others made heroic efforts to look for helium but found none. I conclude that sometimes helium is produced and sometimes it is not. ... We spent the last seven years looking only for light-element transmutations. This leads to a question that has been haunting me for months, ever since I first heard about Mizuno's results. Why did this take so long? Debra Rolison, at the Naval Research Laboratory, reported evidence of heavy-element transmutation years ago. Why was there no follow-up? Mizuno says the other Japanese researchers have completely ignored his and Ohmori's findings.

I hope they pay attention to Miley when he reports his results at ICCF-6. For that matter, why did it take so many years before people did a serious replication of Ni H_2O cold fusion? The pace of this research is still much too slow, and the scope of investigations is still much too cramped by our preconceptions about what must be happening.

In Rothwell's conclusion, he quoted Bockris, who was asked by Miley how to bring respect to the research. Bockris explained that the poor reproducibility of experiments was the major impediment. However, this was true primarily for experiments designed to observe excess heat.

Experiments designed to observe heavy-element transmutations and isotopic shifts were generally more repeatable. The isotopic shifts, in particular, were also more convincing to critics. But scientists focused on trying to prove the existence of tabletop fusion were not interested in such results.

CHAPTER 16

Going in Circles

While George Miley, a nuclear engineer, and Tadahiko Mizuno, a physicist, worked to characterize the products of the reactions in their experiments, electrochemists searched for experimental evidence to support their deuterium-deuterium (D+D) "cold fusion" idea. They pursued this hypothesis even though the experimental evidence against the idea was mounting.

In 1996, Victor J. Stenger (1935-2014), a visiting Fellow in the Department of Philosophy at the University of Colorado, wrote an essay called "ESP and Cold Fusion: Parallels in Pseudoscience."

"Still, with no basis other than faith," Stenger wrote, "the ESP and cold fusion faithful continue to press their case with religious fervor. Like parapsychology, cold fusion seems to have passed on from science to pseudoscience, carried on by a few diehards who have allowed their desires to overcome their reason."

Rather than write that the available data didn't support the hypothesis of room-temperature fusion, Stenger argued that the entire field lacked credible evidence for nuclear reactions. He, like many people after 1989, had stopped paying attention to new developments in the research. The progress was easy to miss: The only regular sources of news on "cold fusion" were the newsletters and magazines of Hal Fox and Eugene Mallove.

As discussed in Chapter 8, D+D nuclear fusion occurs in three specific reaction paths, producing specific products, with specific energies and specific ratios between the products. These characteristics did not fit the "cold fusion" experimental data.

Scientists who believed that the underlying process for the

phenomena was fusion had limited experimental evidence to support their hypothesis. They speculated that, in their experiments, because of the low levels of emitted neutrons and the absence of detectable high-energy gamma-rays that normally occur in fusion, a modified version of the third reaction was occurring. Instead of this reaction $d + d \rightarrow$ *Helium-4 + 23.8 MeV gamma-ray*, which is scientifically verified, they proposed the reaction $d + d \rightarrow$ *Helium-4 + 23.8 MeV (heat)*. Therefore, researchers believed, the dominant nuclear product in the experiments would be helium-4.

Unlike the conference in Texas organized by John Bockris and George Miley in September 1996 that focused on transmutations, the Sixth International Conference on Cold Fusion (ICCF-6), in October 1996 in Hokkaido, Japan, centered on the D+D "cold fusion" idea. At ICCF-6, Rothwell saw a significant difference from the Texas meeting:

> Many workers are stuck in the rut of trying to replicate the 1989 simple palladium-heavy water electrolysis method. This requires high loading and other conditions which are nearly impossible to achieve. Why anyone would still be trying to use this method so many years after better methods have been invented is a mystery to me. The majority of scientists in the field ignores these promising approaches and continues using only [bulk] palladium.
>
> Instead of selecting the easiest and most successful methods, they insist on using the oldest, least effective, and most frustrating technology, as if they were computer scientists who insist on building vacuum-tube machines in the age of transistors. ... Six years of low-level [excess-heat] results have failed to convince mainstream scientists that cold fusion is real. Six more years will not convince anyone, either. (Rothwell, 1996, ICCF-6)

As Rothwell suggested, easier experimental methods included the use of thin-foil cathodes (Debra Rolison, John Dash), thin-film cathodes (James Patterson and George Miley), and electrolytic co-deposition (SPAWAR). All of these methods involved greater reactive surface area.

The Search for Helium-4

Melvin Miles was the first researcher to find a direct correlation between the production of helium-4 and excess heat. He published the results of these electrolysis experiments with deuterium and heavy water in 1991. (Bush et al., 1991; Miles et al., 1991)

Four requirements were necessary to support the $d + d \longrightarrow Helium\text{-}4 + 23.8\ MeV\ (heat)$ idea. The first is that the produced helium-4 concentration must be above the atmospheric level of helium-4 in order to rule out contamination. Miles' results were lower; therefore, critics were not willing to accept such measurements. (Passell, 1994)

The second requirement is that production of helium-4 and excess heat must be simultaneous (temporally correlated). Miles' results showed this.

The third requirement, which Miles achieved within an order of magnitude, was that the helium-4 produced in the cell be quantitatively in the correct proportion to the amount of heat (energy) released. In other words, for every helium-4 atom produced, approximately 24 MeV of excess heat had to be measured. Miles' experiments produced 39, 25, 44, 88, 83, 52, and 62 MeV per helium-4 atom.

The fourth requirement was that the experiments produced no other nuclear products associated with significant heat release. Although the fourth factor may seem obvious in hindsight, it was not evident to me until a 2006 phone call in which theorist Lewis Larsen explained the significance of other nuclear reactions to the total energy balance in any given experiment. Until then, I had never seen any published paper or news report that mentioned this problem.

The issue of other exothermic processes and products observed among the phenomena was at the root of the growing debate in the field about the possible mechanism. In fact, reactions with light water and heavy water revealed both the exothermic production of newly transmuted elements and changes in isotopic ratios. Scientists who believed in the fusion idea defended their hypothesis by one or all of the following:

1. Not analyzing for products of other possible nuclear processes in their own experiments.
2. Ignoring other nuclear products reported in other researchers' experiments.
3. Claiming that the other reported nuclear products were not strongly exothermic and therefore did not contribute to the measured excess heat.
4. Claiming that other reported nuclear products were simply the result of experiential errors.

Tullio Bressani, a professor in the Department of Physics at the University of Torino, Italy, and a well-respected member of the field, was an enthusiastic proponent of the D+D "cold fusion" idea. His summaries of the reported helium-4 evidence at ICCF-6 and ICCF-7 are good records of the growing support for the idea. (Bressani, 1996; 1998)

In his ICCF-7 paper, Bressani dismissed reports of heavy-element transmutations because he could not conceive of a mechanism by which they could occur. (Bressani, 1998)

Two U-Turns for Hagelstein

I had long wondered why and when the consensus within the field had shifted from neutron-based ideas in the 1989-95 period to D+D fusion-based ideas in the late 1990s. In hindsight, I see that 1996 was the turning point. Other researchers did as Bressani had done, selecting only helium-4 data and ignoring or disputing all other data.

Prominent among these was Peter Hagelstein. For the second time in seven years of trying to explain the phenomena, he made a U-turn. Back in April 1989, he had rushed to publish and patent ideas based on his theories of D+D "cold fusion."

By March 1990, he had abandoned his fusion ideas and instead proposed mechanisms based on virtual neutrons and weak-interactions.

In October 1996, he switched back to D+D fusion again. In the coming years, he proposed and abandoned a number of models, each one attempting to explain some portion of the phenomena. None could explain how deuterium nuclei could overcome their electrostatic repulsion from one another at room temperature at sufficiently high rates to explain the experimentally observed phenomena. By 2005, he said he had attempted 150 models.

Hagelstein had been well aware of the broad array of nuclear phenomena in 1996, including heat generation in experiments without deuterium. He had known that, logically, neutron-based ideas made the most sense, but he abandoned his neutron-based model because he couldn't envision a plausible way to make it work. (Hagelstein, 1996)

In 1998, at ICCF-7, Hagelstein returned to the fusion concept, postulating that helium-4 was the sole product of the experiments, despite the fact that he knew that the phenomena included the "production of excess heat, tritium, helium, neutrons, MeV charged particles, X-rays, gamma rays, and induced radioactivity." Yet he continued to pursue theoretical models that depicted helium-4 as the sole product of the experiments. (Hagelstein, 1998)

In this paper, he also incorrectly attributed to Fleischmann and Pons the idea that helium-4 was the main product of their experiments. Hagelstein wrote that "Pons and Fleischmann conjectured that $d + d \rightarrow$ $4He + Q$ *(Lattice)* may account for the excess-heat production." As discussed in Chapter 1, Fleischmann and Pons never made such a conjecture. The conjecture was first proposed in April 1989 by other scientists; one of them was Hagelstein.

After ICCF-6, scientists who believed in the fusion idea perpetuated this myth, boosting the apparent credibility of the idea by creating the impression that it originated with the field's two progenitors.

U.S. Navy's Fruitless Search

Miles' 1991 observation of a correlation between helium-4 and excess heat resulted in a joint U.S. Navy research program that ran from 1992 to 1995. It was called the "Navy Program to Understand Anomalous Effects in Electrochemically Loaded Materials" and was informally known as the "Tri-Navy Program." Three Navy labs participated in the collaboration, the goal of which was to search for excess heat but not helium-4.

The labs were the Naval Research Laboratory (NRL) in Washington, D.C., the Naval Air Weapons Station, in China Lake, California, and the Naval Ocean Systems Center, in San Diego, California, later renamed the SPAWAR.

The program was funded and managed by the Office of Naval Research (ONR). Debra Rolison, the top electrochemist at NRL, had done the only Fleischmann-Pons-type experiments at NRL by then. She presented her positive results at conferences in October 1989 and March 1990. By November 1990, however, Rolison submitted a paper to a journal that cast doubt on her previous results. (Chapter 11)

The idea for the program originated in a conversation among Miles and other scientists during the ICCF-2 conference in Como, Italy, in June/July 1991. Miles told me how it began.

Stan Szpak, an electrochemist at SPAWAR, had presented important co-deposition results. Miles had reported a correlation between excess heat and helium-4 production. When Robert Nowak, an electrochemist and program manager at ONR, returned to Washington, D.C., he got Fred Saalfeld, the head of ONR, interested in a Navy program. David Nagel, the superintendent of the Condensed Matter and Radiation Sciences Division at NRL, also helped put the idea together.

"There were opponents of cold fusion at ONR," Miles wrote, "but Fred Saalfeld's interest in a Navy cold fusion effort sealed the deal. Saalfeld took an active part in the Tri-Navy Program and was present at all the review meetings that took place in Washington, D.C."

Nagel wrote the proposal for the program, with input from the principal investigators, and Saalfeld approved it. Nowak directed the program and distributed the funds. A primary goal was for NRL to replicate the China Lake experimental protocol and observe excess heat, as Miles had reported at ICCF-2. Another objective was to have NRL and China Lake use the SPAWAR co-deposition protocol and observe excess heat. A third objective was for the Navy to make its own palladium cathode materials. By 1992, researchers thought that reproducibility was largely a materials problem.

The plan included testing NRL's cathodes for the production of excess heat at China Lake and at SPAWAR. The program began in early

1992. The cathodes were produced at NRL by Ashraf Imam, a metallurgist. He told me his story:

> Saalfeld was very keen, and he took a special interest in this program and funded it. I was given the task of making the material, so initially the material which we were looking at was pure palladium. From my discussions with other scientists in the field, I learned that the presence of other elements in palladium may play an important role and contribute to generation of excess heat.
>
> I ordered material from Johnson Matthey with two different impurity levels, 99.99 and 99.999. I analyzed the as-received material, but unfortunately I found that both of the batches they sent were five-nine purity, despite the fact that labels indicated the different grades. That was a big setback — at first. But then I thought, "Why don't I start putting different types of elements into the palladium and make alloys?" I made several alloys, including palladium-boron.
>
> At NRL, Dawn Dominguez had been assigned to do the experiments and look for excess heat. Initially, she didn't want to use the alloys because, according to her understanding of the project, her task was to replicate Miles' earlier experiment that had been performed with pure palladium. Miles, however, was willing to test all the alloys.

Imam sent his cathodes to Miles at China Lake and to Szpak and Pamela Mosier-Boss, at SPAWAR. At NRL, Dominguez (b. 1948) also performed in-house tests looking for excess heat using Imam's pure palladium cathodes. In San Diego, the SPAWAR researchers never ran any experiments using the NRL materials as part of the program, according to Mosier-Boss:

> We received some palladium-silver alloy cathodes from NRL, but we didn't do anything with them. We were not equipped to do calorimetry, and we received no funding from ONR for the Tri-Navy project. We were quite busy

doing our own co-deposition experiments, so anything from NRL was on the back burner.

In the first year of the program, before October 1992, Nowak stopped inviting SPAWAR to the program review meetings. At China Lake, Miles tried to use the SPAWAR co-deposition method to search for excess heat but didn't observe any. It wasn't a valid replication of the SPAWAR experimental protocol because the SPAWAR researchers had only analyzed for, and found, other phenomena: tritium, X-rays, unusual surface morphology, and unusual thermal effects. They had not attempted to measure excess heat. In fact, it wasn't a suitable experiment to search for excess heat, as Miles explained to me.

"Much too late," Miles wrote, "I realized that the amount of palladium deposited was much too small for excess heat to be detected with my calorimeter unless there was an exceptionally large power-density result."

For several years, Miles had known that pure palladium was not the best cathode material for the experiments. Palladium with some impurities in it (for reasons that are still unknown) — cerium, lithium and silver — more reliably produced excess heat. Martin Fleischmann, in a Jan. 21, 1992, response to a letter Miles had sent him, told Miles, "The impurities in the D_2O are probably of key importance, and this goes for metals, as well as the borates and silicates. I am sure you know that our favored strategy now is to use alloys, which you have listed."

After three painstaking years of failure after failure with all other cathode materials, when Miles used Imam's palladium-boron cathodes in 1994, eight of the nine produced excess heat, an 89 percent success rate. Imam's palladium-boron alloy was an inter-laboratory success. Even the cathode that didn't produce excess heat revealed useful information; it contained microscopic cracks.

Miles reported the good news at the final planning session in January 1995 in Washington. Nowak seemed pleased. "At a dinner that evening," Miles wrote, "Nowak told me that my results likely saved this Navy program." Dominguez, like Miles, had observed only failures with her pure palladium cathodes and had not yet tried the palladium-boron alloy. The Washington-based proponents of the program were in a bind. To succeed, to justify the funding, and to keep critics at bay, the program had to show independent replication, and because SPAWAR was no longer part of the program, a successful replication had to take place at NRL.

The problem, however, as Miles learned when he visited NRL, was that its senior experts in electrochemistry and calorimetry were reluctant to get involved with the controversial experiments. Staff members were afraid of harming their careers, and they were aware that the entire lab could be put at risk from funding cuts. This risk was not speculative.

According to Scott Chubb, as discussed in Chapter 11, Peter Zimmerman told people at NRL that he would go to Congress and push for a reduction in NRL funding because NRL was doing illegitimate science. Nowak confirmed such threats in an article written by Bennett Daviss in the March 29, 2003, issue of *New Scientist*.

"Fairly prominent individuals within the physics community voiced threats," Nowak said. "They said that they were aware that federal funds were going into cold fusion research and they were going to do what they could to stop it."

Daviss also spoke with Saalfeld, who told him that he had to defend his decision to other scientists and managers at ONR. "I told them that there is a phenomenon here that we don't understand," Saalfeld said, "it might have relevance to naval science, and we're going to explore it."

According to Miles and Imam, Rolison declined to get directly involved in the project. Instead, the person at NRL who performed the electrochemistry experiment was Dominguez. She wasn't an electrochemist — her expertise is in organic chemistry and material science — but, as she told me, she received a lot of excellent support from other electrochemists.

Miles did the best he could to help her. In January 1992, Dominguez spent two weeks with Miles in his lab at China Lake learning about the experiment. In January 1995, the last year of the program, Miles went to NRL at the request of Nowak to help set up experiments there.

According to Miles, instead of following the experimental protocol he discussed with her, Dominguez followed instructions given to her by Rolison and Nowak, both electrochemists. Miles said that, when

Dominguez asked for help from other researchers, they refused or cooperated only minimally.

Michael Ravnitzky, an editor of this book, requested and obtained a copy of Dominguez's final report from NRL. Independently, Miles provided me with the marked-up copy in which he reviewed Dominguez's work.

A crucial problem with Dominguez's experiments was that her calorimetry was 10 times less accurate than that of Miles. Whereas Miles' China Lake calorimeter was accurate to +/- 20 milliwatts, Dominguez' NRL calorimeter was accurate to only +/- 200 milliwatts. As Miles explained to me, Dominguez would have had to obtain 600 milliwatts of excess heat to get a scientifically meaningful result. In Miles' successful experiments, he rarely saw excess heat greater than 600 milliwatts. Thus, excess heat measured in these types of experiments might have been detected for Miles but not for Dominguez. The failure was compounded by Dominguez's late use of the palladium-boron alloy.

Dominguez did not dispute the problem she encountered with her calorimetry. "Toward the very end of the program, we got a new calorimeter," Dominguez said. "Part of the reason why we couldn't see anything during most of the program was because our calorimeters weren't very good. We were trying to replicate Miles' work, and he had this home-made calorimeter with a plastic bottle filled with vermiculite and stuff like that, and a lot of insulation."

Some of the best scientific instruments are often custom-made because they may require highly accurate measurement of a particular parameter. For example, Fleischmann and Pons' "home-made" calorimeter was accurate to +/- 1 milliwatt.

Dominguez confirmed that the larger size of her cell and the additional wires running into it caused her calorimetry to be less precise than that of Miles. She remembered learning about Miles' success with the palladium-boron alloy. "The news came at the very end of the program while we [at NRL] were trying to get the new calorimeter going, and then they terminated the program."

Dominguez also had trouble with the new calorimeter. Miles explained to me that NRL had purchased a commercial Hart Seebeck calorimeter, but the NRL researchers didn't understand how to use it properly. In January 1995, Miles wrote, NRL called in Roger Hart, the founder of Hart Scientific, the company that manufactured the calorimeter. Hart examined three months of the NRL data taken with this calorimeter and told Miles that it was useless because NRL had not determined the baseline for their excess-heat measurements.

Dominguez explained to me that her understanding of the program instructions was to use a pure palladium cathode. "The goal of the program," Dominguez said, "was to do as clean an experiment as we could and replicate the China Lake experiment, which had been done with pure palladium. That's why the plan was to start with high-purity palladium."

Miles had always been perplexed about the cause of the termination of the Tri-Navy program. The final meeting took place at China Lake June 7-9, 1995, according to Miles:

> The first and second day seemed to go okay. But when Nowak and Rolison arrived in the morning of the third day, it was obvious to me that everything had changed. Nowak announced right away that our meeting was over, as was the whole program, and he and Rolison walked out. Most of the rest of us stayed and talked, anyway, but we all wondered what had happened overnight.

The mid-1990s was a difficult time to do this controversial research. Without successful replication of the China Lake result at NRL, the program was vulnerable to internal critics at ONR and NRL, as well as to external critics. Saalfeld and Nowak had worked at NRL and knew the political terrain well. (Nowak declined to speak with me about this history.) Saalfeld confirmed the accuracy of this section by phone, with the exception that he thought that Dominguez was an experienced and well-recognized electrochemist.

Saalfeld's description of the program for Daviss in 2003 for the *New Scientist* article differs somewhat with my account of this history. Yet it was clear to me from my conversation with Saalfeld in February 2016 that, as expected of someone in his position, he was not involved in the day-to-day operation of the program and did not know many of the details. The reason for the variance in the accounts is difficult to determine.

I asked Imam whether he wanted to publish his alloy results in a peer-reviewed journal. Imam explained that the fear of the new science ran all the way to the top of NRL:

> I was not encouraged to publish my results. The report that I, Miles, and Fleischmann wrote was initially rejected by Timothy P. Coffey, the director of NRL. He didn't want to approve the report because he thought it would harm NRL's reputation because it was controversial. Saalfeld disagreed with Coffey and said the results should be published. Saalfeld said that a result is a result and should be published, at least for internal use, but not for public use.

The final reports from the principal investigators (Dominguez, 1996; Miles, 1996; Miles, 2001) are available on the *New Energy Times* Web site. Mosier-Boss, who had never seen a copy of Dominguez' report until we obtained it, explained additional shortcomings of the NRL work. Her summary is also available on the *New Energy Times* Web site.

Despite the failure of NRL's experiments, Imam and Miles were granted two U.S. patents. Imam explained that, when Nagel heard about Miles' success with his alloy at China Lake, Nagel asked them to apply for a patent. Imam told Nagel it would never be approved because it was for "cold fusion." However, he focused on the fact that the alloy produced a very-high-strength material, and he mentioned only secondarily that it could be used to produce excess heat from electrolysis.

"It took five years," Imam said, "but the NRL scientific board and lawyers finally approved the application, and eventually the patent office issued two patents with Miles and me as inventors." (US 6,764,561 and US 7,381,368)

When the Tri-Navy Program ground to a halt in mid-1995, the person perhaps most affected by the termination of the program was Miles. It is a sad story that evoked painful memories for him, even years later. As described in the next chapter, Miles, like Bockris, faced penalties and punishment for thinking outside the box.

CHAPTER 17

Despair, Darkness, and a Glimmer of Hope

T he Tri-Navy Program, the informal name for the cooperative research effort funded by the Office of Naval Research (ONR), came to a lackluster end in 1995. As the anniversary of the first decade of the field approached, prospects for continued research opportunities dwindled.

Researchers at the Naval Research Laboratory (NRL) had failed to accurately repeat the experiment that Melvin Miles had performed at the Naval Air Weapons Station, in China Lake, California. But this was not publicly known until 2016. All that was known in 1995 was that NRL failed to confirm the excess heat reported by Miles.

ONR program manager Robert Nowak was officially in charge of that program. When I spoke with Miles in 2016, he searched his 1995 notebooks to refresh his memory of this history. According to Miles' notes, Nowak was disillusioned with the Navy program and, in February 1995, had turned control of it over to Debra Rolison.

Two more people unofficially managed the program. The first was David Nagel, an NRL division head who wrote the program proposal. The second was Michael Melich, a former NRL physicist.

"Both Melich and Nagel," Miles wrote, "were often involved in decisions about this program and in giving advice. However, neither actually did any research for the program." NRL's failure was bad news for Miles, particularly given the atmosphere of hostility from people outside the field. The end of the Tri-Navy Program marked the end of Miles' U.S. funding for his research, as he explained:

When I was first hired at China Lake in 1978, my supervisors were very supportive and always gave me sufficient internal research funding for my battery work. When I began cold fusion research in 1989, this all changed.

At China Lake, researchers had a salary, but everyone had to struggle to get funding. Equipment, travel and materials were paid out of this funding. If you did not have funding, you were placed on overhead, that is, your salary was taken from funds earmarked for the operation of the base. Alternatively, you had to get permission to use funding from someone else. Both options placed you in a vulnerable situation, and this was my situation from 1995 to 1997.

In February 1997, Miles' colleague Emilio del Giudice (1940-2014), an Italian theoretical physicist, invited Miles to Milan, Italy. They went to visit a laboratory that the Pirelli Research Group, part of the Pirelli Corporation, was considering funding the research. Miles was slated to head the laboratory.

Sent to the Stockroom

On April 23, 1997, while Miles was waiting to hear from Pirelli, he was called into a meeting with Robin Nissan, the head of the China Lake Chemistry Department, and Robert W. Gedridge, one of Nissan's branch heads. Miles' diary tells the story of what happened that day:

I was essentially told to stop writing papers, patents, and preparing for presentations except for directly funded work. I was to report to Juanita Morton to perform stockroom tasks, inventory of chemicals, etc. I was not to attend International Energy Conversion Engineering conferences or any other meetings. My work schedule was to be 0800 to 1800. Cannot change schedule. I refused to sign on this drastic change in my work. Obviously, Gedridge and Nissan want to get rid of me — [to] produce documentation for a bad performance rating — [to] get me out with the next [reduction-in-force].

Very depressed, stressed out about this treatment. I have led the research department for years in terms of publications, presentations. This would reduce me to less than a high-school helper. Stomach hurts, pain in chest. Thought I was going to have a heart attack later that day. Tried to work — but stress made it difficult. Why are they doing this?

Called Dr. Kevin Seymour for an appointment. Married Linda 19 April. Thought there would be something from the office — a card at least. Got nothing except this attack. How can I survive this treatment? I hope I can find a job elsewhere — but this is not easy at age 60.

A few months later, on July 29, 1997, Miles received bad news about Pirelli. "Emilio told me that Pirelli feared loss of respect by becoming involved in cold fusion; therefore, they pulled out," Miles wrote.

He never reported to the stockroom clerk. "At first," Miles wrote, "my response was passive resistance. I started the inventory in my own room working very slowly at this task instead of doing research."

Later that year, Miles received an invitation to do research for the New Energy Development Organization, in Japan, and took a six-month sabbatical from China Lake to work there. When that ended, Miles accepted a variety of teaching positions at universities in the U.S. In August 2001, he formally retired from the Navy.

Among Electrochemists

Electrochemist Robert Nowak was undeterred. Nowak had been a strong ally for many electrochemists in the field. Before helping to fund Miles at China Lake and Dawn Dominguez at NRL, Nowak had helped to fund electrochemists Martin Fleischmann and Stanley Pons at the University of Utah, John Bockris at Texas A&M University and perhaps other like-minded scientists.

In the mid-1990s, Nowak left ONR and went to the Defense Advanced Research Projects Agency (DARPA). There, he continued his support of the research by funding a team at SRI International led by Michael McKubre.

Undoubtedly, McKubre was the best-supported LENR researcher in the U.S. Funding from the Electric Power Research Institute and DARPA also supported other scientists at SRI International who worked with McKubre.

By the late 1990s, researchers like McKubre, who were convinced of the fusion concept, dominated the *ad hoc* International Advisory Committee (IAC) that managed the International Conference on Cold Fusion series.

The Continuing Fissure

The IAC was created in 1991. According to former members I spoke with, it operated without any written rules. The most important *de facto* rules were as follows: 1) Anybody who wanted to be the organizer of the next year's ICCF conference would submit a proposal to the IAC. Membership in the IAC was not required to submit a proposal. 2) Members of the IAC would approve or disapprove of such proposals and, if necessary, vote on competing proposals. 3) Membership on the IAC comprised the founding members as well as the chairmen of each new conference.

As of 2016, no woman had chaired an ICCF meeting. In addition to normal attrition, the chairmen of each year's conference sometimes dismissed members from the committee.

Most IAC members rejected the claims of their peers who reported heavy-element transmutations and light-water excess heat.

At ICCF-7, which took place in Vancouver, Canada, April 19-24, 1998, McKubre was spinning the illusion of scientific consensus and camaraderie in the field. ICCF-7 chairman Fred Jaeger, the president of ENECO, asked McKubre to give the closing comments for ICCF-7. Here is an excerpt:

There is a cohesive spirit in the cold fusion community. ... One of the strengths of this community is the attitude of warmth and friendship which surrounds us. I have a great hope that we will succeed. I think if you were to design a project, a concept, the sole purpose of which was to bring people from several nations together to work, people from several disciplines together to work, and to listen and talk to each other, you couldn't design one better than cold fusion.

It may be that Fleischmann knew this all along and has perpetrated this on us just for the purpose of an electrochemical — materials science — nuclear processes man-on-the-moon project. (McKubre, 1998)

The chairman of each ICCF conference had nearly total control over the program. He had the power, either directly or through delegation, to decide whose papers were accepted. The ICCF scientific review committees often existed only on paper, and the selection process for which scientists would be allowed to present papers and which topics would be discussed was in their control.

Peer Review in Crisis

On May 7, 1999, George Miley, was awarded a $100,000 federal contract for a proposal to use low-energy nuclear reactions to transform radioactive waste into harmless byproducts. Eugene Mallove called it a "stunning milestone in cold fusion history."

Mallove wrote that it was the first overt funding for the research by the Department of Energy (DOE) since 1989. Miley was one of 45 recipients among 308 peer-reviewed proposals sent to the DOE from universities, national laboratories and industry. In his proposal abstract, he discussed his reported transmutation and isotopic shift results and wrote that the transmutations were accompanied by excess heat. Miley's conclusion, as reported by Mallove, was thought-provoking:

This new field of low-energy nuclear reactions (LENRs) is so fundamental that it is impossible to predict where it will lead. Nuclear waste amelioration is one of the more obvious possibilities. Small nuclear power cells may also be feasible. Beyond scientific feasibility, all such applications must stand the tests of engineering and economic feasibility. In any case, the new fundamental knowledge gained from the proposed project could start a movement toward a whole new field of nuclear reaction applications, in addition to waste management. (Mallove, May 1999)

It didn't take long for trouble to start. A news article published in *Science* magazine in May criticized a meeting called Conference on Future Energy (COFE), where "cold fusion" papers had been presented. In response to the article, the organizer of the meeting, Thomas Valone, who is also a U.S. patent examiner, wrote a letter to *Science* in June.

Valone wrote about the political attacks against his conference from influential Washington, D.C.-based science critics, including Peter Zimmerman and Robert Park. In his letter, Valone mentioned that Miley had just won an award from DOE. The *Science* article had quoted Richard Garwin, a key participant in the 1989 DOE-sponsored review of "cold fusion." Mallove also wrote a letter to *Science*, objecting to the tone of the original article and criticizing Garwin for ignoring data, expressing an uninformed opinion, and suffering from paradigm paralysis. (*Science*, June 1999)

A month later, *Science* reported that DOE was considering revoking Miley's grant, which had not yet been funded. *Science* described Miley's award as a "potentially embarrassing stumble" for DOE. (*Science*, July 1999) On Sept. 7, 1999, the backroom politics succeeded: Miley's grant was killed by a secret panel of six independent peer reviewers selected by DOE's Office of Science. (Mallove, December 1999)

People like Zimmerman, Park, and Garwin were philosophically opposed to the idea that a broad new area of nuclear science might be possible. However, the problem Miley faced was much broader, as David Goodstein, then the vice-provost of the California Institute of Technology, explained in a paper. He wrote that, in the latter part of the

20th century, scientific research became detached from its roots, a time when progress was limited primarily by the imagination and creativity of its participants. The institution of anonymous peer review, Goodstein wrote, had become critically flawed and was and still is in critical danger:

> Peer review is quite a good way to identify valid science. It was wonderfully well suited to an earlier era, when progress in science was limited only by the number of good ideas available. Peer review is not at all well suited, however, to adjudicate an intense competition for scarce resources such as research funds or pages in prestigious journals. ...
>
> Editors of scientific journals and program officers at the funding agencies have the most to gain from peer review, and they steadfastly refuse to believe that anything might be wrong with the system. Their jobs are made easier because they have never had to take responsibility for decisions.
>
> They are also never called to account for their choice of referees, who in any case always have the proper credentials. Since the referees perform a professional service, almost always without pay, the primary responsibility of the editor or program officer is to protect the referee. Thus referees are never called to account for what they write in their reviews. As a result, referees are able, with relative impunity, to delay or deny funding or publication to their rivals. When misconduct of this kind occurs, it is the referee who is guilty, but it is the editors and program officers who are responsible for propagating a corrupt system that makes misconduct almost inevitable. (Goodstein, 2006)

I met Miley several years later, and he didn't want to talk about the 1999 incident. He seems not to have written about it. Theorist Robert W. Bass (1930-2013), who was known for his emotional communication style, voiced what Miley and other scientists might have felt. In addition to being a professor of physics and astronomy at Brigham Young University and a Rhodes Scholar, Bass was also the principal inventor of a magnetic confinement fusion design, which he called a Topolotron. He described the attitude of what he called the "scientific establishment" in 1999:

> The taxpayers have not yet realized how they have been betrayed by a selfish scientific oligarchy more concerned with retention of its present privileges than with the traditional scientific goal of the honorable objective search for truth.
>
> In the long run, their ignoble refusal to consider the evidence will do to the scientific establishment what the Inquisition's silencing of Galileo did to the late-medieval Church. The Cardinals who refused to look through Galileo's telescope for fear of jeopardizing vested interests are no different from those in today's high-energy physics and hot fusion research communities who have insisted upon classifying cold fusion a "pathological science" and have intimidated their students from researching it. (Bass, 1999)

---

10 Years Later and Still Struggling

On the 10-year anniversary of Fleischmann and Pons' fateful and mistaken announcement of room-temperature D+D nuclear fusion, the state of the field was gloomier than it had ever been. Major financial support in the U.S. had ended. The Japanese government's New Hydrogen Energy LENR research program ended in the late 1990s.

There had been progress in the 1990s, but with the exception of insiders, the broader scientific community didn't know about it and didn't care. The news media considered the subject dead, and with rare exceptions, editors of scientific journals would not send out submitted manuscripts for peer review.

Eugene Mallove asked for perspectives from the field's participants. (Mallove, April 1999) I have excerpted a few of the responses that are particularly illuminating:

Michael McKubre

It is been a long road, tortuous, often dark and occasionally painful. I would not have chosen any other course and could not have found better companions. The end is not in sight, but the path is clearly marked. Let us walk it with vision and with pride.

George Miley

Unfortunately, research in the area has been much slower than originally anticipated by Pons and Fleischman and other pioneers. No one anticipated the complexity of the problem. The early [idea] that any high school student could build and operate a cold fusion experiment must rank with the statement that nuclear fission would produce electricity "too cheap to meter" as classic mistakes. The polarization and viciousness that developed around the field, both by outsiders and insiders, has been much more severe than could ever have been predicted. Instead of people being drawn into cold fusion research by the exciting new science involved, they were driven away by the attacks on the field and on individuals.

Edmund Storms

Radiochemist Edmund Storms, who retired at 60 from Los Alamos National Laboratory (LANL) in 1991 after 34 years of service, was another contributor who offered a perspective at the 10-year mark. Since he retired from Los Alamos, Storms has worn the hats of both scientist and promoter, conflicting roles which compromised the integrity of each activity.

On June 23, 1989, he bypassed the LANL public affairs department and told the Salt Lake City *Deseret News* that he and his wife, Carol Talcott, had found tritium in their experiments.

A spokesman for LANL quoted in the newspaper came close to denying Storms' claim. Storms was sharply criticized by one of his peers for behavior that was "inappropriate and a breach of the normal standards of conduct for professional scientists." (*Fusion Fiasco*)

After his retirement, Storms became a prolific writer and promoter of "cold fusion," although most of his reviews of the topic were published on his own Web site (http://home.netcom.com/~storms2/) rather than in mainstream journals or encyclopedias. In his 10-year retrospective, Storms spoke about the research using religious terminology:

> The cold fusion debate has struggled on for almost ten years with essentially no progress in converting skeptics to believers. Indeed, the attitude of conventional scientists has hardened. On the other hand, new demonstrations of the claims have appeared, reproducibility has been achieved using several methods, and a relationship between heat production and appearance of a nuclear product has been determined. All of these successes were demanded by skeptics before claims could be accepted. So why has this acceptance been denied? ... Asking critics to carefully examine cold fusion today is rather like asking President [Bill Clinton] to define sex.

In fact, reproducibility of excess heat had not been achieved using any method, and the relationship between excess heat and the production of helium-4 was far from conclusive.

Richard Oriani

Richard Oriani (1920-2015) had worked for 44 years as an industrial chemist. After he earned a Ph.D. in physical chemistry from Princeton University, he worked for 11 years as a research associate at the General Electric Co. research laboratory and then for 21 years as assistant director for research in the Bain Laboratory of the United States Steel Corp.

After retirement, he became a professor in the Chemical Engineering and Materials Science Department of the University of Minnesota and the first director of the Corrosion Research Center at the university. He had early success in replicating Fleischmann and Pons' excess heat, but it didn't come easily, as he told me.

"I worked for three-and-a-half months before getting definite positive results," Oriani wrote. "I also had been researching the problem of hydrogen embrittlement of steels for quite some time." In 1990, Oriani was the second scientist (after Bockris' group) to publish an excess-heat confirmation. (Oriani, 1990)

Richard Oriani (Photo: S.B. Krivit)

In 1999, he offered a more scientifically accurate perspective than Storms did:

> Progress in enhancing the generation of thermal energy and in understanding the underlying mechanisms has been disappointingly low since 1989. In my opinion, several phenomena have been established. These include the possibility of generating more thermal energy than the energy supplied and that the excess cannot be understood by chemical reactions; the possibility of generating helium and tritium; and the possibility of generating elements that were not present in the specimen prior to the experiment. The word "possibility" is carefully employed here because each of these phenomena cannot be reproduced at will. ...
>
> Credibility of the field has been hurt by a large number of poor, uncritically done experiments and by a large number of poorly considered theoretical approaches. ... People who claim that their device or process can produce

excess energy reliably should, after patent application, if desirable, make their recipes or device openly available to others.

Peter Gluck

Another contributor to Mallove's 10-year perspective was Peter Gluck (b. 1937), an accomplished Romanian chemical engineer with 25 patents to his credit. Gluck was not a researcher but a passionate enthusiast and observer. He had a knack for creativity, systems thinking, and the art of problem-solving. He retired in 1999 and began writing and later blogging.

Unfortunately, as time went on, his passion for "cold fusion" turned to desperation and bitterness. In 2011, Gluck, along with Jed Rothwell and several well-recognized scientists in the LENR field, abandoned their sound judgment and began endorsing newcomers, particularly Andrea Rossi, who made fraudulent technological claims. (Chapter 35) Rossi had already been imprisoned for and convicted of fraud in his past energy ventures.

Gluck and Rothwell, however, seemed anxious to believe in Rossi's claim of a commercially viable 1 megawatt thermal fusion reactor. The pair used a blog and an Internet discussion list, respectively, to attack people who disagreed with them.

For many years, Gluck wanted to solve important technological problems in many realms, including energy. He was attracted to big problems, difficult problems, and this research was one such "wicked problem" that consumed his determined mind.

He envisioned a world that that would fulfill the interests of scientists, accommodating the natural level of "infinite complexity and provide an endless and eternal field of thinking and action for" scientists. He defined for himself a set of core values on which he based his actions: "Information sharing, communication, collaboration, coopetition, networking, friendship, collegiality, effectiveness and efficiency, and creativity."

When Gluck later took on a high-profile role of promoting Rossi, friendship and collegiality were no longer as important to Gluck.

Gluck had followed the research since 1989 and wrote several thoughtful essays in the early 1990s. At some point after 2003, he and I struck up a friendship. Gluck was primarily interested in excess heat and, as was natural for him, he looked at the reproducibility problem from a systems perspective.

The experiments performed with solid cathodes (as opposed to co-deposition experiments) were and, as of 2016, still are poorly reproducible. He knew that the rare instances of massive heating in electrolytic cells (such as Fleischmann and Pons in 1985 and Mizuno in 1991) represented an approximation of an engineering target for performance. He labeled cathodes used in these experiments "healthy."

However, he knew that the typical excess heat in the experiments produced only milliwatts-to-1 Watt, and in comparison, the majority of experiments were only marginally effective from an engineering perspective. He labeled these cathodes "sick." He labeled all the other cathodes that produced no excess heat "dead." It was a useful analytical process although it likely was not appreciated by scientists in the field. It reminded them of the rarity of the massive heating events in the face of their continued mediocre excess-heat results, results that rarely convinced anybody who wasn't already convinced.

Gluck had good reason to dream of better times. He had lived much of his life under the rule of Romanian dictator Nicolae Ceausescu, as he explained:

> I have a dream. When, and if, the state gives back the two houses confiscated in 1950 from my parents by the communists, I will be able to afford to establish a cold fusion laboratory at my former workplace. I will try to apply my ideas to create a reproducible gas-phase cold fusion experiment. Unfortunately, the state is represented by ill-willed, monstrous bureaucrats who are trying to delay the process, destroying my nerves, consuming my time and residual energy. Time is working against me.

In 2007, when I was on my way from a science conference in Istanbul, Turkey, to another conference in Russia, I stopped for a night in Romania to visit my former long-distance science pen pal. His town,

Cluj, is in the Transylvania region, best known in the U.S. for the legendary tales of Count Dracula. I was treated with extraordinary graciousness to a simple meal. Gluck introduced me to his family, including his wife, Judith, daughter, Antonia, and three grandchildren, Rudi, Silvia and Nora — who all seemed to have an infinite source of energy. Gluck's comments in 1999 captured his altruism:

> Cold fusion will change the lives of billions of people by assuring them independence and prosperity regarding energy. This belongs to my professional credo. However, it also has changed the lives of a selected group of people, even before it was accepted and applied. I belong to this group. Cold fusion is a unique intellectual and sentimental adventure.
>
> My life was intensified, my feelings of hope, disappointment, love, and hate became more vivid. As a result, I will live probably five years less than projected. But I don't regret it at all because I got the opportunity to live a real peak experience. No other scientific field could do this for me or for my friends. I was there from the very start.
>
> I became totally dedicated to CF. And finally, too, a devilish cycle of great expectation and giant disappointments started; however, after each wave, or ebb and tide, more and more certainties have accumulated, and the light(s) at the end of the tunnel are visible. And the field grew and grew and grew beyond any expectations and prediction. Many diverse systems and ten times more theories appeared in many parts of the world.
>
> It is a Herculean task to find some unity in this diversity. The field is cursed with irreproducibility; however, I think this is actually a blessing because it demonstrates that the multiplicity of phenomena takes place in elusive, strongly localized, active sites. As a marvel of synchronicity, the cold fusion revolution was associated with a political change and with the Internet renaissance, giving us communication beyond geography, collaboration beyond history,

information beyond informatics. All we lack now is a bit more understanding and open-mindedness.

Steven B. Krivit, Rudi, Silvia, Judith Gluck, Peter Gluck

After I investigated Andrea Rossi's claims in 2011 and published my conclusion that Rossi's claims were bogus, Gluck was so distraught and angry (not at Rossi, but at me) that he demanded that I abandon my profession as a journalist for, in Gluck's opinion, my erroneous conclusion about Rossi. Needless to say, I lost a friend.

John Bockris

John Bockris' contribution to the retrospective was much like that of Oriani, and Bockris emphasized the irreproducible production of heat, as well. One of the biggest problems, he wrote, was that the subject had been dominated by chemists, electrochemists, and physical chemists rather than nuclear scientists. He described how he envisioned the long-term future of the field:

> The discovery of low-energy nuclear reactions is on a par with the discovery of nuclear fission by Hahn and Meitner in 1939. It may be that we have the nuclear power source we need. But whether this will be in five or fifty

years, a great deal will depend on the dying-off of the old physicists and their replacement by a group willing to face facts.

Sometime after that, Bockris wrote a letter to his former student Tadahiko Mizuno offering advice on how to proceed and what might be necessary for success in the research. Mizuno was apparently having great difficulties. "My wife believes that you should stay at the university," Bockris wrote, "even if it costs you your life's work in cold fusion."

The first piece of advice he offered was to form an alliance among researchers. The second, easier said than done, was to identify a fully reproducible experiment. Mizuno explained Bockris' third suggestion:

> The third way to success is to decide on a theory, because nobody will accept anything in cold fusion unless they have a working model. By this he means that they must have a way of picturing what happens, whereby they can imagine how the reaction occurs. At this point, we will begin to convince people. Which theory should we settle on? There are hundreds of theories of cold fusion, but for [Bockris], the emphasis is always going to be on neutrons. (Mizuno, 2001)

As the new millennium approached, opportunities for the controversial research disappeared. The Japanese government program had ended. The Electric Power Research Institute's funding of SRI international's program had ended. Still, the scientists and their supporters held onto a glimmer of hope.

CHAPTER 18

The Raging Debate

Despite some progress in the research, most scientists who were pursuing the mysteries of the new science in 2000 experienced one disappointment after another. Funding was at an all-time low. Journal editors friendly to the topic were scarce. The spokesperson for the American Physical Society called the researchers "voodoo scientists." However, new data were emerging that offered potential to explain the phenomena.

By 2000, the majority of researchers were convinced that the primary nuclear product was helium-4 and that each helium-4 atom produced and released 24 MeV of heat. This quantitative correlation, they claimed, proved that the results were really products of a deuterium-deuterium (D+D) fusion process, because it had a distant relationship to the well-known D+D fusion reaction that occurs in super-hot ionized plasmas. Much to their dismay, nobody had measured a definitive correlation between the two in the benchtop experiments.

In 1991, electrochemists Melvin Miles (U.S. Navy-China Lake) and Benjamin Frederick Bush (b. ~ 1957) (University of Texas) were the first researchers to observe a partially quantitative correlation between the production of excess heat and helium-4. They claimed that their results proved the fusion idea. (Bush, Ben, 1991; Miles, 1991) However, their results did not convince even people in the field because the amount of helium-4 they measured was still below the ambient level of helium-4 in the atmosphere.

But helium-4 reported by SRI International electrochemist Michael McKubre in 2000 was higher than the level of atmospheric helium-4. McKubre's result boosted the D+D "cold fusion" idea and elevated his status in the field. In 1998, as McKubre followed the helium-4 and excess-heat trail, he knew what needed to be done.

McKubre had the money from the Defense Advanced Research Projects Agency, thanks to Robert Nowak. McKubre just needed a good experimental method that would reliably produce helium-4. In 1998, McKubre heard about an experiment that, rather than using deuterium in heavy-water electrolysis, used deuterium in a gas-loading experiment. It was the invention of Leslie Catron Case (1930-2010), who had three degrees in chemical engineering and a degree in business administration, all from MIT.

Case had worked for DuPont Chemical for a number of years, then switched to teaching chemistry. He started his own company and private laboratory in Nashua, New Hampshire, always hoping for but never achieving commercial success.

When the "cold fusion" news broke in 1989, it grabbed his attention. In October 1992, he went to the Third International Conference on Cold Fusion (ICCF-3), in Nagoya, Japan. There, he heard Eiichi Yamaguchi (b. 1955), a researcher at Nippon Telegraph & Telephone Corp. who had a doctorate in science, talk about his results.

Yamaguchi spoke at a press conference that coincided with the ICCF-3 meeting. "We now have evidence of the reality of cold fusion," Yamaguchi said. "Only nuclear fusion could have created the helium atoms." (*Los Angeles Times*, 1992)

It was true that only nuclear reactions could produce helium, but Yamaguchi's helium-4 looked nothing like the evidence expected for D+D fusion. Therefore, Yamaguchi didn't convince the broader scientific community. However, Les Case was impressed.

Yamaguchi had performed an experiment based on the titanium-chip and deuterium-gas experiment developed by Francesco Scaramuzzi, in April 1989. Scaramuzzi, the head of the cryogenics laboratory at Energia Nucleare e Energia Alternative (ENEA), the Italian equivalent of the U.S. Department of Energy, had cycled the deuterium-loaded titanium to liquid nitrogen temperature and back up to room-temperature. The Scaramuzzi group analyzed for neutrons and found them (Vol. 2, *Fusion Fiasco*), but the researchers didn't look for other products.

Later in 1989, researchers at the Bhabha Atomic Research Centre

(BARC), in India, repeated a version of the Scaramuzzi experiments. Rather than induce a temperature change from liquid nitrogen temperature to room temperature, the BARC researchers went in the other direction. They ran the cells through cycles of heating to 600° C and back down to room temperature. Yamaguchi started his experiments with similar temperature cycling.

In 1989, Yamaguchi reported seeing a "gigantic neutron burst ... and explosive release of deuterium gas." (Yamaguchi, 1990) By 1992, he also analyzed for and found helium-4. However, he didn't measure a precise quantitative correlation between excess heat and helium-4.

Case's experiment seemed to come closer. As a chemical engineer, Case recognized the advantages of avoiding the complexities associated with electrochemistry. He also realized that gas-phase systems would allow higher operating temperatures, compared with electrolysis experiments, which are limited by the boiling temperature of water. But he needed a nuclear laboratory to perform his research. He explained his search in an article in *Infinite Energy* magazine:

> There was no laboratory in the United States that I could find that would work with me. After all, it was cold fusion or something related to cold fusion, and most scientists wouldn't touch it — even for money. I finally determined that, because all Eastern Europe is known to be on a very low wage scale — low price scale—there were some Eastern European neutron laboratories that were of possible interest. So I got myself a plane ticket to Berlin and took the train going east to Warsaw. (Case, 1999)

He ended up at the Department of Physics and Mathematics at Charles University, in Prague, the Czech Republic, which welcomed him (and his money) with open arms. There, he began testing a variety of catalysts, placing palladium on a variety of other substances, eventually finding his best success with a palladium-on-carbon catalyst. Once he began seeing excess-heat results, he began filing foreign patent applications. He was disappointed by the lack of reaction to the publication of his first application in November 1996.

"I expected that there would be a very big response when this was published," Case wrote, "but there was no response whatever. Nobody was paying any attention." Finally, he decided to "take the bull by the horns," and he appeared in Vancouver, Canada, at the ICCF-7 conference in April 1998 unannounced and without submitting an abstract. Nevertheless, he was permitted to give a brief talk, at which he told people that he had developed a scalable, reproducible, experimental procedure for generating excess heat and helium-4. He claimed that his experiments were producing excess temperatures of 35° C, far greater than had been observed in any electrolytic experiments. (Case, 1998; 1999)

He said that samples he had sent to Oak Ridge National Laboratory measured helium-4 concentrations at 91 ppm, a scientifically significant number considering that background concentration of helium-4 in air is 5.2 ppm. Case thought that in a few years he could scale up his experiment to a self-sustaining commercial device that produced 100 megawatts of heat. It was rather optimistic.

At ICCF-7, unsurprisingly, he received a lot of attention, including an offer by McKubre to replicate the experiment. By the summer of 1998, SRI researchers had begun their attempts to repeat and confirm Case's experiment. McKubre reported the preliminary results on March 26, 1999, at the American Physical Society meeting in Atlanta, Georgia.

The meeting had been organized by Scott Chubb. Nagel wrote that Chubb's manager, Phillip R. Schwartz, the superintendent of the Naval Research Laboratory Remote Sensing Division, "stopped Chubb from working on 'cold fusion,' and restricted Chubb from studying and publishing in the area as part of his NRL duties." Chubb said that "the topic bothered Schwartz and he did not want to be involved."

For more than a decade, Chubb, without using his NRL affiliation, organized "cold fusion" sessions at the annual APS March meeting, not without some personal risk to his career.

In no way did the APS officially recognize the research, but it had an open policy of allowing its members to present short papers on any topic of their choosing. The APS did its best to obscure the "cold fusion" session and scheduled the 1999 meeting for 2 p.m. on Friday, by which time the weeklong meeting was mostly over. Exhibitors had packed up on Thursday, and most attendees had left by Friday morning. It was the first such session at APS since the 1989 fusion controversy.

McKubre and his colleagues reported two sets of experiments in 1999. The first was a set of electrochemical experiments performed by his team in 1994. McKubre said that his group came within 4% of measuring the correct amount of helium that would validate the proposed reaction $d + d \rightarrow Helium\text{-}4 + 23.8\ MeV\ (heat)$. Jed Rothwell, who reported on the meeting, did not explain why McKubre hadn't presented this data five years earlier, except to say that it was part of a larger EPRI-sponsored research program, the results of which had only recently published.

The EPRI publication was effectively an electric industry secret; it was available to members of the EPRI consortium of electric power companies, but to anyone else, it was priced at $20,000.

The second set of experiments was the SRI replications of the Case experiment. At SRI, the experiments produced only 11 ppm, not 91 ppm, as Case had reported. The APS procedures allowed only for 10-minute presentations, and proceedings were not normally published. In anticipation of the next ICCF conference, McKubre began preparing what may have been his most significant paper.

As the attendees of the ICCF-8 conference in May 2000 arrived in Lerici, Italy, McKubre approached researcher John Alfred Thompson (1954-2010), who told me the story many years later. Thompson said McKubre was exuberant. "Alf, I got it! I got it!" McKubre said, as Thompson remembered. Rothwell and Eugene Mallove captured the excitement in *Infinite Energy* magazine a few months later:

> The best news from ICCF-8 is that spectacular progress has been made in correlating helium-4 and helium-3 with excess heat, especially at SRI International. ... McKubre has outdone himself with the Case and Arata replications, which are far better than the originals. The helium issue is now closed. Cold fusion produces helium commensurate with heat, meaning the helium-4 production rate is comparable to a hot fusion deuterium-deuterium (D+D) reaction yielding helium-4 plus a 23.8 MeV gamma ray, only there is no

gamma ray, just equivalent energy in the form of heat.

The helium can only be a product of the reaction, not contamination, for several reasons: mainly because, in 4 out of 16 cases, it was measured at levels far above atmospheric concentration. In other words, the helium may have leaked out of the cell, but under no circumstances could it have leaked in. ... Helium has been detected for many years by different researchers, but never in such large amounts, at such high concentration. It has never been observed with such large signal-to-noise ratios or produced so consistently and reliably. ...

After the lecture, McKubre was asked about evidence of transmutations or other changes to the cathode material. He said it was too early to comment on this subject. (Rothwell and Mallove, 2000)

In fact, it was always too early for McKubre to comment about possible heavy-element transmutations. He had never looked for them, and he wasn't about to start now. Heavy-element transmutations would involve energetic reactions and would, therefore, invalidate the D+D "cold fusion" idea. McKubre had to have understood why. Rothwell knew why in 1999, when he reported on Benjamin Bush's presentation at the American Chemical Society national meeting in Anaheim, California:

[Benjamin] Bush dismisses reports of heavy-element transmutations. He saw no evidence for them in neutron activation analysis of his cathodes. ... He is adamant about transmutations; he told me, "I just walk away when the topic comes up." This reminds me of the anti-cold-fusion skeptics who refuse to look at data. Bush's main reason for [lack of] doubt is that the energy balance can be explained by the D+D reaction alone; there is no room in the equations for energy from heavy-element transmutation. This argument is weak. Edmund Storms [claims] that heavy-element transmutations produce much less energy than deuterium fusion and they would contribute little to the energy balance. Bush's energy balance data is crude, but McKubre has more refined statistics: 105% +/-10% of the expected helium balance. (Rothwell, 1999)

There are three important points to note about Rothwell's quote. First, he revealed that Benjamin Bush's reasoning was backward. If properly measured data provided evidence of heavy-element transmutation products, and the D+D "cold fusion" equation had no room for such products, then the equation, not the experimentally measured data, was wrong. The second point is that Storms, like Bush, allowed his belief to override the scientific method. The third point is that Rothwell, unlike Storms, had not yet begun to believe in the D+D fusion idea.

In the SRI replication of the Case experiment, McKubre reported that his group measured 31 MeV of excess heat per helium-4 atom produced. (An alternate calculation gives 32 MeV.) As for the 1994 experiment, McKubre never reported that measurement with an absolute value, like he did for the Case experiment. Rather, he obfuscated the facts and reported that the single helium measurement "results in a number that is 104 ± 10% of the number of atoms quantitatively correlated with the observed heat via the $d + d \rightarrow 4He + 23.82 MeV (lattice)$ reaction."

Nobody at the time apparently asked McKubre what the absolute value was for that measurement or why he took six years to report what was apparently, in 2000, the most important experiment that had ever been reported in the field.

Furthermore, Bush as well as McKubre knew that there was evidence for heavy-element transmutations in heavy-water experiments.

The Outlier

Thomas Passell (b. 1929) studied nuclear chemistry under the well-known nuclear chemist Glenn Seaborg at the University of California, Berkeley. Passell worked at Phillips Petroleum Atomic Energy Division and several other industrial companies before going to the Electric Power Research Institute (EPRI). There, as a program manager, he was responsible for managing most of the $10 million that EPRI spent to investigate the palladium-deuterium experiments.

I met Passell in 2003. He graciously sent me a copy of an unpublished 1993 report sent to the Pentagon from Richard Garwin and Nathan Lewis. As part of a JASON evaluation, Garwin and Lewis had been contracted to visit SRI and inspect the experiments. Both men had been very public in their dismissal of the research in 1989.

Garwin and Lewis reported that they had looked at experiments at SRI International and found that some had shown strong evidence of excess heat. They wrote that "such an excess could not possibly be of chemical origin." (In 2010, I discovered a 1998 EPRI report in which McKubre and his colleagues reported an error that invalidated the apparent excess heat measured for those experiments.)

Thomas Passell (Photo: S.B. Krivit)

Passell told me in 2004 that two public critics in 1989 asked him within a year for funds to study room-temperature fusion. Darleane Hoffmann (b. 1926), a nuclear chemist and professor at the University of California, Berkeley, was listed on a formal grant proposal. She had been a member of the 1989 Department of Energy Cold Fusion Panel that had dismissed the new research. (Vol. 2, *Fusion Fiasco*)

The other person who asked Passell for funds was Ronald Ballinger (b. 1945), a professor in the departments of Nuclear Engineering and Materials Science and Engineering at MIT. Ballinger testified in the April 26, 1989, congressional hearing on "cold fusion." Two days later, Ballinger and Ronald R. Parker, the director of the MIT Plasma Fusion Center, planted a story in the *Boston Herald* that accused Martin Fleischmann and Stanley Pons of fraud.

"Ballinger wanted $50,000 to do electrochemistry on the subject," Passell wrote, "as he told me during breakfast at the ICCF-1 conference in Salt Lake City in March 1990." Passel commented further about this in Mallove's "Cold Fusion: Fire From Water" video. "It goes to prove that the search for money in research is a very big thing," Passell said, "and sometimes takes precedence over the search for what we call pure truth." Passell wasn't convinced that helium-4 was the dominant nuclear product in the reactions or that its production was responsible for most of the excess heat measured in the experiments.

McKubre's Hidden Data

Passell was intrigued by possible isotopic changes in materials used in the experiments. He realized that evidence for abnormal isotopic abundances would give a) unambiguous proof of a nuclear reaction and b) information about the underlying process.

Passell produced five papers on this topic. One of the advantages of isotopic analyses was that he could perform them anytime after the experiments had completed, at least for stable nuclides. Unlike excess heat, the evidence didn't disappear immediately. Unlike helium-4, the evidence wasn't hard to capture.

Passell began his search in 1995, with a cathode used by McKubre in his palladium/deuterium electrolysis experiment #C-2 and its virgin material counterpart. The thermal data for this experiment is difficult to find; it is in EPRI report TR-104195 on PDF page 85. (McKubre, 1994)

That experiment produced a maximum absolute excess heat of 3 Watts and a total excess energy of 0.56 megajoules. The relative maximum excess heat was 8.6% greater than the input electrical power and 3.88% greater than the input electrical energy. Of 25 experiments in that series, C-2, which ran for 356 hours, produced the highest level of excess heat. Passell explained why he used the method of prompt gamma activation analysis (PGAA) for his analysis.

"The main advantage of PGAA is its ability to observe the entire contents of a sample cathode nondestructively," Passell wrote. "Most other methods observe a small volume of material near the surface or are destructive, in that they consume the sample. Thus, [with PGAA, researchers] need not wonder whether the sampled part of the cathode is representative of the whole or a local artifact."

Passell specifically looked for changes in boron-10. Isotopic abundances for most elements vary no more than a few thousandths of a percent among the various natural sources. Passell found an 18 percent reduction of the boron-10 isotope.

Passell calculated that, if the excess heat was produced uniformly in the cathode, it amounted to 66 megajoules per gram-molecular-weight of palladium. With that information, he made some calculations based on the nuclear binding energy necessary to make such isotopic changes:

> This work is predicated on the hypothesis that some reaction OTHER than D+D ["cold fusion"] is the likely heat- and helium-4-producing nuclear reaction. One such hypothesized nuclear reaction addressed by this work is *boron-10 + d —> helium-4 + beryllium-8* followed by the breakup of beryllium-8 into two more nuclei of helium-4. The heat produced per reaction is 17.81 MeV. ... It is encouraging to note that the 18% depletion in boron-10 corresponds almost exactly to the number of reactions ... needed to explain the 0.56 megajoules of excess heat observed. (Passell, 1996)

The isotopic changes showed that D+D fusion inadequately explained McKubre's C-2 experiment and that other reaction mechanisms were possible, if not probable. Passell was perplexed why his analyses never revealed unstable nuclides. He had difficulty reconciling this data with his training in physics and nuclear chemistry.

"The most troubling observation," Passell wrote, "is the almost complete absence of radioactivity in palladium and nickel cathodes which have apparently produced excess heat. That is, 'Why should nuclear reactions producing only stable isotopes be the heat producers?'"

From a technology and environmental perspective, it was a wonderful problem to have. He had no answer, but he did know that, if the reactions could be scaled up, deuterium, assuming it was the primary fuel in the experiments, offered an energy source more plentiful than any existing raw fuel source. Passell was determined to search for understanding of the excess heat because the power industry's major business is the conversion of heat to electricity.

Isotopic Shifts Everywhere

In 1999, Passell reported the results of another pair of isotopic analyses. (Machiels and Passell, 1999) This set included a cathode used in an excess-heat-producing experiment performed by Stanley Pons and its virgin counterpart. Passell had funded researchers at the University of Texas, Austin, to perform neutron activation analysis (NAA) to determine the isotopic ratios. The senior principal investigator there was Joseph J. Lagowski (1930-2014). The other principal investigator was Benjamin Bush, the same scientist who, at the American Chemical Society meeting that year, acrimoniously dismissed all reports of heavy-element transmutations.

The University of Texas researchers (and staff members who performed the NAA) measured a 7% depletion of palladium-110 atoms. They also observed a variety of transmutations on the Pons cathode. Compared with the virgin material, they measured increases in six elements, including 56 times the amount of iron and 11 times the amount of zinc.

The amount of heat produced in the experiment, according to Passell's calculation of nuclear binding energy released by the isotope change, was in agreement with the excess heat Pons measured.

Passell performed a similar analysis on three samples from a palladium-deuterium gas-electrolysis experiment performed by Yoshiaki

Arata (b. 1924) and Yue-Chang Zhang, at Osaka University. Arata, a Japanese pioneer in thermonuclear fusion, developed what he called a double-structure cathode.

The cathode contained a hollow core in which finely divided palladium, also called palladium-black, was inserted before the experiment began. After insertion of the palladium-black, the core was welded shut. The palladium-black material was protected from the electrolyte inside the gas-pressure-tight core of the cathode. As deuterium gas separated from oxygen in the heavy-water electrolysis, it permeated the outer structure and entered the hollow core.

In all likelihood, Arata knew about the 1926 experiment by two German scientists, Fritz Paneth and Kurt Peters. They reported the transmutation of normal hydrogen to helium using a method of hydrogen-gas absorption into finely divided palladium. In 1927, Paneth tried to discredit his and Peters' experiment, but his reasoning does not hold up to close scrutiny. (Vol. 3, *Lost History*)

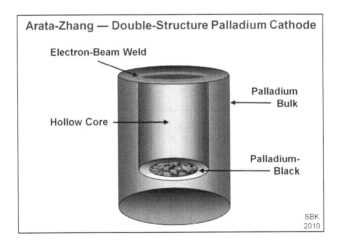

Arata-Zhang — Double-Structure Palladium Cathode

Electron-Beam Weld

Palladium Bulk

Hollow Core

Palladium-Black

SBK 2010

Each sample of Arata's material revealed significant isotopic increases in zinc-64 and in palladium-110. (Passell, 2000) The samples also revealed significant changes in the amount of gold and iridium. Here is the full list of changes measured:

- 7-15 times the zinc-64 by weight
- 8 times the iridium content by weight from sample A
- 0.4 times (decrease) the iridium content by weight from sample B
- 6 times the iridium content by weight from sample C
- 5.5 times the gold content by weight from sample A
- 0.1 times the gold content by weight from sample B
- 0.7 times the gold content by weight from sample C
- 24% increase in Pd-110/Pd-102 ratio over virgin palladium from sample A
- 6% increase in Pd-110/Pd-102 ratio over virgin palladium from sample B
- 21% increase in Pd-110/Pd-102 ratio over virgin palladium from sample C

Passell estimated that the nuclear binding energy released by the production of the excess zinc-64 would be 20 megajoules, which agrees approximately with the 30-to-40 megajoules of excess heat Arata and Zhang measured in similar experiments. (Passell, 2000)

Two years later, at the ICCF-9 conference, Passell showed his colleagues that a variety of other possible nuclear reactions besides D+D fusion could produce helium-4. (Passell, 2002)

At ICCF-10, Passell published more details about his analysis of the Arata-Zhang material. He gently suggested that the greatest revelation of the data is that D+D fusion was "not necessarily the only path forward in understanding these phenomena." (Passell, 2003) But this wasn't information that the fusion proponents wanted to hear, so they ignored it.

Helium Hypothesis Disproved

Certainly, Passell did not intend to embarrass prominent fusion-oriented scientists like McKubre and Arata. However, materials from their own experiments, as Passell showed, disproved their hypothesis that the sole nuclear product was helium-4 and that no other exothermic reactions took place inside the experiments. Although Passell's alternate proposed mechanisms to create helium-4 were as unlikely as room-temperature D+D fusion, he listed other possible explanations.

Two more considerations are important. The first is that, for each of the five samples Passell analyzed from experiments that produced excess heat in McKubre's, Pons' and Arata and Zhang's labs, every one showed isotopic anomalies. This is a 100% success ratio, better than can be said about sets of experiments that produced helium-4.

The second consideration is that researchers who were trying to explain the experimental data as D+D fusion rarely, if ever, made the effort to search for isotopic anomalies. This evaluation applies to McKubre and Arata but not necessarily to Pons, who was clearly open-minded about light-water excess heat.

Revolution-in-Hiding?

In May 1994, Mallove had enthusiastically reported spectacular news about results reported from Siena, Italy. The data was from the experiment developed by Francesco Piantelli and his colleagues Sergio Focardi and Roberto Habel, in which normal hydrogen gas was loaded into a nickel rod. It had produced a record level of excess heat.

In 2000, at ICCF-8 in Lerici, Italy, a member of the group, Focardi, presented the group's results for the first time at an ICCF meeting. The researchers had new results: an experiment that reached a maximum excess power of 70 Watts, with an average of 40 Watts sustained over 10 months. They also found a variety of new elements that were undetectable on the nickel rod before the experiment.

In their extensive review of ICCF-8, Mallove and Rothwell made no mention of the Piantelli group's paper. Several factors may account for this omission: the news of McKubre's heavy-hydrogen experiment took center stage; no independent researchers had replicated Piantelli's normal-hydrogen idea yet; and more researchers thought that normal-hydrogen experiments, as asserted by the D+D fusion proponents, produced no excess heat.

I learned more about Piantelli when I visited him at his laboratory in 2007.

Rocking the Boat

Tadahiko Mizuno and Tadayoshi Ohmori had also reported light-water excess-heat results at the May 2000 ICCF-8 meeting in Italy. They reported results from a high-voltage plasma electrolysis experiment using a tungsten cathode and potassium carbonate electrolyte.

Mizuno told me they chose tungsten because its melting point is 3,422° C, the highest of all metals, and more than double the melting point of palladium. They tried several other metals; all but tungsten melted quickly and made it difficult to continue the plasma electrolysis.

They had reported their first results with this method two years earlier, at ICCF-7. In that paper, they had reported excess heat and few heavy-element transmutations. They did not present clearly their before- and after-electrolysis values of the detected elements on the cathode. They suggested a very general mechanism, based not on fusion or fission but on neutrons and weak interactions. (Ohmori, 1998)

At ICCF-8, they presented a curious array of transmutations, along with the production of excess heat. (Mizuno, 2000) The shaded elements were detected after the experiment with a large amount of excess heat. The numbers below each element show how many times greater its abundance was compared to the virgin material.

20 Ca	21 Sc	22 Ti	23 V	24 Cr	25 Mn	26 Fe	27 Co	28 Ni	29 Cu	30 Zn	31 Ga	32 Ge
7.5x	ND	800x	ND	1400x	ND	104x	ND	20x	<1x	750x	ND	120x
								46 Pd	47 Ag	48 Cd	49 In	
ND = Not Detected								50x	12x	ND	110x	SBK 2016

Some of the elemental changes reported by Mizuno and Ohmori in 2000

They didn't search for helium-4 because they didn't have adequate measurement devices. Four points are worth noting. First, the abundance of seven elements they measured after electrolysis was more than 100 times greater than before electrolysis (Al, 200x, is not shown above). Second, they showed a direct comparison of the elemental abundances between an experiment that produced a large amount of excess heat and another that produced a small amount of excess heat. With the exception of copper, all of the detected elements were more abundant in the experiment with the larger excess heat. Third, for unknown reasons, the newly detected elements revealed a pattern based on their atomic weights. Fourth, the published data revealed there were no newly detected elements that did not exist at or greater than trace levels before the experiment.

Raging Debate

A few weeks before Mallove went to ICCF-8, he prepared his next editorial: "A debate has raged since 1991 on whether these ordinary hydrogen systems are giving rise to such mind-boggling occurrences as the low-energy transmutation of heavy elements." (Mallove, 2000)

Theorist Lewis Larsen, who had begun learning about the research a few years earlier, remembered the intense reaction from the fusion faction. "I had been talking with all the well-known American researchers, such as Michael McKubre, Peter Hagelstein, Edmund Storms," Larsen said. "They said that you couldn't get excess heat or transmutations with light water and that Miley and the Japanese people were crazy because there was no deuterium in the system, and therefore they couldn't explain the data with their D+D fusion idea."

CHAPTER 19

Insights and Gun Sights

A creative, open-minded Japanese researcher who was willing to look for heavy-element transmutations and isotopic shifts provided a new and valuable window into the study of these experimental phenomena.

In 2001, Yasuhiro Iwamura (b. 1961), one of the few relatively young researchers in the field, was gaining recognition. Researcher Takehiko Itoh, one of his close colleagues, had worked alongside Iwamura for a decade. Along with other researchers at Mitsubishi Heavy Industries Ltd., they measured some of the most convincing experimental evidence in the field.

Iwamura earned his Ph.D. in nuclear engineering in 1990, then worked in the Applied Physics Group for Mitsubishi at the company's Advanced Technology Research Center, in Yokohama, Japan. His primary areas of research were the technology used for the company's nuclear power plants and neutron irradiation technology for the detection of explosives. By 2015, Iwamura managed a technology intelligence group at Mitsubishi. In March 2015, he left Mitsubishi and took a leadership role in the newly formed Condensed Matter Nuclear Science group at Tohoku University.

Iwamura started his low-energy nuclear reaction research in 1993, with a deuterium-gas experiment heated to 126° C. At the time, the typical nuclear products expected by researchers were helium, neutrons and tritium. Iwamura analyzed for neutrons and tritium and found both, but the reproducibility of the results was poor. (Iwamura, 1993)

The following year, he started electrochemical experiments with a palladium cathode (atomic number 46) and a platinum anode (atomic number 78). He observed X-ray and neutron emissions and reported his first evidence of a heavy-element transmutation, a quantity of atoms of lead (atomic number 82) on the palladium cathode after electrolysis. (Iwamura, 1995)

He continued his experiments, and in 1998, at the ICCF-7 meeting, reported results from a new experimental method he had developed.

The experiment consisted of two chambers separated by an opening: an upper chamber filled with heavy-water electrolyte and a bottom chamber that was evacuated. The cathode, a thin plate, was mounted to the bottom of the top chamber with an O-ring gasket and covered the opening. The anode was suspended above the cathode. During the experiment, deuterons migrated from the electrolyte, through the cathode, and into the vacuum chamber.

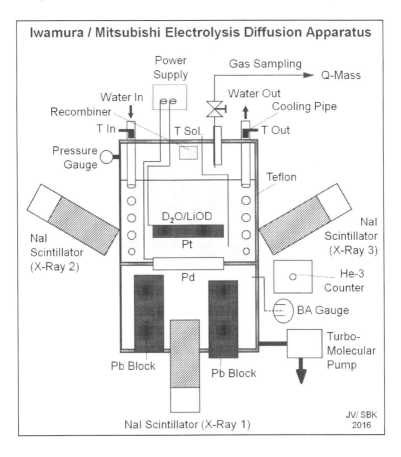

Iwamura / Mitsubishi Electrolysis Diffusion Apparatus

Iwamura and his colleagues measured excess heat from their system, but the most surprising result, as they wrote in a paper they submitted to *Fusion Technology* in 1997, was the detection of various elements that did not exist before electrolysis. (Iwamura, 1998) They found titanium as well as an unusual ratio of iron isotopes on the cathodes. They also found half a dozen elements that had not existed on the cathode before electrolysis, but not at sufficient levels that they could exclude the possibility of contamination in the electrolytic cell.

Yasuhiro Iwamura (Photo: D. Tran)

Researchers knew by 1998 that a high deuterium-palladium ratio was needed to produce positive results in the experiments. But Iwamura focused on creating a high ratio on the palladium surface rather than in the bulk of the palladium. He also thought that rapid diffusion of deuterium through the palladium cathode was important. Last, Iwamura came up with the idea that a second material, such as calcium oxide, sandwiched between thin layers of palladium would help induce successful reactions. Using these three concepts, he was able to dramatically increase his success rate and observe heavy-element transmutations and isotopic changes.

In the *Fusion Technology* paper, Iwamura and his colleagues also described a weak-interaction, neutron-based theoretical model that they thought was most likely to explain the phenomena. They called it the "electron-induced nuclear reaction model" and described it with the equation $d + e \rightarrow 2n + v$. In other words, a deuteron reacted with an electron and formed two neutrons and a neutrino.

Their model also depicted a reaction path for the creation of tritium, $p + 2n \rightarrow t \rightarrow He\text{-}3$, as well as a reaction path for the creation of helium-4, $d + 2n \rightarrow H\text{-}4 \rightarrow He\text{-}4$. The authors did not offer a detailed explanation of the model, and they admitted that it was just an assumption to explain the observed experimental results. In fact, it turned out to be a prescient assumption.

In 1999, Japanese researchers created the Japan CF Research Society, and began organizing their own local conferences. The society held its second annual meeting at Hokkaido University, Japan, in October 2000. Jed Rothwell attended, and he reported the news in *Infinite Energy* magazine. Iwamura's team members, Rothwell wrote, "were the stars of the conference." Rothwell described Mitsubishi researcher Mitsuru Sakano's presentation:

> Sakano described a new set of experiments that expand the Mitsubishi research program and confirm the original work. In the new experiments, they use similar detection equipment and multilayer membrane palladium plates, as before, but they employ gas-loading instead of electrolysis. In the original experiment, which is still continuing, they use liquid electrolysis, and they observed excess heat and massive transmutations in every run. .... Spectroscopy is performed on the cathode material before and after the run.
>
> In the new experiment, the top chamber is loaded with low-pressure (1 atm) deuterium gas. (Higher pressure might load the metal faster, but it might rupture the thin membrane.) The deuterium is absorbed into the top surface of the palladium membrane, and it is gradually sucked through the metal into the vacuum chamber below. The palladium undergoes transmutations similar to those observed with electrolysis. Gas loading pushes the deuterons in more slowly than electrolysis, so the reaction rate is

thousands of times slower. They assume it would be impossible to detect the excess heat at such a low rate, and they do not perform calorimetry. The reaction is slower, but contamination is also reduced by several orders of magnitude.

Another major advantage to the gas-loading experiment is that it allows them to watch the transmutations as they occur, in the cell, without removing or disturbing the palladium sample, which might introduce contamination. An on-line, X-ray Photoelectron Spectrometry (XPS) apparatus is installed in the top chamber.

At the beginning of the run, they evacuate the top chamber, and they examine the surface of the palladium and inventory the elements and isotopes in it. As the experiment proceeds, every week they evacuate the deuterium gas again and take another reading with the XPS apparatus. They see surface transmutations gradually develop and change. ...

After the run, various methods of off-line spectroscopy confirm new element production and isotope shifts. The total mass of some of the new elements is much larger than all sources of contamination can account for. In a few cases, it is so large that the products are visible, forming a small pile of material about the size of the graphite in a pencil point, as shown in a photograph. ...

The Iwamura research program is a masterpiece. This is the most dramatic and clear-cut proof yet revealed that cold fusion transmutes metal, and it is one of the most important experiments in the history of the field. The quality of the data, the care, precision, and high signal-to-noise ratio is rivaled only by the results obtained by McKubre. These are also the most expensive experiments in the history of this field. The dedicated, in situ spectrometer alone costs an estimated $700,000, and it is bigger than Mizuno's entire laboratory. The entire experiment is performed in a clean room, which must cost millions.

[In May this year, as he reported at ICCF-8,] McKubre

proved once and for all that palladium cold fusion produces helium. Iwamura has proved it also produces transmutation. ... I have no idea how to reconcile these two findings. ... McKubre has concluded the principal reaction is definitely fusion: a D+D reaction resulting in helium and occasionally tritium. He hedged when I asked him whether he has observed transmutations in the cathode. He said that, years ago, they checked palladium cathodes for signs of transmutations. They found confusing, unconvincing evidence of foreign material with possible isotope shifts. ...

I asked Iwamura whether these experiments have been independently replicated or otherwise confirmed. They have not been replicated, because of the expertise required and the astronomical cost. However, the results have been independently confirmed. Used palladium cathodes have been sent to laboratories in Japan and Europe for specialized spectroscopy and other tests. These independent tests with different instrument types confirm the isotope shifts, which prove the new elements cannot be contamination. The shifts are orders of magnitude greater than changes induced by things like SIMS [secondary ion mass-spectroscopy] or by electrochemical isotope separation, which, in any case, cannot occur in the gas-loading experiments. ... I asked Iwamura if there is opposition within Mitsubishi toward this research. He said that he has the support of his own upper management, but other people in the company think the research is controversial and even embarrassing, and they want him to keep a low profile. ...

In conversation, Iwamura laughed and said, "You know, people have accused us of faking isotope shifts. I don't know how anyone could fake such a thing, but if we have, it is an amazing accomplishment!" (Rothwell, 2001)

A decade later, in October 2010, the Mitsubishi researchers switched back to the electrolytic method and increased the magnitude of their transmutation yields, on average, by a factor of 100. (Iwamura, 2015)

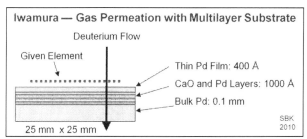

In the course of the experiments, the given element, placed on top of the multilayer substrate, gradually changes to the target element.

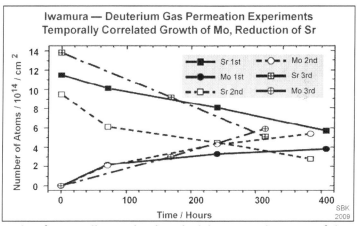

Examples of temporally correlated gradual decrease and increase of elements.

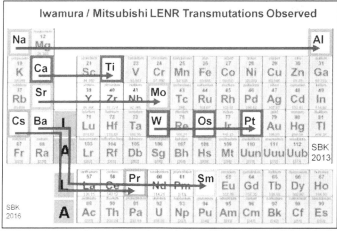

Graphical summary of transmutations observed by Iwamura/Mitsubishi group

Iwamura Transmutations Summary

$$^{133}_{55}Cs \xrightarrow[Z+4]{M+8} {}^{141}_{59}Pr$$

$$^{88}_{38}Sr \xrightarrow[Z+4]{M+8} {}^{96}_{42}Mo$$

$$^{138}_{56}Ba \xrightarrow[Z+6]{M+12} {}^{150}_{62}Sm$$

$$^{137}_{56}Ba \xrightarrow[Z+6]{M+12} {}^{149}_{62}Sm$$

$$^{44}_{20}Ca \xrightarrow[Z+2]{M+4} {}^{48}_{22}Ti$$

$$^{184}_{74}W \xrightarrow[Z+2]{M+4} {}^{188}_{76}Os$$

MITSUBISHI DATA   SBK 2012

Numerical summary of transmutations observed by Iwamura/Mitsubishi group

From Bolivia to Texas

Two unconventional American researchers also brought innovative ideas to the subject. Dennis Letts isn't a typical researcher. Nor is he a typical Texas oilman. In fact, there is probably very little that's typical about Letts, who performs his experiments in a small shed at the back of his Austin, Texas, property.

His longtime colleague Dennis Cravens, of Cloudcroft, New Mexico, a retired chemistry and physics professor, told me that, in his spare time, Letts enjoys going to Pacific islands and beaches with his wife, Kathy. "Letts has a warped mind," Cravens wrote. "He takes his physics books and reads instead of looking at the scenery." Not only does Letts own a few oil wells in Oklahoma, but he's also spent time mining gold in Bolivia. Letts told me his story:

> I was operating two mechanized gold mines in the middle of the jungle in Bolivia in 1989-1991. I called home once a week from the site. My wife, Kathy, told me about the cold fusion announcement in March 1989. I immediately decided to study the discovery.
>
> As soon as I returned to Texas in December 1989, I

went directly to the University of Texas chemistry library and made a copy of Pons and Fleischmann's historic paper. From that beginning, I never dreamed that one day I would have a beer with Martin Fleischmann in Cambridge.

Time passed, and whenever I came back to Texas from Bolivia, I went to the chemistry library to make copies of more papers. I read them with great passion, to the point that I couldn't sleep at night. Every spare moment I had was devoted to understanding the physics and electrochemistry. I carried 50 pounds of physics books and cold fusion scientific papers in all of my subsequent travels in Bolivia. Here are a couple of photos from that period: I'm on the left, and a partner from my oil and gas days, Frank Cameron, is on the right.

Dennis Letts, Frank Cameron and the 1991 Rio Solacama Mining Camp military police.

Letts had a license to carry firearms from the Bolivian military police for mine security and had earned the friendship of a detachment of military police at the mine site.

Once he was done with gold mining, he began palladium-deuterium experiments full time on July 10, 1992, for the most part self-funded. Very early in the development of the topic, researchers knew that the excess-heat effect required some form of external triggering. Even if researchers got lucky and used a cathode that had the correct (yet still-unknown) metallurgical properties, their experiments rarely produced excess heat without some kind of disruption to the electrolytic equilibrium. Some form of triggering was required to cause instability, disequilibrium, in the cell. A sudden increase in current would do the job; even a sudden decrease would work, as would external stimuli.

In 1993, Letts performed experiments with John Bockris at Texas A&M University. They tried radio-frequency (RF) triggering and found that it worked but was difficult to reproduce.

To reduce costs, instead of platinum, Letts used a gold wire as the anode. Unfortunately, the gold went into solution and transferred to the cathode, ruining the RF experiment. The gold coated the cathode with a dark material (gold oxide), and it seemed to Letts that the RF would not get through the thick deposit. Letts remembered his moment of discovery, experiment #500, on September 29, 2000, at 8:13 p.m., dutifully noted in his laboratory notebook:

> I stopped the RF experiment, opened my tool drawer and saw a 1 mW red laser pointer. Several years earlier, I had seen excess-power results with infrared diodes, so I thought I'd give visible light a try. On a whim, and to avoid abandoning the experiment completely, I decided to try shining the laser on the cathode. A few moments later, the temperature of the electrolyte shot up by several degrees.

The Shotgun and the Station Wagon

Eventually, word about Letts' discovery traveled, first to Peter Hagelstein at MIT, then to Michael McKubre at SRI International. McKubre was fascinated and, two years later, invited Letts to bring his experiment to McKubre's lab in Menlo Park, California.

Letts didn't volunteer this story to me, but an interesting man named Bill Harrington, who heard it from Letts, and whom I later met, did. Afterward, I cross-checked the story with Letts, and he reluctantly confirmed it.

Harrington is a wild character; he looks and sounds like a combination of a backwoods conspiracy theorist, a physicist, and a crazed science wizard. Here's the story, as told by Harrington:

> Letts wanted to get the experiment going in his lab first, make sure it was generating excess heat, before bringing it to McKubre's lab. He didn't want to take a chance that it wouldn't work when he got to California.
>
> So in April 2002, he took his little "cold fusion" cell, which is about the size of a large grapefruit, and stuck it in a wine-chiller cabinet that he bought from the Sharper Image catalog. So he had a nice, cool, temperature-controlled portable box for the cell. He put it in the back of his car and attached it to a backup-battery power supply. He plugged the whole apparatus into the cigarette lighter, and he and his wife proceeded to drive cross-country.
>
> He and Cathy had their operation planned meticulously. They would pull into a motel; Cathy would get out and check in and made sure to get a room on the ground floor. She would get back into the car, and they would drive over to the room. Dennis would back the station wagon up to the doorway of the room. Cathy would open the door to the room while Dennis opened the back of the car.
>
> Dennis would flip the device onto its back-up battery, carry it into the room, plug it into the wall outlet, and let it run overnight. The reactor was running continuously from Texas to California.
>
> Meanwhile, Dennis and Cathy were enjoying their cross-country trip, stopping at various tourist places and having a good time during the five days with this nuclear reactor — and a shotgun under the back seat.
>
> They got to SRI, and McKubre and his buddies in their white coats came down to the parking lot, and they're looking at this thing. They saw this guy with a Texas twang and license plate who's come halfway across the country with a nuclear reactor running in the back of his car.

The purpose of the shotgun, Dennis told me, was "to proactively eliminate any misunderstandings."

Nothing, and nobody, apparently, was going to come between Dennis Letts and his electrolytic cell. His proclivity for firearms notwithstanding, Letts is one of the most gentlemanly and considerate researchers I've met in the field. He is a peaceful man with an encyclopedic knowledge of self-learned science.

When I interviewed him in 2003, he had performed more than 6,000 experiments. He is a true Edisonian-style researcher: He allows his creative mind the freedom to explore and fail. His experiments have failed many more times than they have succeeded, but this has never discouraged him. However, his cross-country trip in 2002 was one of his most memorable experiments:

> We made a detour to the south rim of the Grand Canyon for a two-hour visit. I guess that makes me the only cold fusion researcher to have run a Fleischmann-Pons cell at the Grand Canyon. When we got to SRI International, I'm pleased to say that it was still producing excess heat.

Cravens, who collaborated with Letts on the laser-triggering discovery, is the kind of guy you'd want with you if you were stranded on a desert island. In fact, when the Turner Broadcasting System produced *The Real Gilligan's Island*, a reality-show remake of the classic 1960's television sitcom, Craven's was a finalist for role of the professor. The producers invited him to a hotel in Beverly Hills, California, for an interview.

"I showed them how to get the latitude and longitude of the hotel with a string and watch and was within 100 feet," Cravens wrote. He did not end up getting the permanent role, but he appeared briefly in one of the first episodes. "I was the crazy professor with the bamboo flashlight made from lemons and limes — electrochemistry is good for something! The trick, of course, is the two kinds of metal I used for the battery, platinum and magnesium."

CHAPTER 20

American Revival

Ten years had passed since an International Conference on Cold Fusion (ICCF) meeting had taken place on U.S. soil. ICCF-10 was held at the Royal Sonesta Hotel, in Cambridge, Massachusetts, Aug. 24-29, 2003. By then, most of the original researchers had reached retirement age. It was the first ICCF meeting I attended, and I met most of the remaining researchers there.

I had conducted my first interview with a scientist, Edmund Storms, earlier that year, on Feb. 22, 2003. He appeared to have a broad understanding of the research. He generously spent a full day teaching me about the science in his home laboratory in Santa Fe, New Mexico. I had the impression that he was an objective and thoughtfully critical authority on the subject.

Michael McKubre was my second interviewee. I met him on Aug. 8, 2003, at his office at SRI International, in Menlo Park, California. He showed me the SRI lab, which contained several exciting-looking experiments and cool analytical devices. I had never been in a scientific laboratory before, and it was impressive, as was the chance to meet a scientist who was so well-regarded in the field.

In 2003, the knowledge gap between people who thought the science was completely wrong and people who thought it was completely valid was wider than it had ever been. For this reason, I wanted to talk directly to the researchers — and their critics — and form my own opinions and conclusions.

I asked Peter Hagelstein, the chairman of the ICCF-10 meeting, for permission to register as a news media representative. He reviewed my Web site and told me that my basic idea was good but that he had concerns about some of my content. He wrote to me that Mallove, in his *Infinite Energy* magazine, had expanded his scope beyond fusion.

"Much of what he has included in the scope is pretty far-out stuff," Hagelstein wrote, "and the credibility factor has dropped. I am wondering whether you are going to go the same route. There is more than enough solid material on cold fusion available to make for an interesting Web site. You undercut your website by including things that are either wrong or way off topic."

He added that the competing theoretical ideas proposed by Talbot Chubb (1923-2011) and his nephew Scott Chubb were wrong. This seemed peculiar because Hagelstein's technical chairman for the conference was the younger Chubb.

"The article by Talbot," Hagelstein wrote, "includes some material which is in textbooks and some material that is consistent with the theoretical models that he and Scott have been pursuing. Unfortunately, Talbot does not distinguish between what is accepted and what he and Scott are proposing. Their models, for better or for worse, are not widely accepted by anyone at this point. Hence, when the ideas that Scott and Talbot present are given equal weight and not distinguished from the material covering accepted physics, it does real damage."

The criticism was remarkable; Hagelstein's theoretical models were no more widely accepted than any other model that purported to explain how fusion could occur at room temperature.

I was eager to receive a media credential for the conference and gain access to sources for my reporting, so I reassured Hagelstein that I would keep my focus on "cold fusion," review the content of my Web site, and make appropriate changes. The meeting was useful for me. I was able to meet and exchange contact information with many researchers from around the world. Equally significant was my opportunity to learn about the science, the various experimental methods, the instruments, and the analytical procedures.

It was there that I met Fleischmann, the patriarch of this subculture. I was told that I could find him in the bar, and I did, with a glass of beer in his hand. I was thrilled and felt honored to meet him. He was patient and tolerant of my enthusiasm, despite his somber mood. I knew nothing about what he had gone through in his years as a scientific outcast; I was mystified at his melancholy. It was heartbreaking, as I learned through the course of our conversation, that so many doors had closed for him after his and Stanley Pons' public announcement in 1989.

The complete story of how he and Pons were pressured into their premature announcement as a result of the conflict with Steven Jones and the influence of the University of Utah was not widely known until 2016. (Vol. 2, *Fusion Fiasco*). My interview with Fleischmann marked the beginning of a professional friendship that lasted for many years.

In the Corner

Something strange happened to me on the second day of the conference. It bears significantly on my relationship with my reporting and sources during the following years. A business colleague of Hagelstein and McKubre's, Matthew Trevithick, who worked for a Silicon Valley venture capital firm, told me that a film crew was arriving the next day, and he asked whether I'd be willing to interview the scientists he selected. I contained my surprise that I, a rookie in the field, would be chosen for this, and I eagerly accepted the opportunity and the task. I interviewed scientists for 10 hours the next day.

At one point, while we were filming in the hotel café, I noticed that Mallove was sitting at a table in the opposite corner — silently watching the activity. I wondered why I was asked to do the interviews instead of Mallove, a credentialed scientist who had known everyone in the field for many years.

The topic of Mallove came up again, in a Nov. 19, 2003, e-mail I received from Storms. "Gene likes to generate conflict and then takes the result personally," Storms wrote. "What is worse, he will not take advice from people who know more about the subjects that he thinks he understands. As a result, he has harmed the field by giving the skeptics ammunition. This is especially tragic because his heart is in the right place and he has the talent to be a success."

It struck me as odd that people like Hagelstein and Storms, who were considered respected scientists and authorities in the field, were trying to discredit and ostracize Mallove.

"Cold Fusion" Theory Questioned

As I learned only in hindsight, the thorn in Hagelstein and Storms' side was that Mallove, — the man who had dominated news reporting of the research since 1994 — had begun to doubt the fusion concept.

George Miley had been discussing it with Mallove on July 29, 2003, on a group e-mail list. "If the cold fusion community," Miley wrote, "is overlooking some basic concepts that should be known to aid cold fusion research, then that needs to be brought out strongly in ICCF-10." "It's going to be," Mallove replied. "I'm working on the paper right now for the poster session."

Mallove was no lightweight when it came to science. He had a master's degree and a bachelor's degree in Aeronautical and Astronautical Engineering from MIT and a doctorate in environmental health sciences from Harvard University. He had been the chief science writer at the MIT News Office. Whereas, in 2000, Mallove had heralded what he called the spectacular progress made by McKubre in correlating helium-4 with excess heat, in 2003 Mallove had begun to see the limitations of that data. In his review of ICCF-10, Mallove took direct aim at Hagelstein, the Chubbs, and a dozen other theorists.

"The primary theorists in the field that is properly designated Cold Fusion/LENR," Mallove wrote, "have generally assumed that the excess-heat phenomenon is commensurate with nuclear ash such as helium-4." Mallove called this the Mainstream Cold Fusion Hypothesis (MCFH). He wrote that the idea was an excellent place to start but that it had not been proved and appeared to be approximately correct in only a few experiments. (Mallove, 2003)

The following day, Storms argued with Mallove, who, in turn, reminded Storms that nobody, aside from McKubre, had shown a precise quantitatively correlated measurement of helium-4 and excess heat. "It is disappointing," Mallove wrote, "that there has not developed more firm evidence to support the MCFH. There is a dearth of other results like [McKubre's] data. There are also prominent experiments within CF/LENR that cast doubt on the rigor of the MCFH — but not, I repeat, against the basic validity of the large-magnitude excess-heat and nuclear changes and emissions." (Mallove, 2003)

Just one year earlier, at the ICCF-9 meeting in Tsinghua, China, none other than McKubre had expounded a similar broad perspective in his closing address. Most significant, McKubre could see that the variety of experimentally observed phenomena in hydrogen as well as in deuterium systems likely was rooted in the same fundamental process, which, therefore, would have excluded D+D fusion:

> I have resisted the expansion of the field. I resisted the extension of the field into biological nuclear effects, into ... the possibility of heat from nickel/light-water experiments — I have resisted this. And I have resisted the concept of transmutation, that somehow we can change higher mass elements from one isotope into another. It isn't that I think these effects are not well-observed or well-disclosed by able people. My resistance really is — I resented the diversion of focus of attention from what was already a very difficult problem. In general, those effects are just too easy for our critics to attack, to use as sticks to beat us with. At least for the heat effect, possibly also tritium production from nickel/light-water experiments with small additions of deuterium, and for the evidence of new nuclear isotopes, ... the time has come to change. My prejudice must change.
>
> I have to abandon my objections and pay much more attention to what is being said about the yielding of new isotopes and the possibility that nickel/hydrogen systems are representative of the same phenomenon that was observed by so many [researchers] in the deuterium/palladium system. These effects seem to me to be quite clearly exposed in experiments that are scientifically defensible. The work has been performed by very studious researchers with very clear exposition of the systematic errors that might possibly be associated with such experiments. This work has been going on for a number of years and has been accumulating a weight of evidence that I think we must just accept. More particularly, I think that these effects are physically

consistent. By this, I mean consistent in terms of their fundamental physics with the phenomena of heat production and helium-4 production that many of us have observed in the deuterium-palladium system. We will have common theory to describe these effects.

McKubre had never publicly expressed a more accurate summary of the broad collection of experimental research in the field. But, in 2003 at ICCF-10, McKubre's developing vision of a unified common mechanism was overshadowed by a growing false distinction between the two sets of phenomena. At ICCF-10, the well-recognized scientists in the field told me that experiments performed with deuterium involved a separate and distinct reaction branch from those performed with ordinary hydrogen. James Corey, a non-scientist analyst and senior member of the technical staff at Sandia National Laboratories, got the same impression, as evidenced by his presentation about LENRs at the Energetic Materials Intelligence Symposium in Chantilly, Virginia, on Sept. 10-12, 2003.

In 2003, I did not know the history and research in the previous decade. I had no idea that, for several years in the early 1990s, Hagelstein unequivocally stated that the experimental data was inconsistent with the D+D "cold fusion" idea. In 2003, I had no idea that such a strong collection of light-hydrogen research and heavy-element transmutation results existed.

As a result, when some of the well-known and apparent authorities — Hagelstein, Scott Chubb, and Storms — told me at ICCF-10 that George Miley and John Dash's light-water work was not reliable, I accepted their point of view. I didn't know enough about the research to question them, or defend myself in arguments with them.

But people in the field could not completely ignore and dismiss the heavy-element transmutation results reported at ICCF-10: The Mitsubishi results had been confirmed. A group of scientists at Osaka University, Japan, announced that they had replicated the Iwamura transmutation results: a reduction of cesium and an increase of praseodymium. Moreover, each of their three attempts was successful, an enviable success rate.

Even "cold fusion" theorist Talbot Chubb could not deny the transmutation results. Chubb had spent a long career in nuclear science that went back to his days working on the Manhattan Project at Oak Ridge National Laboratory during World War II. He sent an e-mail on Sept. 12, 2003, expressing his excitement about the recent meeting:

> The results presented at this meeting seem destined to affect the course of solid-state and nuclear science. Probably the most important of the results were those concerned with a unique form of nuclear transmutation reported a year ago by Iwamura et al. of Mitsubishi Heavy Industries. ... The new discoveries remind one of the beginnings of neutron-capture physics [when], in 1932, Chadwick discovered the neutron.
>
> It seems likely that the larger international community will build on the Japanese work. Further attempts to replicate the Mitsubishi protocol are in progress. Graham Hubler, at the U.S. Naval Research Laboratory, announced plans for replication testing in consultation with the Mitsubishi scientists.

The transmutation evidence confirmed the broader set of nuclear products resulting from palladium-deuterium experiments. The belief that helium-4 was the primary reaction product was at risk. "We have entered the age of modern alchemy, and we know it," Mallove wrote. (Mallove, 2003)

They all knew it. That included the most resistant person among them, Michael Melich, who went to great lengths to try to vindicate Fleischmann and Pons and, in 2008, unsuccessfully tried to name the entire body of research the Fleischmann-Pons Effect. (Melich, 1992; Melich, 1993; Hansen and Melich, 1993; Nagel and Melich, 2008)

Melich, who had informally managed NRL LENR research from a distance in the early 1990s, was the mastermind behind the later NRL attempt to repeat the Mitsubishi transmutation results. Melich directed and obtained funding from NRL and DARPA for the two-part program, which he called Project NUMONKI. (NPS, 2006) According to Hubler,

Melich also obtained funding for the project from the Defense Threat Reduction Agency (DTRA). The details of the NRL replication attempt took several years to emerge and are discussed later in Chapter 31.

SPAWAR Steals the Show

The U.S. Department of Defense had a significant presence at ICCF-10. Several representatives from the Naval Research Laboratory (NRL) in Washington, D.C., were there, as was Jim Corey, from Sandia National Laboratories, and Michael Staker, from the Army Research Laboratory at Aberdeen Proving Ground, in Maryland. But the work of the Navy's SPAWAR San Diego group stole the show.

The initial core SPAWAR LENR group consisted of Stanislaw Szpak (b. 1920), an electrochemist, Pamela Mosier-Boss (b. 1957), and Jerry Smith (b. 1939), a chemist who worked for the Department of Energy (DOE). In the first two decades of the field, the SPAWAR group was the only U.S. government group to consistently publish the research in the open, peer-reviewed literature. A primary reason for the continuation of the SPAWAR work and the publication of the group's results was the support of Frank Gordon (b. 1944), who, in 1989, was the head of the Department of Anti-Submarine Warfare at the San Diego laboratory, later renamed SPAWAR.

Szpak approached Gordon soon after the Fleischmann-Pons announcement and asked for his support to work on experiments. Gordon recognized the value and potential importance of the research and provided discretionary funding to support Szpak's research group. Gordon also had enough clout to fend off naysayers who occasionally challenged the legitimacy of the group's work on what many people at the time thought was bogus science.

According to Gordon, Szpak was born in the U.S., but when he was two, his parents moved back to Poland. As a young man, he was studying chemistry when World War II broke out. At the time, Hitler's troops were everywhere, confiscating everything they could to use for the war effort, including leather. Normally, the process of tanning leather took many days and was very smelly. It was virtually impossible for private leather industries to evade Hitler's troops. Szpak used his chemistry know-how to develop a way to tan leather very quickly and thus evade detection. Szpak survived the era but at one point suffered a partial loss of sight in one eye and hearing in one ear when a nearby land mine exploded.

In 2013, Gordon spoke glowingly about Mosier-Boss. "If you count her peer-reviewed papers, conference proceedings, and book chapters, it's over 160 in total," Gordon said. "She holds more patents than any other woman in the history of the SPAWAR lab, at 17, with five more pending." When the Fleischmann-Pons news broke in 1989, she had been doing research on high-energy-density batteries for torpedo propulsion.

The SPAWAR researchers in 1989-1990 succeeded where others had failed. Unlike the failures, such as those announced in April 1989 — in press releases no less — by Nathan Lewis at the California Institute of Technology and Mark Wrighton at MIT, the SPAWAR scientists were patient and open-minded. They also had an innovative idea, as Mosier-Boss explained to me in 2006:

> In 1989, I was working in a chemistry research group at the Naval Ocean Systems Center in San Diego, which later became the Space and Naval Warfare Systems Center, under Stanislaw Szpak's direction. A week before the Fleischmann-Pons announcement, Szpak came to me and asked if I thought electrochemically induced "fission" was possible. I asked physicists at our lab, and they said yes.
>
> Szpak had the idea to use the co-deposition loading method. It was different from the method Fleischmann and Pons used because it allowed the necessary prerequisite conditions for nuclear reactions on the cathode to develop very quickly. A few weeks later, Szpak talked with Frank Gordon, the head of the Navigation and Applied Sciences Department, and got his approval. The next thing I knew, I was ordering heavy water and palladium chloride to do experiments.
>
> This method became the foundation of our group's

work, which allowed us to consistently and reproducibly observe the production of charged particles, tritium, neutrons, heavy-element transmutations, cathode heating that was inconsistent with Joule heating, unusual surface morphology, melted metal that was inconsistent with the low applied input power, and other nuclear phenomena for more than two decades.

Pamela Mosier-Boss (Photo: S.B. Krivit)

I asked Mosier-Boss why they persisted in the research beyond the first few weeks after Fleischmann and Pons' March 23, 1989, announcement, particularly in light of the discouraging negative backlash. Here's what she wrote:

> When I was doing spectroelectrochemistry, I found out that Pons and Fleischmann had been the first to demonstrate it. When I was looking at ultramicroelectrodes for gas sensor research, I learned that Pons and Fleischmann had been the first to demonstrate it. When I began doing surface-enhanced Raman spectroscopy, using optical fibers to obtain Raman spectra, and X-ray diffraction of electrodes during electrolysis, the literature showed that Fleischmann had done it first.
>
> If Fleischmann was right about all that, I was willing to give him the benefit of the doubt. I knew that both Pons and

Fleischmann were top-notch scientists. They knew how to conduct experiments.

Smith, the third researcher in the early core group, was a program manager for condensed-matter physics in the Division of Material Sciences at DOE. He, like other program managers, was encouraged to regularly engage with active research projects. He had known Szpak and, in fact, had known Pons. From 1978 to 1985, Smith had been in charge of the electrochemistry program at the Office of Naval Research and had supported Pons' ultramicroelectrode research. From 1991 to 1998, Smith was a co-author on eight papers with Szpak and Mosier-Boss.

Szpak didn't like the idea of waiting weeks, let alone months, for the necessary amount of deuterium to load into palladium, as Fleischmann and Pons had done. High ratios of deuterium and palladium atoms were a prerequisite before anything unusual would happen in the cells. Szpak's Ph.D. dissertation had been on electrodeposition, an electrolytic process in which a thin layer of a metal is deposited on top of another metal. It was natural for him to think of using this method.

He and Mosier-Boss used electrodeposition to co-deposit atoms of palladium from a palladium-chloride solution onto cathodes (substrates) made of copper foil or nickel screens, metals that did not absorb deuterium. The result was that deuterium and palladium atoms from the solution deposited onto the substrates at a high deuterium-to-palladium atomic ratio, layer by layer, right from the beginning of the experiments.

Szpak showed little interest in trying to prove that the experiments produced excess heat or in trying to vindicate Fleischmann and Pons. Instead, Szpak was primarily interested in characterizing the reaction and investigating the possible factors responsible for initiating and producing the anomalous effects.

Szpak didn't think the underlying mechanism was fusion. After Lewis Larsen sent a preprint of his and Widom's fourth paper to the SPAWAR researchers on Aug. 24, 2006, and after a conference call between Larsen and SPAWAR on Sept. 13, 2006, Szpak wrote in a 2007 paper that a weak interaction was the source of low-energy neutrons. Szpak, however, had not published his ideas before speaking with Larsen. (Szpak, 2007)

"During the conference call," Larsen wrote in an e-mail, "Szpak mentioned nothing about any earlier ideas he might have had about the weak interaction in LENR processes. When I later saw his paper — he submitted a revision to the journal after our conference call — I was incensed when he made a weak-interaction claim based on his own 'unpublished' work. That's when I called Frank Gordon with smoke coming out of my ears."

The SPAWAR group's research slowed in 1993 because of a base relocation and closure program. Gordon was reassigned as the executive director of a Navy in-service engineering center for command, control and communications throughout the Pacific. Szpak retired in 1995. In 1996, as a result of another reorganization, Gordon returned to the San Diego lab, this time as the head of the Navigation and Applied Sciences Department. His enthusiasm for the research was still strong, as Gordon told me in 2015.

"I called Szpak and invited him to come back to work as an emeritus researcher," Gordon said. "I told him that he wouldn't have to write any proposals — because I wasn't going to pay him! He was thrilled."

Gordon again helped Szpak and Mosier-Boss with discretionary funds to pay for their supplies. The research continued, and they were occasionally joined by other researchers. The co-deposition method had many advantages and became the foundation of the group's work.

By effectively creating the cathode surface during the experiment, the method also minimized the possibility of interfering defects in the physical structure of the cathode. The process enabled the deposition of very pure palladium onto the cathode, and by plating onto foils and mesh, the researchers could create cathodes with relatively large surface areas, which are known to amplify the reactions. It also allowed them to embed temperature sensors directly within layers of copper-foil cathodes.

The co-deposition method enabled the researchers to consistently and reproducibly observe anomalous effects, something that few other chemists in the field could claim. Over the years, the group reported the production of charged particles, tritium, neutrons, heavy-element transmutations, cathode heating that was inconsistent with Joule heating, unusual surface morphology, melted metal that was inconsistent with the low applied input power, and other nuclear phenomena. (Mosier-Boss, 2008; 2009)

But until ICCF-10, even though they were publishing in mainstream journals, few of their colleagues had noticed their work. Gordon often said of the team's work that they were hiding in plain sight. He and Szpak had gone to the ICCF-2 meeting, Szpak had gone to the ICCF-3 meeting, but they didn't attend later meetings until ICCF-10.

The team came out of hiding in February 2002 with a two-volume compendium of the research, *A Decade of Research at Navy Laboratories*. (Szpak, 2002) The reports included the successful experiments performed by Melvin Miles, who had recently retired from the Navy's China Lake laboratory. Ashraf Imam, the metallurgist at NRL who made the working cathodes for Miles, was a contributing author, as was NRL physicist Scott Chubb.

A year later, on March 29, 2003, the group appeared on the cover of *New Scientist* magazine and in a favorable news story by Bennett Daviss, absent the vitriol that had been so prevalent for the last decade. Then, in August 2003, Gordon presented the group's work to the community for the first time in a decade. Szpak wasn't well enough to travel from San Diego, and Mosier-Boss was off on another assignment.

Gordon displayed one of the most stunning and informative measurements in the history of the field: temperature readings taken by an infrared video camera directed at the broad face of a nickel-mesh cathode during a co-deposition experiment. The video provided the public with its first glimpse of the low-energy nuclear reactions in action.

It had long been suspected, thanks in particular to the research of John Dash, Tadahiko Mizuno and Tadayoshi Ohmori, that the reactions took place in highly localized spots on the cathode surfaces. They had seen craters and other evidence of rapid heating and cooling. Now there was visual proof: tiny flashes momentarily appearing and disappearing in milliseconds. A copy of the video is available for viewing on the *New Energy Times* YouTube Internet channel.

The thermal measurements revealed by the video show that the cathode was much hotter than the surrounding electrolyte. As shown in the images below, the white spots represent the hottest temperatures.

Measured temperatures progressively decrease from the center of the electrode outward.

Joule heating from the electric current would normally show the reverse, hotter temperatures in the electrolyte than on the electrodes. This was not textbook science.

The measurements were conducted with the help of Massoud Simnad, a professor at the University of California San Diego, and Todd Evans, a researcher at General Atomics Inc. who provided the IR camera. The researchers had made this fascinating video in 1994, but it was not seen publicly until 2003.

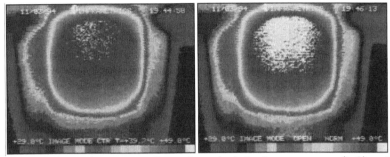

SPAWAR thermal imaging video as the cathode begins to warm up (left) and a few minutes later (right)

In another of their creative approaches to characterize the reactions, the SPAWAR researchers took advantage of their access to military technology used for underwater detection devices. They suspected that the flashes they were seeing with the infrared video camera suggested miniature explosions. They designed an experiment to search for acoustic and pressure waves that would accompany such explosions. The co-deposition process made it possible. They took a piezoelectric transducer and, after preparing it to minimize interference, used it for the cathode, onto which they co-deposited palladium.

In the same way that the infrared video visually displayed small, discreet events, so too did the transducers, which picked up discrete temperature and pressure spikes, confirming the SPAWAR researchers' idea of mini-explosions. This kind of instrumentation also offered them the possibility of real-time feedback while the anomalous heating events were occurring, potentially enabling them to make many modifications to refine and narrow the working parameters.

The SPAWAR researchers saw other evidence of high-temperature events in response to both external electric and magnetic fields.

Formation of circular and square rods (top), a folded thin film, (bottom)
(Photos: P. Mosier-Boss)

In co-deposition experiments performed in the presence of an external electric or magnetic field, scanning electron microscope (SEM) images revealed unusual morphological changes on the surfaces of the cathodes, suggesting highly energetic events. Experiments performed in the absence of such fields showed uniform structures throughout the electrode.

Moreover, they saw, as had Mizuno and Ohmori, evidence of molten palladium that had cooled quickly, giving evidence of possible flash boiling, reaching a temperature of at least 2,963° C. The image below is of palladium that was deposited on the surface of a gold foil and used as a cathode. It suggests that a very-high-temperature event occurred, closely followed by fast quenching from immersion in the electrolyte.

Evidence of possible flash boiling of palladium that had solidified quickly in a 2003 experiment. Photo: Charlie Young

"Cold Fusion" Proponents Try Again

One day during the ICCF-10 meeting, Randall Hekman, a retired judge from Michigan who was interested in developing LENR research commercially, was having lunch with David Nagel, a longtime proponent of the research who retired from NRL in 1998. Hekman had an idea. The new evidence presented at ICCF-10, he told Nagel, seemed strong enough to warrant DOE's revisiting the topic. The DOE hadn't looked at the subject since 1989, when it appointed a panel of independent scientists to advise the government.

The members of that panel (not DOE staff members) found no convincing evidence that the research would ever produce significant amounts of energy. However, there were deep methodological flaws in the 1989 panel's assessment, as described in Vol. 2, *Fusion Fiasco*. More significant, that conclusion was adopted by DOE as the official position on the topic. From that, the opinion formed worldwide that nothing was real about the science.

Hekman wanted to move the field forward. The researchers hoped that DOE would take a second look and consider the progress that had taken place.

CHAPTER 21

From Death to Rebirth

R andy Hekman was eager to advance the science into a practical energy technology and wanted the Department of Energy to take a fresh look at it.

In August 2003, at the Tenth International Conference on Cold Fusion (ICCF-10), as Hekman told me, he and David Nagel took the idea of a new DOE review to Peter Hagelstein and Michael McKubre, who they believed had the most credibility in the field. But Hagelstein said he didn't think there was any way it would work. Hekman argued that he knew how to lobby and that he personally knew U.S. Secretary of Energy Spencer Abraham.

On Sept. 12, 2003, Hagelstein wrote to Abraham saying that scientists had made much progress and that the weak evidence available in 1989 was now superseded by much stronger results. The lobbying was fruitful. The first meeting took place a few months later. On Nov. 5, 2003, Nagel, Hekman, McKubre and Hagelstein met with James F. Decker, principal deputy director of DOE's Office of Science.

Meanwhile, after I returned home from the ICCF-10 meeting, I dropped everything else I was doing and wrote my first white paper on the subject, "The 2004 Cold Fusion Report." In late 2003, I was unaware of the proposed DOE review.

On Dec. 13, 2003, journalist Bob McDonald, of Canadian Broadcasting Corp.'s Radio One, produced a short program on the subject. He chose three people to represent the field: Hagelstein, McKubre and Edmund Storms. McKubre told McDonald that he thought he would have a working reactor within two years. Hagelstein told McDonald that he had equations for a predictive theory.

I asked Hagelstein about this theory. He replied that he had "examined more than 100 models and variants before arriving at the model currently under investigation." Although Hagelstein hadn't specifically claimed that he had figured out the theory for "cold fusion," he suggested that he had done so. Even though Hagelstein, along with his colleagues, had initiated a request for DOE to reconsider "the evidence for nuclear emissions from metal deuterides," theorist Lewis Larsen had already been talking privately with DOE.

Behind the Scenes

Larsen, a quantitative investment analyst who had studied biophysics, had begun developing his theoretical ideas in the late 1990s. By 2000, Larsen had most of the basic concepts worked out. (Chapters 25-27 describe the development of his and Allan Widom's theory.)

Larsen, armed with his first and only theoretical model for LENRs, wanted to raise money for commercial research and development. But the DOE panel's negative 1989 conclusion cast a shadow on the subject, and he, like Hekman, thought that it would be useful if DOE could re-evaluate the topic more favorably to help overcome public resistance.

Starting around 2000, behind the scenes, Larsen began cultivating relationships with people at DOE headquarters, Sandia National Laboratories and Lawrence Livermore National Laboratories. Larsen told me that he spoke with people at other government agencies, but he declined to provide details. In his outreach, he explained that the observed phenomena were not created by a fusion process.

The only person who vaguely knew that Larsen was working on a theory at the time, least of all that it proposed a non-fusion explanation, was George Miley. In the summer of 2000, Larsen had begun collaborating with Miley and let him in on his secret. To others in the field, Larsen simply said that he had a company, Lattice Energy LLC, and was hoping to develop the research commercially. I did not learn anything about Larsen until 2005.

One of the people Larsen met was John Fairbanks, a program manager with DOE's Office of Heavy Vehicle Technologies. Fairbanks was interested in ways to capture waste heat from vehicles and was specifically interested in thermoelectric devices to provide additional power for onboard electrical devices. There had been a long association between thermoelectric devices and nuclear heat sources, for example to power spacecraft, so Fairbanks was interested in what Larsen had to say. Fairbanks invited Larsen to speak at a workshop co-sponsored by DOE and the Electric Power Research Institute on high-efficiency thermoelectric devices. The workshop took place Feb. 17-20, 2004, in San Diego, California.

Mildred Dresselhaus, a professor at the Massachusetts Institute of Technology (MIT), was one of the two keynote speakers, and Larsen was warned by many people that she might be hostile to the subject of LENRs. Dresselhaus was all-too-familiar with "cold fusion." She had served on the 1989 DOE review panel and was a former head of DOE's Office of Science. She also was a member of the same department at MIT as Hagelstein, and she was not known as a supporter of Hagelstein's theoretical ideas.

"Many people thought Dresselhaus might attack me," Larsen said, "but I spoke with her before my talk, and, far from attacking, she was friendly, listened politely, and showed a mild interest in my topic."

Larsen identified his topic as low-energy nuclear reactions and explained that the phenomenon had nothing to do with fusion. Larsen said nothing about the theory that he was working on. In his slides, Larsen discussed Yasuhiro Iwamura's heavy-element transmutation data as an example of the best research in the field.

"You can argue about excess-heat measurements and ponder near-absence of 'normal' nuclear products," Larsen wrote, "but transmutation experiments involving LENRs are irrefutable."

"Even though the whole topic of 'cold fusion' was still regarded very skeptically," Larsen said, "during and after my presentation there wasn't one snicker in the room. In addition to Dresselhaus, Thomas Passell, a close colleague of McKubre's, was there. Most likely, McKubre and Hagelstein knew through Passell that I spoke at this meeting."

Big News From DOE

By late February 2004, my "cold fusion" white paper, co-written with Nadine Winocur, my wife at the time, was nearly finished. I began hearing rumors that DOE was going to take a second look at "cold fusion." Confirmation of the rumor arrived on March 3, when Toni Feder, a writer for *Physics Today*, distributed an advance draft of her article about "cold fusion researchers' attempts to have a new review carried out." The impetus behind the review was not DOE; rather, it was the researchers who had pushed for it.

Feder had sent her draft to the sources in her article and other involved parties. *Physics Today* is a monthly magazine published by the American Institute of Physics, the umbrella organization that includes the American Physical Society. Feder's article didn't appear until April.

On March 17, a researcher forwarded to me an e-mail from Hagelstein that tipped me off about the impending review, and I immediately checked with DOE spokeswoman Jacqueline Johnson, who confirmed the rumors: DOE was going to conduct a second review. Both Feder and I were scooped, however, on March 20, 2004, by Bennett Daviss, writing for the U.K.-based *New Scientist* weekly magazine. Our white paper published on March 22, 2004.

For most people who knew nothing about the status of the research at the time, the DOE news was hard to fathom. By that time, the popular view of the subject was that it had no scientific merit and perhaps was even the product of fraud and incompetence.

The other thing that happened on March 17 was that Eugene Mallove sent me his critique of a draft of my forthcoming white paper. I realized only years later that the most important critique he offered was about my sentence "It is now known that that the amounts of excess heat in cold fusion are consistent with the change in energy that results when heavy hydrogen is converted into helium-4."

Mallove did not mince words with me. "You're on VERY thin ice in stating that. There is only ONE experiment in which such a fact has been even approximately proved, and that is the SRI international reproduction of the Case catalytic fusion work."

I showed Mallove's critique to Storms, and Storms convinced me to keep my text, writing that there had been other similar replications. It took me several years to learn that Mallove was correct and that I had been misled by Storms.

In April, Mallove wrote his next editorial, which published two months later. He expressed his concern about how the proposers of the review, Hagelstein and his colleagues, would present the matter of theory to DOE.

"No such satisfactory, comprehensive theory exists in the LENR community," Mallove wrote, "nor is one likely to be framed in the next few years. But theory-driven optimism is rampant in the LENR field, so there is a real danger that one or more theorists will try to push their theory too far and will put the field in a position from which it would be difficult to recover, by handing the opposition another straw man that could be shot down." (Mallove, 2004)

On April 22, 2004, McKubre and Hagelstein formed Spindletop Corp., perhaps in the hope that the department might decide to fund LENR research. The first planning meeting for the DOE review occurred on May 13, 2004, according to a document listing the participants' names and the agenda for a conference call.

In Cold Blood

At 11 p.m. on May 14, 2004, paramedics in Norwich, Connecticut, arrived at Eugene Mallove's mother's house. She was no longer living there, and Mallove leased the house to renters. Paramedic Daniel Miles found Mallove lying on his back, in the driveway, at the rear of the residence, near a large dumpster. Mallove was dead.

The police report described the evidence of a violent struggle, with multiple pools of blood at separate areas on the pavement and blood drops on the side of the dwelling and on a nearby vehicle. The police report described the grisly autopsy:

> Dr. Carver noted numerous injuries to the victim's face, head and neck area. The victim also had numerous cuts and

abrasions to his extremities and a knife wound to his right forearm. The victim had sustained thirty-two lacerations to his facial area which were caused by a blunt instrument. Dr. Carver also stated the victim's tracheal area had been crushed, and this was determined to be the actual cause of death. Dr. Carver ruled the manner of death a homicide.

In June and July 2005, two suspects were charged, but the case against them was dismissed for lack of evidence in November 2008. In April 2010, two different suspects, Candace L. Foster and Chad Schaffer, were charged with the crime. During testimony, they identified a third suspect, Schaffer's cousin, Mozelle Brown.

On May 25, 2010, Karen Florin, a reporter with *The Day*, a New London, Connecticut, newspaper, described the court hearings. Foster and Schaffer had lived at the house with Schaffer's parents. Mallove had traveled down from his home in New Hampshire to clean out the property to prepare the house for the next tenant. According to the testimony, Schaffer and Brown carried out the initial assault, left, then returned to the scene of the crime with Foster, Schaffer's girlfriend. Mallove was still alive. They turned him over, blood spurted from his mouth, and he cried out for help. The two men told Foster that they had to make the scene "look like a robbery," according to Foster. "They stomped him in the face and suffocated him," Foster said. "They put, like, a bag over his face."

Brown was charged in May 2011. On April 20, 2012, the trial for Schaffer came to an abrupt end when he accepted a plea offer from the state. Reporter Florin wrote that no forensic evidence tied Schaffer to the crime. On April 4, 2014, according to Florin, Judge Hillary B. Strackbein ruled that the state had enough evidence to prosecute Brown. He was sentenced to 58 years in prison.

The most perplexing aspect of this case is the seeming lack of motive for such a brutal crime. According to Florin, the motive offered by police was that Brown and Schaffer "attacked Mallove because he was throwing out items that belonged to Schaffer's parents, who recently had been evicted from the home."

In May 2004, the news of Mallove's tragic death had a powerful impact on many people, including me. I was now the only journalist regularly covering the topic. I received letters from researchers I knew expressing their hope that I would continue writing.

DOE Organizes Review of LENRs

The participants in the May 13, 2004, conference call to plan the second DOE "cold fusion" review included proposers Nagel, Hagelstein, McKubre and Hekman. Representing DOE were staff members Patricia Dehmer, Dennis Kovar, Gene Henry and James Horwitz. The outcome of the meeting was an agreement about the structure and process of the review. DOE, however, identified it as a review of low-energy nuclear reactions.

DOE, aided by suggestions from the proposers, selected 18 reviewers who, at the time, were anonymous. DOE set up the general structure for the review, but they left the details of the science to the proposers. The proposers provided eight papers to the DOE staff, who in turn gave them to the reviewers to study. DOE also asked the proposers to produce a paper summarizing the most significant work in the field as well as recommending directions for research. Hagelstein and McKubre did most of the work on the paper, and they were assisted by Nagel, Talbot Chubb, Graham Hubler, and Hekman. (Hagelstein, 2004; McKubre, 2004) According to McKubre's presentation at ICCF-11, one more person worked behind the scenes on the project: Michael Melich. (McKubre, 2004)

The proposers were criticized by other scientists in the field because the majority of the research presented in the paper was McKubre's. Ten of the 15 pages of the paper were dedicated to McKubre's palladium-deuterium replication of Fleischmann and Pons' electrolytic experiments and McKubre's replication of Leslie Case's deuterium-gas experiment.

McKubre's measurements of excess heat and helium-4 in those experiments provided excellent data; however, he and Hagelstein, insisted that the data proved the idea of room-temperature fusion.

McKubre, Hagelstein and their co-authors steered clear of the research showing evidence of heavy-element transmutations, isotopic shifts, and experiments performed with nickel-hydrogen systems. In a presentation on Nov. 1, 2004, at ICCF-11, in Marseilles, France, McKubre tried to explain the omissions. According to McKubre, a) DOE changed the terms of the agreement and b) the 15-page limit imposed by DOE was unreasonable:

> We wanted a comprehensive review of the field, of the whole field, of everything that had been done since the original ERAB report. So we were faced with a bit of a dilemma. What we had asked for and what had been agreed to was very different from what was on the table. We had to trust ourselves and trust, in part, the progress of this whole field. The results of this deliberation [would] affect, in some way, everybody in this room. The brief format did not allow us to review the field. In 15 pages, how can you review the field of cold fusion?

In fact, writing a review of the field in 15 pages was possible. I did that in the *Journal of Environmental Monitoring* in 2009 at the request of Harpal Minhas, executive editor at the Royal Society of Chemistry. (Krivit, 2009).

I was curious about what instructions the DOE staff members had given to the proposers, particularly concerning the scope. On my request, Patricia Dehmer wrote back to me explaining how she and Kovar conducted the review and what they requested of the summary document:

> DOE did not prescribe what topical areas were to be included (or excluded) from the review document. My notes of an initial conference call with the principal investigators show that the following was the DOE position regarding the review document.
>
> "Principals will provide a summary of the status of the field which articulates what are considered to be the most recent significant experimental observations and publications, and identifies those areas where additional

work would appear to be warranted based upon what has been learned from progress in this area."

The reviewers did not make any laboratory visits, but on Aug. 23, 2004, 11 of the 18 reviewers attended a daylong meeting with the proposers and other LENR researchers. The meeting took place at the Doubletree Hotel in Rockville, Maryland. Hekman told me about the key moments:

> The individual panel members seemed very cordial and eager to delve into the question at hand. ... During the morning, various members of our team and others who had been invited to assist us gave their perspectives. My read on the review panel, based on questions they asked as well as their body language, was they were generally supportive and impressed with the testimony. ...
>
> But then, at the start of the afternoon session, things took a decided turn for the worse. Our team began to suggest theories to explain the research data. Hagelstein stands out in my mind as being one who offered his theory. As he talked, I clearly remember the eyes of many panelists begin to roll, the looks on their faces expressing incredulity. Their questions and comments began to take on a derisive tone. Little by little, we lost the crowd at this point. ...
>
> I believe our failure to convey a cohesive theory of LENR caused us to lose the argument and, hence, waste much time and resources.

One of the additional researchers the team invited to give a presentation during the oral review was Vittorio Violante, a chemical engineer at the Italian National Agency for New Technologies, Energy and the Environment (ENEA), in Frascati, Italy, and a close associate of McKubre.

Violante told the reviewers that, in a recent experiment, he too had measured just the right amounts of excess heat and helium-4 to confirm the idea of D+D "cold fusion."

For 15 years, many researchers had struggled to obtain such confirmatory data. Thus, it was a remarkable coincidence that, just two months before the DOE panel meeting, Violante performed the only experiment that confirmed McKubre's experiment.

On Dec. 1, 2004, DOE released a five-page conclusion of the review on its Web site. I also asked for and received a 45-page document from DOE containing the reviewers' comments. DOE wrote that "the conclusions reached by the reviewers today are similar to those found in the 1989 review." In other words, the science was not real, and it did not indicate a potentially new source of energy.

DOE had asked the proposers to identify the most significant experimental observations in the field. They did not. Instead, they promoted the "cold fusion" concept, which even Hagelstein, in 1993, knew was a poorer match for the experimental data than were neutron and weak-interaction processes. (Hagelstein, 1993)

Instead, they highlighted excess-heat data, which were only indirect evidence of nuclear reactions. The proposers irresponsibly omitted a vast body of research performed by other scientists in the field that provided direct evidence of nuclear reactions, such as heavy-element transmutations and isotopic shifts.

Fifteen years earlier, in 1989, the reviewers selected by DOE, with very little guidance and direction from DOE and no guidance from researchers in the field, came up with reasons to ignore and dismiss direct nuclear evidence shown by the experiments: production of tritium from U.S. and Indian national laboratories, low fluxes of neutrons from U.S., Italian and Indian national laboratories, and isotopic shifts from U.S. national laboratories.

In 2004, LENR scientists provided inadequate information to DOE's independent reviewers and tried to convince them of the erroneous fusion idea. Based on that information, the reviewers made appropriate recommendations and DOE found no reason to fund and give its stamp of approval to the research.

Not only did Hagelstein, McKubre, Nagel, Chubb, Hubler and Melich adversely affect the progress of the field in the U.S. by promoting their "cold fusion" hypothesis, Germany and likely other nations had deferred to the U.S. and were waiting to hear the conclusion from DOE.

CHAPTER 22

The New Generation

D uring the summer of 2004, while preparations for the Department of Energy review of low-energy nuclear reactions were under way, I began expanding my white paper, "The 2004 Cold Fusion Report," into a book. I wanted to make the information available to a wider audience.

Nadine Winocur and I finished writing that book, *The Rebirth of Cold Fusion*, on Sept. 20, 2004, and published it in November. The book provides a historical snapshot of the subject, based on my experience at that time. Any part of *The Rebirth of Cold Fusion* that discusses theory, however, is obsolete.

The book is also heavily one-sided, and here's why. While writing, I asked the researchers for evidence to support their claims. I then contacted as many of the well-known critics of the field as possible and sought their views. But the critics knew nothing about any of the research that had been reported in the past decade and were not in a position to offer a substantive critique.

The Baker Sends Some Dough

In 2004, while writing that book, I met a philanthropist who was interested in the research. I was looking for financial support to help with the costs of producing the book. He prefers to remain anonymous, so I will call him "Sam." He funded the printing costs based on our mutual understanding that, if he had been involved or was going to get involved in the science, I would not use him as a source. When the book was done, Sam was pleased. "Keep up the good work," he said. I told

Sam, "Fine, keep sending money." Thus began a relationship that lasted seven years.

Sam's wealth came from his family's commercial baking company. For much of his life, he was also keenly aware of the environmental risks facing the planet, and he had followed thermonuclear fusion research for many decades, later becoming interested in "cold fusion" research. Because my book was so one-sided, it was popular in the "cold fusion" field, as I was referring to it in 2004.

Dinner With Stan and Sheila Pons

Stanley Pons, who, with Martin Fleischmann, announced in 1989 that they had created the first controlled and sustained nuclear fusion reaction on Earth, had remained out of the spotlight since the late 1990s. By 2004, Pons, unlike Fleischmann, no longer attended the International Conference on Cold Fusion (ICCF) series, and he had avoided all news media.

I had no contact with Pons, but I knew that a French researcher, Jean-Paul Biberian, who, like Pons, lived in southern France, did. In August 2003, I sent a few paragraphs from the draft of my book to Pons, by way of Biberian, for fact-checking. Pons e-mailed me and confirmed what I sent him. I told him that I would be in Marseille at the end of October for ICCF-11 and asked whether I could meet him. He agreed enthusiastically.

We agreed ahead of time that the meeting would be off the record. Pons, his wife, Sheila, and I met at one of his favorite restaurants, and it was a memorable evening for me. Although he did tell me a few things about the past, his primary reason for not discussing the history in detail was that it brought him too much pain. He had finally regained the happiness he once had, before the University of Utah fusion press conference. What I can say about him is that he was one of the most gracious men I've ever met. He humbly revealed to me things he would have done differently, with the benefit of hindsight.

After dinner, he told me I could contact him again as long as I didn't bring up the past. In later years, while I was trying to map out certain historic details, I encountered questions that only Pons could answer. I broke my agreement with him and sent him questions. Soon after that, the phone number and e-mail I had for him no longer worked.

Fleischmann Concedes

The big news the year before, at ICCF-10, had been the confirmation of Yasuhiro Iwamura's heavy-element transmutations. In 2004, at the ICCF-11 conference, transmutation results continued to capture the attention of the researchers.

Fleischmann, the grandfather of the field, finally dropped his resistance and, during his closing comments for the conference, accepted the heavy-element transmutation results as valid.

"This conference is notable for the results of transmutation experiments which have been presented," Fleischmann said, "and I would agree with Francesco [Celani] that these are now much more solid. I must say, this is an aspect of the field which I viewed with intense skepticism originally. ... The transmutation experiments now seem quite believable."

1922 Transmutation Results Confirmed

More than eight decades earlier, on March 11, 1922, two chemists at the University of Chicago made a bold announcement at an American Chemical Society meeting. The next day, the news went nationwide, probably worldwide. It was a first: Chemists at the Kent Chemical Laboratory of the University of Chicago, according to news reports, said that they had transmuted a metal, tungsten, into another element, helium.

The senior chemist was Gerald L. Wendt (1891-1973), 31 at the time; the junior chemist was Clarence E. Irion (1896-1976), 26 at the time. They didn't use the word transmutation, but they reported their results as the artificial decomposition of tungsten into helium.

The Wendt-Irion experiment used the exploding electrical conductor method: running a high-voltage discharge inside an evacuated bulb through a tungsten wire. Rather than transmuting tungsten, they may have transmuted hydrogen into helium because it is almost always present as a monolayer on the surfaces of metals. Despite doubts expressed by critics at the time, including Sir Ernest Rutherford, no one unambiguously identified any error in Wendt and Irion's 21 successful experiments. (Vol. 3, *Lost History*)

Rutherford, a pioneer of modern physics, led the charge against Wendt and Irion and aggressively depicted their results as invalid long before the duo's paper published. Understandably, the public accepted Rutherford's position rather than the evidence of the two unknown chemists. The public and scientific community assumed that the transmutation results were wrong, and the experiment was soon largely forgotten.

Evidence of nuclear reactions from exploding electrical conductor experiments was confirmed 80 years later by Leonid Urutskoev (b. 1953) and his colleagues at the Kurchatov Institute, Russia's top nuclear energy research institution. Urutskoev began his career at Kurchatov, in the Plasma Physics Department, and earned his Ph.D. in physics there in 1979. In 1987, he was promoted to the head of the laboratory in the Plasma Physics Department.

On April 26, 1986, disaster struck at the Chernobyl nuclear power plant in the Ukraine. Responsibility for the plant fell on authorities in the Soviet Union, who called on the Kurchatov Institute for help.

Urutskoev was asked to lead a team to develop and test a gamma-ray detector to be suspended beneath a helicopter on a 300-foot cable to map out radioactive hot spots. It was successful at Chernobyl and was later used at operating nuclear power plants in Russia and Europe.

His personal memoir of being in the Chernobyl area within weeks of the accident is dramatic. (Urutskoev, 2015) Here is an excerpt:

> The place was teeming with military jeeps and buses full of uniformed servicemen and civilians in white overalls. All the people there wore similar standard work clothes and had respirator masks on, which made them look sort of faceless. Yet no traces of panic were detectable; everybody was busy doing their work with purposeful determination.

Military deactivation vehicles kept incessantly washing the radioactive dust off the streets and pavements. The whole picture resembled a stirred anthill.

There was one more obvious sign which could not but implant people with an intuitive feeling of alarm: the total absence of birds. None of them could be seen anywhere around, neither dead nor alive. Sensing the trouble, they all had just flown away.

Urutskoev's LENR story began in 1998. He knew nothing about the Wendt-Irion experiment at the time. He had been studying the well-known phenomenon in which rapid, high-voltage electrical pulses sent through thin wires could produce an intense explosion sufficient to fracture concrete.

Serendipitously, after performing such an explosion with the titanium wire, he and his colleagues performed a mass spectrometric analysis on the titanium (Ti) powder after the explosion and noticed unusual changes in the isotopic composition of the titanium.

The Ti-48 isotope had dropped significantly — by 5% — compared to a measurement error of +/- 0.4%. To their amazement, none of the other titanium isotopes increased significantly. Instead, simultaneously with the disappearance of the Ti-48, they measured a sharp (10-fold) increase in other elements detected in the samples. The net percentage increase of the new elements corresponded to the fraction of lost Ti-48. (Urutskoev, 2002)

According to the authors, the results were independently verified by another Russian research group, at the Dubna Joint Institute for Nuclear Research. (Kunznetsov, 2003)

Urutskoev and his colleagues wanted to learn whether the strange nuclear changes were inherent only in titanium. When they repeated the experiments with several other metals, they again detected isotopic shifts. They also detected a very low flux of neutrons and surmised, based on the detection instruments they used, that those neutrons must be ultracold, that is, possessing very low energy. Urutskoev did not rush to publish and demanded unusually high repeatability compared to other researchers in the field. He explained his philosophy to me:

I am a professional researcher. My teachers used to say "a measurement should be repeated at least 7-10 times, results should be plotted, the mean value should be calculated, and error bars should be determined. Only then are you allowed to start sharing results with others." I always followed this sacred rule.

Our first paper, published in 2000, was based on nearly 400 experiments; 200 of them confirmed the reality of isotopic changes and transformations of chemical elements.

Urutskoev group's exploding electrical conductor chamber (Photo: Kirill Alabin)

Urutskoev and his colleague Georges Lochak attended the ICCF-11 meeting, their first participation in that conference series. In their paper, they wrote that the transformations did not involve strong nuclear interactions. Also, they wrote that some sort of collective process was responsible for the transformations because single atoms did not have enough energy to make such changes. (Lochak, 2004)

They drew three conclusions from their experimental findings:

1. Contrary to the opinion of the majority of physicists, the possibility of low-energy transformation does not contradict the conservation laws.

2. This process is collective in principle and can be simulated within the framework of processes based on weak interactions.
3. Since weak [nuclear] interactions are characterized by small cross-sections, a catalyst is needed.

Urutskoev's plasma physics group in 2011 at Sokhumi Institute of Physics and Technology: Leonid Urutskoev, far left in blue shirt; Kirill Alabin, in green shirt; Valery Mizhiritsky, far right; Gennady Ostapenko, front left. Other students, left to right: Timofey Shpakovskii, Georgy Steshenko, Almaskhan Markolia, Pavel Belous, Saida (last name unknown), Aleksey Levanov, Gleb Zhotikov, and Artyom Biryukov (not pictured, photographer)

In 2007, Urutskoev told me, the management of the Kurchatov Center issued a written directive forbidding him to continue research in LENRs. Management did not believe his results, and a dispute developed. Urutskoev accepted professorships at the Russian Presidential Academy of National Economy and Public Administration as well as at the Moscow Institute of Physics and Technology and began teaching. LENR research was only marginally better received at the universities, and Urutskoev searched for a well-equipped laboratory that was immune to the conservative Russian science politics.

He found what he needed at the Sokhumi Institute of Physics and Technology, in Abkhazia, a partially recognized state south of Moscow and between Russia and Georgia. One of his students, Kirill Alabin, who earned a Ph.D. under Urutskoev in LENRs, explained why:

> Our group was initially funded by the Russian organization ROSATOM (State Atomic Energy Corporation). We chose the Sokhumi lab because its distance from the capital provided us with greater scientific freedom from the government bureaucracy.
>
> In the former USSR, Sokhumi was one of the most advanced labs in the nation. It was home to the first two thermonuclear fusion reactors in the USSR: a stellarator-type and a Tokamak-type. We borrowed most of our parts from installations there that were no longer being used.
>
> The Abkhazia government didn't have much money for our research, but they had the space we needed and plenty of old machinery, devices and diagnostic tools. In order to pursue this research, I had to leave my whole life in Moscow and move down to work in Abkhazia for an indefinite period among many uncertainties.

Alabin said that Urutskoev attracted a large team of experts, most from Abkhazia, who made enormous contributions to the group's work. Sometime in 2007, Urutskoev's friend Henri Lehn, a French scientist, told him about the Wendt-Irion experiment and sent him a copy of the 1922 scientific paper. (Wendt, 1922) Urutskoev and his colleagues began new exploding-conductor experiments and looked for anomalous products in the residual gas.

In 2012, they reported preliminary results in the Russian journal *Prikladnaya Fizika (Applied Physics)*. (Urutskoev, 2012) In numerous experiments, they detected particles with a mass of four, which they presumed to be helium-4 atoms. Based on the precautions they took and the magnitude of their results, they were able to conclude definitively that the helium detected in the chamber was not the result of inward leaks from the atmosphere. Eighty years after Wendt and Irion's claim, Urutskoev's group confirmed that nuclear reactions could be induced in a similar type of experiment. An additional 10 years later, they confirmed the production of helium.

Cirillo and Iorio's Transmutation Results

In 2003, Domenico Cirillo (b. 1977) a member of the younger generation of LENR researchers, followed his curiosity and began doing experiments in his home laboratory in Italy. He worked with his friend Vincenzo Iorio (b. 1958), replicating a high-voltage electrolytic experimental method developed by Tadahiko Mizuno.

Domenico Cirillo (Photo: S.B. Krivit)

The typical input power for such experiments ranges from 100 Watts to 800 Watts. The experiments can last a few minutes or a few hours. Researchers typically use light-water electrolyte. Tungsten is a favored cathode because it has the highest melting temperature of any metal. Once the experiment heats up, a glowing ball of light — an underwater plasma — surrounds the cathode. The brilliant visual effect is accompanied by a continuous loud roar, like a miniature jet engine. It's one of the most exciting types of benchtop experiments to witness; however, the light and sound do not necessarily mean that nuclear reactions are taking place. The presence of newly detected chemical elements, however, does.

Cirillo first reported a portion of his and Iorio's results in April 2004 at an Italian meeting sponsored by the National Organization for New Energy, an organization that looks for new scientific applications.

In November 2004, he presented his work at ICCF-11. It went largely unnoticed because he was given the opportunity to present only during the poster session rather than the oral sessions. In their experiments with tungsten cathodes, ultra-pure light water, and potassium carbonate, they observed the presence of eight elements on the cathode after the experiment, shown below, that were not detected before the experiment. (Cirillo, 2004)

Cirillo wrote that neutrons were at the heart of the reactions, and in his ICCF-11 poster, he depicted a weak interaction, $e + p \longrightarrow n + v$, as the basic idea behind the underlying nuclear process, rather than any kind of fusion process. The general concept wasn't new, but it had been forgotten. Several researchers in the 1990s, including Peter Hagelstein, Tadahiko Mizuno, and Yasuhiro Iwamura, proposed such a reaction. It was first mentioned by Larry A. Hull, a scientist who wrote a letter to *Chemical & Engineering News* on May 15, 1989.

Cirillo also displayed several slides depicting the complex nucleosynthetic reaction networks that would take place if there were a source of neutrons in the experiment.

68	69	70	71	72	73	74	75	76	77	78	79	80	81
Er	Tm	Yb	Lu	Hf	Ta	**W**	Re	Os	Ir	Pt	Au	Hg	Tl

- W (Tungsten): Probable starting material.
- Dy (Dysprosium, atomic number 66): If present, may have existed as another starting material.
- Shaded elements: Detected after experiment.
- Lanthanide series elements Er-Lu are normally depicted in another row in the Periodic Table.

Elements detected on tungsten cathode after Cirillo's electrolysis experiment

When Cirillo submitted his paper to Edmund Storms, who was serving as an editor for the ICCF-11 proceedings, he ran into some problems. Storms wrote to him on Nov. 17, 2004, rejecting his paper as submitted:

> While I think this is excellent work and important to make available to a wide readership ... I found it difficult to read. Consequently, I did some editing to the first part of the paper. Later in the paper, I found sections about which I had some concerns, and I inserted comments and questions. Upon reaching page 12, I realized that you will have to shorten the paper considerably before submitting it to the proceedings. At that point, I stopped making changes. If you welcome my changes and comments, I would be pleased to examine your shortened version.

I compared Cirillo's preprint with the published version, after Cirillo acceded to Storm's requirements. I found two major differences in the versions. Page 7 in Cirillo's preprint discusses his perspective on the significance of his experimental data to theory — a non-fusion theory. Cirillo took out that text but managed to keep several crucial sentences in his published version.

"We propose that, as the temperature increases," Cirillo wrote, "electrons in the metal start to oscillate in a coherent way. ... A considerable number of electrons are available to the surface region. We believe this condition is important to initiating the observed transmutation reactions."

The idea of coherent oscillations of electrons on the surfaces of metals is very important and will be discussed in Chapter 26. Cirillo also removed an entire section about possible reaction mechanisms. He depicted and proposed the weak-interaction idea $e + p \longrightarrow n + v$, rather than the "cold fusion" idea $d + d \longrightarrow Helium\text{-}4 + 23.8\ MeV\ (heat)$ as the explanation for his results. With this removal, Storms accepted Cirillo's paper for the proceedings.

Jed Rothwell openly admitted to his and Edmund Storms' censorship in an e-mail he posted to an Internet discussion list in early 2004.

"At LENR-CANR.org," Rothwell wrote, "we have censored out some of the controversial claims related to cold fusion, such as transmuting macroscopic amounts of gold, or biological transmutations, along with some of the extremely unconventional theories."

Undaunted by the censorship, Cirillo continued to follow his passion for the research and, in 2007, completed LENR research for his thesis in fulfillment of the requirements for a master's degree in mechanical engineering. In 2012, Cirillo and several other authors published evidence for low-level neutron emission from their experiments. (Cirillo, 2012)

Without Cirillo and Iorio's knowing it, their transmutation of tungsten to gold, among other elements, may have confirmed another experiment from the previous century. In 1925, Hantaro Nagaoka, a well-known Japanese scientist, had performed experiments with tungsten cathodes. Rather than use a solution of water, Nagaoka applied a high electric field to mercury atoms in transformer oil, which provided an abundant source of hydrogen. (Vol. 3, *Lost History*)

Nagaoka thought the transmutation was going downward from mercury (that is, from atomic number 80 to 79), but according to what theorist Lewis Larsen told me in a telephone interview, it was going up from tungsten (from atomic number 74 to 79):

> Using only limited knowledge about nuclear physics that was available at that time, because neutrons and fission had not been discovered yet, Nagaoka mistakenly believed that gold was produced as a result of high electric fields disintegrating mercury atoms into lighter gold atoms and other constituents of atomic nuclei.
>
> This made sense in the context of his knowledge because mercury was known to be heavier than gold and found in the same row of the Periodic Table of Elements. What Nagaoka did not know back then was that, during the electric discharges, ultra-low-momentum neutrons were being created from some surface plasmon electrons, present on tungsten electrodes, that reacted directly with hydrogen protons in the transformer oil. Those neutrons were captured by tungsten atoms and transmuted it into gold.

The present was catching up with the past.

Flashes in the Pan

For Tadahiko Mizuno, 2005 began with a bang — literally. At 3:51 p.m. on Jan. 24, he and a guest, Tomio Gotoh, were observing one of his high-voltage electrolysis experiments: a tungsten cathode in light water with potassium carbonate electrolyte. They were nearly killed.

Normally in such experiments, as the researcher slowly increases the electrical power, the glowing plasma gradually develops. It begins as a small, faint glow and ends as a large, brilliant, turbulent lightning-in-a-bottle effect. Normally, researchers adjust the power to control the size and turbulence of the plasma and keep it from getting too large and too wild, which might cause the glass cell to break. That was not the case on this day, as Mizuno explained:

> The event occurred in the early stage of the experiment before a plasma normally forms. Soon after ordinary electrolysis began, voltage was increased to 20 volts and current to 1.5 amps. Five or six seconds later, a bright white flash was seen on the lower portion of the cathode. The light expanded, and at the same instant the cell exploded.
>
> The explosion blew open the Plexiglas safety door and spread shards of Pyrex glass and electrolyte up to 5 or 6 meters into the surrounding area. When the explosion occurred, we were observing the cell from about 1 meter away. I was wounded in the face, neck, arms and chest by shards of glass 1 to 5 cm long. Fortunately, there were no injuries to my or my guest's eyes. Our injuries are light, and

we are expected to recover in a week. However, the explosion made such a tremendous noise that both of us are temporarily completely deaf. (Edited for clarity) (Mizuno, 2005)

Mizuno safely removed a large piece of glass next to his carotid artery. He told me that he removed the glass himself because he doesn't like going to hospitals. Within 24 hours, long before his wounds had healed, Mizuno wrote up an accident report, and Gotoh took photographs. After a week, their hearing returned.

Mizuno's most plausible speculation was that hydrogen and oxygen in the off-gases in the cell headspace mixed and sparked. The problem with this scenario is that, as he observed, the flash preceding the explosion began in the submerged portion of the lower part of the cathode. Also, the cell was open, and only small amounts of gas were present. Mizuno wrote that the voltage, current and electrolyte temperature were below the levels needed to produce a plasma.

Bottom of the glass cell after the explosion (Photo: Tomio Gotoh)

Theorist Lewis Larsen speculated that a low-energy nuclear reaction began to cause rapid heating on the tungsten cathode. The heating occurred so rapidly, he thought, that the electrolyte flash-boiled, immediately vaporizing the water into steam. When water vaporizes into steam, it expands 1,600 times in volume with tremendous force. The rapid production of that much steam caused the surrounding electrolyte to expand in all directions simultaneously.

Despite the trauma Mizuno and Gotoh suffered, and the damage to the lab, Mizuno recorded useful data. Later that year, at the Twelfth International Conference on Cold Fusion (ICCF-12), he reported an estimated heat output 800 times higher than the electrical input from the event. (Mizuno and Toriyabe, 2005)

He recorded a rise in temperature from 25° C to at least 70° C within 10 seconds, and he reported that the cell was shattered by the sharp increase of inner pressure. Whatever caused the rapid heating, it did not occur in a split-second, as it would have with a hydrogen-oxygen combustion reaction. Rather, the event took place in the 20 seconds preceding the explosion. As Mizuno remembered, the tip of the cathode, near the bottom of the cell, began to glow first, followed by a glowing ball of white light surrounding the cathode tip.

The moment he saw the rapidly developing glow, he was transfixed — staring at the cell with his full attention — and watching it explode in front of his eyes. He was able to remember and convey to his sister, who drew sketches for him, several phases of the explosion. He remembered time going by very slowly, as if he could play a videotape in his mind and see every frame of the film.

Mizuno Temperature Data for Jan. 24, 2005, Event					
RTD	**Cm Below Electrolyte Surface**	**Seconds Before the Explosion**			
		20	**15**	**10**	**5**
RTD-1	2 cm	25.124° C	49.982° C	84.985° C	79.998° C
RTD-2	6 cm	25.114° C	50.234° C	88.128° C	79.234° C
RTD-3	10 cm	25.234° C	50.241° C	79.982° C	70.412° C

Cripples, Lepers, and Idiots

In the U.S., January 2005 began with a thud. Journalist Toni Feder wrote in the lead of her *Physics Today* news story "Cold Fusion Gets

Chilly Encore" that "claims of cold fusion are no more convincing today than they were 15 years ago." She wrote, "That's the conclusion of the Department of Energy's fresh look at advances in extracting energy from low-energy nuclear reactions."

Feder spoke with LENR researchers, who put a rosier spin on the report. "The greatest vindication for the cold fusion community," McKubre said, "was that, instead of being treated like cripples, lepers, and idiots, we were treated like normal scientists, in the handling of this review."

Although the adjectives used by McKubre seem extreme, they accurately described how many of the researchers regarded their struggles and their disappointments. On Feb. 27, 2005, McKubre sent me an e-mail expressing his contempt for the DOE panel members' responses. He also tipped me off to a developing news story:

> The DOE review has stimulated all manner of interest and money from the commercial sector (and governmental if not DOE). You will be able to discern the evidence within the next month. ... The more DOE postures in the negative, the more ridiculous they are going to be and seem. The first significant development will happen outside the U.S. but not very far away. Have no doubt. We are going to win this war, after losing every battle.

Hot News From Canada

On June 5, 2005, in the Vancouver, Canada, airport, I answered the customs officer's questions carefully. He wanted to know where I was going with my video and photography gear. I delicately avoided using the word "nuclear." I had tracked down McKubre's tip.

Patrick Cochrane, the chief executive officer and member of the board of directors of Innovative Energy Solutions Inc. (IESI), had accepted my request to observe a demonstration of his company's claimed fusion energy device in Edmonton, Alberta, Canada. The company was incorporated in the U.S., and, as I later learned, its registered corporate office was a private mailbox in Henderson, Nevada.

McKubre had led me to believe that the long-awaited technological breakthrough that would demonstrate, once and for all, that "cold fusion" was real, had arrived. If true, after 16 years of struggle, the research had jumped overnight from a poorly understood, poorly reproducible scientific curiosity to a controllable, revolutionary, clean-energy technology steadily producing thousands of Watts of excess heat.

It was something that everyone involved in the research had imagined and expected would someday happen. After countless arguments with skeptics, cynics and deniers, the cold fusioneers' satisfaction and vindication seemed imminent this day. In addition to inviting McKubre and Hagelstein to the demonstration, Cochrane flew Martin Fleischmann from England to celebrate the auspicious occasion.

I arrived at the industrial park and, after parking my rental car, walked into a large, brightly lit room with bare concrete floors and a high ceiling. A few tables and chairs and a couch were just inside the doorway. With the exception of a large sign with the company's logo, there was nothing on the barren walls. A workbench in the far left corner was covered with mechanical and electrical tools. Two men in jeans and T-shirts, one of them smoking a cigarette, were fiddling with a device the size of a small gasoline generator.

In the center of the room stood a Rube Goldberg-like device with stainless-steel pipes connected by high-pressure fittings to a heavy-duty motor, a tank, and a heat exchanger. On the right side of the room, against the wall, was an array of large stainless steel tanks, connected in a series to each other. I later learned that these were for purifying water for another device. These did not look like science experiments. They looked like engineered technology — and they looked *cool*.

I was the only journalist there, and I appreciated the opportunity to have the scoop. Cochrane allowed me there on the condition that I embargo the news for several months. A half-dozen scientific guests mingled with another half-dozen people whom Cochrane identified as investors. Cochrane, a businessman who seemed to have some understanding of science, looked and sounded at ease while speaking with the visiting scientists. He appeared to have no doubt that, as one of his technicians said, "this is the first invention that replaces fire."

The systems were like no other experimental apparatus I had ever seen. No expensive heavy-water or palladium was required. No messy electrochemistry was needed. In the mid-size device, which was about 6 feet in its largest dimension, machine oil was pumped through a recirculating loop. The vital component was a clear acrylic cylinder about half a foot long and 4 inches around. that was located midstream in the loop. The center of the cylinder contained a special orifice that forced the oil through a narrow opening at high speed and high pressure. The proprietary secret, I was told, had to do with the design of the orifice. I was also told that the recirculating fluid was machine oil doped with boron.

The result, we were told, was that the hydraulic mechanism created electrostatic conditions, which in turn caused cavitation and some new form of nuclear fusion, without strong radiation, to take place. The large system in the room worked on the same concept but, instead of using machine oil, used pure water and produced hydrogen gas.

The demonstrations were thrilling to observe. After a few minutes of a whirring motor, the system warmed up, and a consecutive series of three types of effects were visible: tiny sparks, large arcs and then a luminous glow. The scene had the excitement of a Hollywood movie. "It looks better at night with the lights turned off," Cochrane said. And indeed, as we watched the device that evening, a shower of sparks a few millimeters long burst from the orifice into the flowing stream of oil. As the technician increased the speed and pressure, the sparks gave way to large arcs, about 100 millimeters long. After another increase in power to the pump motor, the arcs dissipated and were replaced by a spectacular glowing purple-blue plasma-like corona. The large-screen display connected to the device displayed the temperature measurements, converted to Watts — hundreds of Watts — several times the amount of input electricity.

Cochrane told me that the devices were based on research by a Korean man, Hyunik Yang, whom Cochrane identified as the co-founder and chief technical officer of IESI. However, during my visit, when I interviewed Yang and asked him to tell me about his discovery, he evaded my question. Instead, he told me about related research that had been done in Russian government rocket science laboratories.

Afterward, I learned that Yang had little to do with the development of the concept but that the idea originated in 1972 from a Russian scientist named Alexander Ivanovitch Koldamasov.

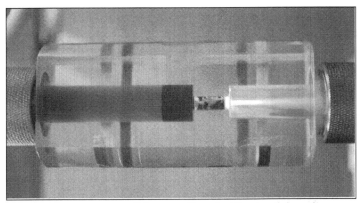

Koldamasov-type cell in operation at a Russian laboratory.

I had the opportunity to interview Koldamasov, a retired scientist from the Volgodonsk branch of the Russian National Research Institute of Atomic Engineering in 2007. He explained the convoluted chain of information and technology transfer (none with his consent): first to other Russians, then to Korea, and eventually to Canada. In one of these steps, Ukrainian researcher Vladimir Vysotskii and a Russian researcher, Alla Kornilova, began doing Koldamasov-type experiments. In Vysotskii's presentation at ICCF-12, in Yokohama, Japan, he reported experiments that had been performed at Moscow State University and later at Keldysh Research Center, part of the Russian Space Agency.

Vysotskii provided me with a photo from one of these experiments, shown above. The experiments he described were visibly different from the ones I observed in Canada. In the Russian experiment Vysotskii described, the machine oil, brownish in color, approaches the orifice at high speed and high pressure. As it exits downstream from the orifice, a purple laser-like beam is emitted from the orifice in the direction of the oil flow. At other speeds and pressures, a bright green glow also emanates from the upstream side of the orifice. No Russian scientists I spoke to were able to definitively explain, using conventional science, the visual phenomena.

In 2005 and during the first day of demonstrations in Edmonton, McKubre and Fleischmann looked at the measurements on the computer display screen. On such a short visit, they were unable to perform in-depth due diligence on the heat measurements. They assumed that the power measurements were correct and that they were displayed honestly.

One other test occurred while I was there. A technician used a handheld radiation detector to look for gamma-ray emissions. Fleischmann seemed to know more than the technician and had him place the detector a few feet away from the device to see whether the level of radiation dropped accordingly. It did, and for the first and only time, I saw a great big smile break out on Fleischmann's face, and he chuckled. Everyone began posing for pictures with their cameras, and the feeling of winning the "cold fusion" war, as McKubre had called it, and witnessing a part of history was infectious.

"From my perspective," McKubre said, "if I had known it was going to take 18 years, I sure as hell wouldn't have started on it." Fleischmann responded with an incomplete sentence and walked away from my video camera as his smile faded. "If I had known it would take that long," Fleischmann said, "I would have — my timetable was 1995."

But What About the Heat?

Despite the elation, two concerns troubled me. First, why had they done a demonstration? What was the purpose? If they had something of this magnitude that was close to a viable technology, why would they want to draw attention to it before they were ready to launch a product and thereby tip off potential competitors?

The second concern was the lack of scientific documentation, particularly about the heat claim. Sometime later, I received a copy of a Feb. 17, 2005, report written by Hagelstein, a theorist. He had visited IESI on Feb. 13, 2005, and written a report for IESI to use to solicit investments. However, there is no evidence in the report that Hagelstein did any of his own tests or brought any of his own instrumentation. He simply restated the information about power measurements given to him by IESI. He wrote that he could imagine how the results could have been faked, and he wrote that he didn't think they had been. He wrote that the best way to perform due diligence would be to arrange for an independent test in a respected laboratory. Nevertheless, he gave IESI a useful document for its next set of investors.

"It was possible to verify by touch the presence of a sizable temperature increase," Hagelstein wrote. "My present view is that they are probably seeing a very large excess-heat effect." That was the extent to which Hagelstein personally tested for excess-heat production.

After my visit to Edmonton, I began looking into the business aspects of IESI. On Nov. 15, 2004, the company had issued a press release announcing that it owned commercial technology and, within a year, would have a device that produced 12 megawatts of electricity from an input of two megawatts of electricity. The company didn't state whether that electricity would come from the conversion of heat or would come directly from the device. The IESI people claimed that their device could also produce electricity directly. They had one unconnected cell, sitting on the workbench, that had electrical taps going into it. They said it could capture electrons from the electric arcs produced in the flowing stream of oil. This would have been a simple demonstration, but they gave an obscure reason why they couldn't show it to us.

In hindsight, I was relieved that I had agreed to an embargo. Without it, I might have helped to promote a swindle or a fraud. I never saw any IESI evidence that even remotely resembled scientific data. I never learned of any independent tests performed on IESI devices. My assessment is that the real power and energy balance, either from heat or direct electricity, was not as claimed.

Cochrane and the two other principals in the company, Ronald Foster and Frederick Dornan, took a total of several million dollars from several hundred investors in the U.S., Canada and the U.K. When it became clear to the shareholders that IESI had massively overstated its projections, the shareholders sued, and they won. The three principals were slapped with $15 million in compensatory damages and $45 million in punitive damages. Before the court passed the judgment against them, they had filed for bankruptcy. The investors never saw their money again. During the discovery phase of the lawsuit, IESI's computers were stolen; therefore, the directors were unable to produce company records. The Alberta Securities Commission issued a decision and sanctions against the company and personally against Cochrane, for illegal conduct and securities violations.

Throughout the years, people have often called me asking for more technical details. However, I have published everything I know about this. I've not had any further contact with Koldamasov or Yang. I don't believe that Yang ever knew as much about the mechanism and device as the Russians did.

In the course of my investigation, I found a paper by a student named Anna Alexandrovna Alimuhambetova, at Volgodonsk University, who researched the history of Koldamasov's invention. Her paper appears to be a useful resource for people to explore this experiment. Perhaps more important are the references she provides in her paper for earlier and related research. I do not know more about this effect; however, I have placed links to all related documents on the IESI investigation page on the *New Energy Times* Web site for readers who may be interested.

The Man Behind the Curtain

Earlier in 2005, on March 24, I was eating lunch with several scientists in the cafeteria at the Los Angeles Convention Center. We were attending the American Physical Society March meeting. A half-dozen of us were at the table. One was Darcy Russ George (b. 1949), a scientist without a college degree but possessing an excellent understanding of nuclear science.

George had made short visits to many well-known companies and government laboratories as a guest researcher. There is no record that he was ever a permanent staff member of any recognized scientific organization. George does not appear to have had any regular employment history, although he has formed and abandoned dozen of small companies, often after selling stock in them or seeking donations.

When George first heard about "cold fusion" in 1989, he wanted to make a documentary film about it, but that never happened. In 1992, he got involved in a commercial startup with a scientist named Roger

Stringham, who had invented an experimental method that used ultrasonic waves to stimulate cavitation and nuclear effects.

During lunch in Los Angeles, George casually mentioned that he had sold his newly formed fusion company for $2 million. The other scientists at the table were astounded and thrilled. There was something about George at the time that made me cautious about trusting him, and I refrained from reporting the news. Seven months later, at the ICCF-12 meeting, the hot topic was the sale of George's company to a larger company, which had retained him as its chief scientist. It was the first publicly traded "cold fusion" company. Xing Zhong Li, a well-known Chinese researcher and physicist, enthusiastically announced the news.

"Our senior pioneer, Tom Passell, is involved in a new startup called D2Fusion," Li said. "This company has received a total of $2 million of investment so far, and [Passell] used part of his investment retirement account to buy 10,000 shares. When I get back to China, I will log in to this Web site and put in a little money to show my respect."

Passell, representing D2Fusion, presented the results of his best experiments: 300 milliwatts of excess heat. Passell was an old friend of George's and had done the experiments with another mutual friend of theirs, Thomas Benson. It was still not news.

After the meeting, however, Nobel Laureate Brian Josephson contacted me. Josephson wrote that George had told him that D2Fusion was going to be selling 1-kilowatt fusion-powered heaters commercially within a few months, and he was considering investing. I later learned that Josephson was also known for his interest in paranormal phenomena, which caused people to doubt his scientific objectivity.

"There's a reasonable chance," Josephson wrote, "that $1 could turn rapidly into $10." Actually, it was extremely unlikely. George's heat production claim was 3,000 times greater than what his scientific evidence showed. It was like an automobile manufacturer of a 40-mile-per-gallon car claiming its car would soon get 120,000 miles per gallon. Now this doubtful claim was news, and I began investigating.

In my first telephone interview with George, he told me things that were not true, evaded important questions, and obfuscated his responses to other questions. Eventually, after George had sold stock to his colleagues and convinced some of them to work for him, I learned that there had been no actual $2 million sale. Instead, George had received only a worthless document promising such a payment from Nelson Skalbania, a businessman who had been convicted of investment theft. Moreover, George's 1-kilowatt heater had been a fantasy, or worse.

While investigating George's activities. I found that he regularly made statements that were factually inaccurate, fabricated at least one document, attempted to take credit for other scientists' work, participated in corporate shell games, and engaged in other deceptive practices.

In 2012, George, along with a new set of business partners, took $3 million from a First Nation (Canadian aboriginal) group in Canada to perform an unauthorized and illicit geo-engineering experiment designed to capture carbon from the atmosphere and cash in on lucrative carbon credits. In 2013, the Canadian Broadcasting Corp. asked me to appear on television to discuss what I found in my earlier investigations of George.

Two months after the show aired, George's partners fired him. He sued his ex-partners. They counter-sued him and claimed that George had made false and misleading claims and did not have the expertise he claimed. In the suit, his former partners wrote that George "exhibited a tendency to behave in a manner that was irrational, unprofessional and offensive to others, and engaged in certain inappropriate conduct including a physical assault upon the project leader."

What I found most surprising in the course of my multiple investigations of George's activities was that none of his peers in the field publicly admonished him. None of them seem to recognize or, if they did, care about the damage he was causing to the image of the field, for which they had worked so hard to bring credibility. To the contrary, on May 10, 2006, many of the researchers were shocked when I published such a hard-hitting exposé of one of their flock. Some of the researchers praised me; some condemned me. As events unfolded, some who condemned me later changed their minds. George called me many unfriendly names, but he never identified any substantive factual error in my investigations.

CHAPTER 24

Cold Fusion Is Neither

D uring the Tenth International Conference on Cold Fusion (ICCF-10), in August 2003, I met and interviewed almost all of the remaining major figures in the field. In our conversations, I asked each of them whether any theory explained "cold fusion." Not one experimentalist suggested a technically plausible theory. On the other hand, more than a dozen theorists each told me that their theory, and only their own theory, could explain it.

Overcoming Miracles

On May 2, 2005, the paper "Ultra-Low-Momentum Neutron-Catalyzed Nuclear Reactions on Metallic Hydride Surfaces" appeared on arXiv, a popular physics preprint server. The authors were Allan Widom, a condensed-matter physics professor at Northeastern University, and Lewis Larsen, chief executive officer of Lattice Energy LLC, in Chicago.

Two days later, David Nagel, a retired NRL physicist teaching at George Washington University, contacted me by e-mail about the Widom-Larsen theory. I asked Nagel why he was interested in it. "I believe that this paper describes what may prove to be a viable mechanism for cold fusion," Nagel responded.

This was the first time in my experience that anyone other than the author of a theory had suggested that any LENR theory might have merit, but I was still reluctant to invest time thinking about it. "I can't make heads or tails of it," I told Nagel. "I don't have the background. But I do understand Huizenga's 'three miracles,' and I would be interested to know if it answers them."

John Robert Huizenga (1921-2014) was a professor of chemistry and physics at the University of Rochester who chaired the 1989 Department of Energy "cold fusion" review panel. (Vol. 2, *Fusion Fiasco)* In 1993, Huizenga wrote a scathing book denouncing the work. He dismissed the entire body of research because he didn't see any theoretical basis for nuclear reactions at room temperatures. He denigrated the anomalous phenomena as requiring three "miracles."

Miracle #1 was the mystery of how the Coulomb barrier, the electrostatic repulsion that surrounds positively charged nuclei, was overcome to allow high rates of nuclear reactions to occur at room temperature. Miracle #2 was the lack of strong neutron emissions expected from deuterium-deuterium (D+D) fusion reactions. Miracle #3 was the lack of strong emission of gamma rays or X-rays expected from D+D fusion reactions.

In June 2005, Nagel sent me slides in which he tried to simplify the explanation of the Widom-Larsen theory. Nagel's slides described how the Widom-Larsen theory resolved the Coulomb barrier problem. More precisely, the Coulomb barrier was not relevant because the Widom-Larsen theory involved uncharged particles: neutrons.

This was an eye-opener; I had never heard of any "cold fusion" theory based on neutrons. Everyone I had spoken to had told me that the underlying process was some novel variation of D+D fusion. Nobody mentioned to me any type of neutron-based idea. By 2000, all such ideas had been forgotten or ignored by most people in the field.

Nevertheless, months went by, and I paid little attention to the theory. Nagel's excitement, however, continued unabated. He sent me a copy of an e-mail that he sent on Sept. 30 to Akito Takahashi, the organizer of the forthcoming ICCF-12 conference in Japan. "The meeting will be very interesting," Nagel wrote, "given the appearance of the papers by Widom and Larsen this year. If you have not seen them, I will send them to you." Takahashi, a staunch supporter of the D+D "cold fusion" idea, was unlikely to have shared Nagel's enthusiasm.

On Nov. 5, 2005, Lino Daddi, a physics professor at the Italian Naval Academy in Leghorn, Italy, distributed an e-mail suggesting that the Widom-Larsen theory might be correct. "Perhaps we have a theory that explains all the anomalous phenomena," Daddi wrote. "The transmutations observed from Yasuhiro Iwamura are explained without the problematic multiple reactions of Takahashi."

At the time, I did not know that, on May 27, 2005, Daddi had presented a paper at a colloquium at the University of Rome about weak interactions and the $e + p \longrightarrow n + v$ reaction. (Daddi, 2005) Daddi had been thinking about the possible role of neutrons in LENRs at least since 1995. (Daddi, 1995)

The main difficulty with all the room-temperature D+D fusion-based theories was that nobody had identified a feasible way to overcome the Coulomb barrier. Invariably, the explanations proposed for room-temperature fusion required some sort of physics miracle. In 1977, cartoonist Sidney Harris drew a cartoon for *American Scientist* and the *New Yorker* magazine republished it in 1989 following the Fleischmann-Pons news.

Sidney Harris cartoon, courtesy ScienceCartoonsPlus.com

Widom and Larsen had published the first LENR theory that, to my knowledge, did not rely on a physics "miracle."

In October 1989, as discussed in Chapter 11, the renowned physicist

Edward Teller likely was the first person to glimpse the logic of neutron-based reactions as an explanation for LENRs.

A New Paradigm

I published a brief report on the theory on Nov. 10, 2005. I tried to interview Widom and Larsen, but neither was willing to speak with me at that time. On March 9, 2006, the first Widom-Larsen theory paper finally published, in the *European Physical Journal C - Particles and Fields*, a respected, mainstream, peer-reviewed journal. Such mainstream publication was unprecedented for a theory in this field.

Nagel called me again and implored me to take a close look at the theory. I thanked him politely and told him that I wasn't interested. Nagel persisted and broke through my resistance. He offered to spend an hour with me on the phone, teaching me the basic concepts. I finally accepted his offer. As we approached the end of that hour, a light bulb went off in my head, and I could see the new paradigm.

I had known nothing about weak interactions or neutron-capture processes before the phone call. This was all news to me. Practical applications of weak interactions were rare; they were generally limited to beta emitters for medical diagnostics and cancer therapy, and use in beta-voltaic batteries. I had had no idea that Cirillo, Mizuno, Iwamura, Hagelstein, Daddi and Teller had already suggested neutron- or neutral-particle based ideas. Teller's idea had been lost in workshop proceedings that were not publicly available until I located one of the rare copies and sent it to Jed Rothwell in March 2010, so he could publish it on his LENR-CANR Web site. Thanks to Nagel, I understood how neutron-based concepts, such as the Widom-Larsen theory, could explain the experimental data far better than any fusion theory.

With Nagel's tutoring, I could then see that the Widom-Larsen theory resolved Huizenga's three miracles because 1) the Coulomb barrier does not apply to neutrons, 2) the theory does not describe a fusion reaction, so high-energy neutrons are not expected, and 3) according to the theory, dangerous gamma rays are prevented from emission beyond the local reaction surface and converted to useful but harmless infra-red radiation. I didn't know whether the Widom-Larsen theory was correct, incorrect, or somewhere in between. Nevertheless, I immediately understood the capability of the theory to advance the research.

The Death of the Rebirth of Cold Fusion

After I hung up the phone with Nagel, I felt a sense of relief that at least one viable explanation for the experimental phenomena existed. Yet I faced some personal discomfort if the Widom-Larsen theory did turn out to be correct. I still had 3,000 unsold copies of *The Rebirth of Cold Fusion*. In that 60- minute phone call, I got the sense that those books had just become obsolete and I considered that several thousand Web pages on the *New Energy Times* Web site might need updating.

Beginning with my editorial in the Sept. 10, 2006, issue of *New Energy Times*, I started to distinguish between "cold fusion" and LENRs. In October 2006, I received an encouraging and favorable analysis of the Widom-Larsen theory written by David Rees, a particle physicist with the U.S. Navy SPAWAR San Diego laboratory.

Among the hundreds of professional and amateur scientists who have attempted to explain LENRs, Widom and Larsen were the first theorists to describe a sequence of events that could explain the process from start to finish. Their theory was supported by detailed physics and mathematics, yet they could explain their concept in plain English.

It was a strange feeling for me to begin thinking, six years into my investigation, that "cold fusion" might not be fusion.

A Gun, a Dog, and a Boat

As a child, Lewis Larsen was inquisitive about nature and largely self-taught. He was born in 1947 in Menominee, Michigan, to parents with modest careers. His father was the methods manager at his grandfather's furniture manufacturing plant, and his mother was the assistant manager of a J.C. Penney department store.

She was a voracious reader, and Larsen remembered making many trips with her to the library. He got his first library card when he was 6 and quickly became an avid student of the natural world. "We had a summer home on a lake in central Wisconsin. We went out there every year, and I studied nature," Larsen said. "I know that sounds boring, but by the time I was 9 or 10, I could name every animal, insect and plant in North America. My parents were very supportive of my appetite for knowledge. By age 10, I had a complete set of Peterson's Field guides."

When he was 5, Larsen became friends with two boys who were five years older than he. One was an avid mineral collector; the other was fascinated by biology, dissecting animals and hunting. The friendships accelerated Larsen's appetite for knowledge.

"I learned from books and directly from nature," Larsen said. "I spent my summers at the lake. I had my gun, my dog, my boat, and my friend Roger. It was idyllic. I'm still friends with these men to this day."

His guidance counselor in high school knew that the University of Chicago had a special early-entrance program. In the 10th grade, Larsen took the College Board exams and did very well on them. After interviews at the university, he was selected for early-entrance and offered a four-year full-tuition scholarship. He skipped the last two years of high school. In the summer, before beginning his freshman year, he was selected to attend a National Science Foundation summer science training program for college-bound high school seniors at Carnegie Institute of Technology in Pittsburgh, Pennsylvania (now Carnegie-Mellon University). That summer, he studied plasma physics.

Larsen's father took measures to keep his son's ego in check. For three summers during college, Larsen worked on a garbage truck. He remembered what his father told him: "You're not gonna be some smart-ass college punk. I'm gonna get you a job where you can get some humility." Larsen also had a sense of humor about his summer job. "The job was $1.65 an hour and all you can eat," Larsen said.

Prediction of Technology Revolution

When Larsen began his undergraduate studies at the University of Chicago at 16, his initial aspiration was to study astrophysics because he was fascinated by the physics of stars. But he learned that scientists at the time were making all kinds of fundamental discoveries in the field of biochemistry, so he chose that path. During his undergraduate studies, he audited a graduate course in theoretical astrophysics taught by the famous scientist Subrahmanyan Chandrasekhar (1910-1995), who later received a Nobel Prize in Physics in 1983 for his theoretical studies of the physical processes of importance to the structure and evolution of the stars. Larsen also attended Astronomy Department open seminars where abundances of the elements and nucleosynthesis — the formation of new atoms and isotopes by nuclear processes — were discussed. What Larsen learned as a teenager in those seminars was later helpful.

In 1968, at 22, he graduated with a bachelor's degree in biochemistry. He continued at the University of Chicago in the Graduate School of Business. At the same time, rather than wait to be drafted and sent to Vietnam as a private, he volunteered and enlisted in the U.S. Army Reserve Officers' Training Corps so he could serve as an officer. He never went to Vietnam because, while he was in summer training at Ft. Benning, Georgia, he contracted a series of severe ear infections that destroyed much of his hearing, a disability for which he received an Honorable Medical Discharge.

In 1970, he graduated with a master's degree in business administration, specializing in operations research and finance. Following his passion for biology, he went to the University of Miami in Florida to pursue a Ph.D. in theoretical biophysics. He was working on a theory of energy flows in biological systems, specifically using bacteria. He was funded by the Atomic Energy Commission under the auspices of a block grant. He completed his coursework, but federal budget cuts occurred before he finished his dissertation, and he was unable to complete his Ph.D. He taught a variety of subjects as a relief instructor for one year at the University of Wisconsin and in 1973 began a business career.

He started out doing systems analysis, then progressed to professional commodities trading and technology investment banking. In his first job in the investment world, as a physical commodities trader for Louis Dreyfus Corp., he ran the international export and trading program in sorghum, oats and soybean oil. He was featured three times in *Barron's* magazine for his ability to predict patterns in stock indexes, commodity prices and interest rates. (Laing, 1986, 1988, 1999) In 1999, journalist Jonathan Laing wrote that, a decade earlier, Larsen had a "dead-on prediction that a coming technology revolution would vault the U.S. decisively ahead of the Japanese in international economic competitiveness."

Looking for a Energy Wild Card

In 1997, Larsen was running a technology consulting company specializing in energy and information management and control systems. One of his clients asked him, "Are there any wild cards in energy?" Larsen remembered the fusion controversy from 1989.

When he heard about the announcement from Martin Fleischmann and Stanley Pons in 1989, he looked into their backgrounds and publication history. "I saw that Fleischmann had a long track record of doing competent and honest research," Larsen said. "Pons was from a wealthy family; he didn't need money, and he was doing research that he loved. I couldn't believe that they had made as many measurement errors as people had accused them of doing. It didn't make any sense to me that these guys would lie and blow their careers away."

The fusion part of Fleischmann and Pons' claim never made sense to Larsen, but because he wasn't working in the fusion field, it didn't bother him so much. Larsen also appreciated calorimetry because, while he was studying biophysics, he designed an experiment with very sensitive calorimetry to measure heat from a bacterial culture over a long period. Larsen searched in 1997 for a credible scientist who was at a major university, had a decent reputation in thermonuclear fusion research and was working in LENRs. Through the early Internet search engines, he found George Miley, at the University of Illinois at Urbana-Champaign. Larsen's most important adventure in science had begun.

CHAPTER 25

Insight From the Stars

When Lewis Larsen and George Miley met in 1997, Miley showed Larsen a five-peak graph of various transmutation products measured in Miley's experiments. The graph showed the abundances of the many elements — all across the atomic spectrum — that were present on Miley's cathodes after the experiments. The pattern revealed by the five peaks was distinct.

As soon as Larsen saw the graph, he recalled a chart he had seen 30 years earlier when he had audited Subrahmanyan Chandrasekhar's astrophysics class at the University of Chicago. "I recognized similar peaks and abundances, and I began to suspect that LENRs were neutron-catalyzed reactions, just like similar processes in stars," Larsen said.

In 1919, Ernest Rutherford proposed that the atom contained a neutral particle. He also suggested that a proton and an electron might combine to form a neutral particle. In 1932, James Chadwick experimentally confirmed the existence of the neutron.

In 1946, Fred Hoyle, an English astronomer and cosmologist, proposed that neutron creation in the hot cores of collapsing stars could be explained by the direct reaction of an electron and a free proton, to form a neutron and a neutrino, expressed as $e + p \rightarrow n + v$. (Hoyle, 1946) Astrophysicists call this process neutronization.

This equation appears identical to (and is often confused with) a different process known as K-shell electron capture. Neutronization takes place among free particles at billion-degree temperatures. Electron capture takes places inside of an atomic nucleus at temperatures and pressures available on Earth. In that instance, the equation is referred to as inverse beta decay. According to Larsen, the term inverse beta decay is a narrowly correct but conceptually misleading way to describe neutronization as well as the Widom-Larsen process. Decays are normally spontaneous and do not require the addition of energy to trigger; neutronization reactions do require energy.

As Larsen explained to me, Hoyle's idea was that the conditions in a dying star are hot enough and dense enough so that electrons have enough energy to react directly with protons to transform into a neutron and a neutrino. This stellar reaction is a simple two-body process. On Earth, according to the Widom-Larsen theory, many-body collective effects provide the requisite amount of energy to enable electrons and free protons — outside of a nucleus — to make a neutron and a neutrino.

For competitive reasons, Larsen did not explain his insights to Miley at the time. In 1998, Miley showed Larsen transmutation results from Tadahiko Mizuno's experiments. To his astonishment, Larsen saw another similar multi-peak spectrum. Although Miley's experiments used nickel and light hydrogen, and Mizuno's experiments used palladium and heavy hydrogen, the spectra were essentially the same.

As described earlier, most researchers in 1998 thought that the experiments required palladium and heavy hydrogen and that experiments with nickel and light hydrogen were not credible. Larsen, however, could see that the two sets of data suggested that the same mechanism was responsible for the results, despite the differences between the experiments.

"Miley's transmutation experiments and Mizuno's experiments showed different relative sizes of the peaks in different parts of the mass spectrum due to different seed elements initially present in the electrolyte," Larsen said. "However, the most amazing thing was that Mizuno's experiments were with heavy water and Miley's were with light water. I knew the coincidence could not be just accidental. I knew then that the heavy-water and light-water results were part of the same phenomenon."

Where Mizuno identified a range for the peaks he observed, they equate to mass numbers 65, 120 and 194. Where Miley identified a range for the peaks he observed, they equate to mass numbers 25, 65, 120 and 200. Widom and Larsen calculated the peak abundances of the

Atomic Mass Peaks of Elemental Abundances					
Source	Peak 1	Peak 2	Peak 3	Peak 4	Peak 5
Mizuno (Measured) [ICCF6, 1996]			65	120	194
Miley (Measured) [2000]		25	65	120	200
WLT (Calculated) [2006]	12	31	62	117	206

transmutations, based on their theoretical model, to occur near mass numbers 12, 31, 62, 117 and 206.

The graph below shows a) measured data, b) an interpreted curve of that data, and c) a predicted curve for such data. The numerous small points on the graph indicate cumulative quantities (y-axis) of elements, as measured by their atomic masses (x-axis) from six nickel-hydrogen electrolysis experiments reported by Miley in 1996.

The upper curves were drawn by Miley in 1996. These curves represent the pattern he interpreted from his experimental data (Miley and Patterson, 1996). The lower continuous curve was drawn by Larsen in 2006 with no fitting, based on calculations he and Allan Widom made using their theory. (Larsen, 2009)

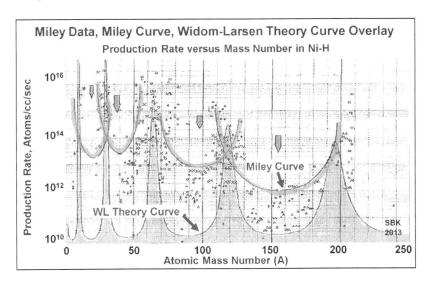

The visual characteristics of the Widom-Larsen optical model reveal a) production of a five-peak atomic mass spectrum, b) peaks near mass numbers 12, 31, 62, 117 and 206, c) increased spacing between peaks as mass increases, and d) declining amplitude of peaks as mass increases.

In another graph, below, the numerous small points indicate cumulative quantities (y-axis) of elements, as measured by their atomic masses (x-axis) from palladium-deuterium electrolysis LENR experiments performed in 1991 by Mizuno. (Mizuno, 2009) As in the previous graph, the upper curves were drawn by Miley, and the lower, continuous curve, depicting five peaks, was drawn by Larsen.

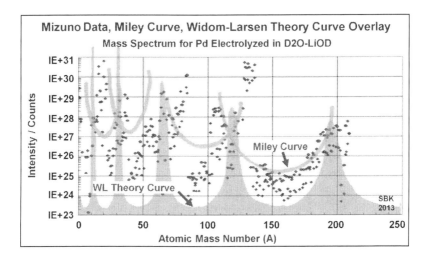

There is an interesting story related to these two sets of peaks. In 2008, Michael McKubre and I were in South India on a lecture tour. Mahadeva Srinivasan had invited us. Our trip was sponsored partially by the Indian government. On our first stop, at the International Conference on Systemics, Cybernetics and Informatics, in Hyderabad, McKubre, Srinivasan and I gave lectures on LENR. McKubre was first. His talk, as requested and advertised, was supposed to be a broad review of the LENR field. After his talk, Srinivasan was clearly upset and was the first to ask a question. "Mike, thank you for the great presentation," Srinivasan said, "but why didn't you say anything about transmutations?" McKubre explained his reasoning to the attendees:

> My friends in the cold fusion community chastise me
> very severely, but I'll say it again since we're all friends. I
> know what to do with heat; I know how to use heat. In the

transmutation business, I don't know what to do with the ability to turn expensive elements into cheap ones. I don't have a use for that.

Now, in turns of a demonstration of a nuclear effect, I think it's useful, and other people have done that. But my interest is a little bit beyond demonstrating that there is a new nuclear effect here. I know that there's a new nuclear effect here. What I'm interested in is, What is it good for?

A few days later, McKubre spoke at the Bhabha Atomic Research Centre and mentioned transmutation data, and even though he saw the same pattern that Larsen saw, McKubre didn't seem to recognize its significance. "Mizuno and Miley," McKubre said, "you can overlap them and see the same thing."

Back in 2004, theorist Talbot Chubb explained the value of transmutations at the Eleventh International Conference on Cold Fusion (ICCF-11), but, like McKubre, he wasn't particularly interested in them. "I agree that the heat production is a primary concern," Chubb said, "and I think that transmutation and ... producing energetic particles are part of the picture, but the real core is the heat. The other things are experiments that teach us something about what process is going on. But I think the heat observations are going to make the difference."

The Ultra-Low-Momentum Neutron

After seeing the Miley and Mizuno data, Larsen knew that neutrons were the key and, moreover, that they had to have ultra-low momentum. "I had a conceptual understanding of the theory worked out by then but not the precise details of how the neutrons were formed," Larsen said. "I realized they had to have ultra-low momentum because researchers saw the results of transmutations but they never saw the neutrons, aside from, possibly, spallation neutrons — and they never saw deadly, high-energy gamma-rays. Therefore, I also knew there had to be some kind of gamma-conversion mechanism. Somehow, the electrons were suppressing the gamma radiation."

In 1999, Larsen consulted with two of the nation's top nuclear physicists, who, at the time, were working at Lawrence Livermore National Laboratory (LLNL). He wanted to know whether Miley's data appeared to be accurate and whether they reflected a real nuclear process. Without explaining the source of the data, Larsen asked them, "Is this bullshit or not?" Both affirmed that the data looked real.

Larsen and his partners formed Lattice Energy LLC in 2000. Through a prior relationship he had with Boeing Corp., Larsen signed a small research contract to produce LENR electrodes. Larsen hired Miley to fabricate them, but Miley never delivered any. Also that year, Larsen began a broad outreach program, as mentioned in Chapter 21, not only to LLNL but also to the Department of Energy headquarters and Sandia National Laboratories. His company got its first seed capital in 2001.

Starting in July 2005, Larsen and Widom participated in numerous invitation-only federal government meetings and briefings to discuss the theory. Larsen said the various federal agencies and groups were interested in the theory for different reasons, but he declined to elaborate on why or identify to me all of the places where he and Widom were invited to speak. Federal employees frequently suggested to Larsen and Widom that they might obtain federal grants for research. "They sucked all kinds of technical information out of us, but they never gave us a dime and never proposed any significant funding," Larsen said.

Around July 2002, Larsen and Miley were at Sandia National Laboratories giving a presentation with senior technical staff about LENRs as well as work being done in Larsen's company for Boeing. At one point during the meeting, the Sandia staff members left Larsen and Miley alone in the room.

"Miley had just returned from the ICCF-9 conference in China," Larsen said. "He pulled out a copy of a paper he had received from Yasuhiro Iwamura that had just published in the *Japanese Journal of Applied Physics* which showed new LENR transmutation results. Miley asked me if I believed it and if I thought the data was correct. I responded, 'Yes, I'm very confident.' Without explaining why, I told Miley that not only did I believe it but also I understood it. I told Miley that I thought that Iwamura and his team had done fantastic experimental work." Larsen went to work to figure out the puzzle.

CHAPTER 26

Larsen's Nucleosynthetic Reaction Networks

In the spring of 2003, Lewis Larsen began to visualize what could be occurring in LENR experiments if ultra-low-momentum neutrons were created in the systems. The hand-drawn diagram below is an excerpt from a larger drawing that depicts the nuclear reactions Larsen

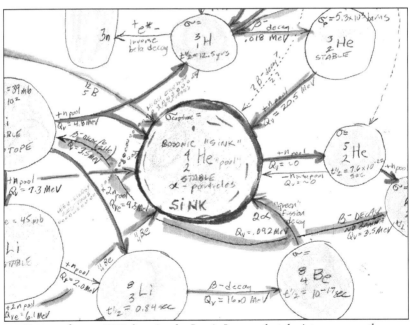

Excerpt from a 2003 drawing by Lewis Larsen that depicts, among other reactions, nuclear processes that produce helium-4.

knew were possible in a palladium-deuterium electrolytic system with a lithium-based electrolyte, given the availability of such neutrons.

Within a few years, Larsen began creating computer-drawn versions of his nucleosynthetic reaction network concepts and publishing them online. The end of Chapter 11 features one such illustration depicting several pathways in which palladium-102 — given a source of ultra-low-momentum neutrons — can transmute to silver-109.

Clues on the Nanosurface

In July 2003, Larsen hired Edmund Storms as a consultant to perform LENR experiments and publish some of the results. Larsen's company, Lattice Energy LLC, sent Storms a scanning electron microscope with an attached energy-dispersive X-ray spectrometer for Storms to use in his home laboratory.

Larsen asked Storms to reproduce the laser-triggering experiment developed by Dennis Letts, in Austin, Texas, and his colleague Dennis Cravens, in Cloudcroft, New Mexico, as discussed in Chapter 19. The Letts-Cravens discovery in 2000 led to what Larsen considered one of his most significant insights. It was the first part of the answer to the question of how neutronization, denoted as $e + p \longrightarrow n + v$, which takes place among free particles in the cores of collapsing stars, could take place among free particles, outside of nuclei, on Earth.

In 1998, George Miley had sent Larsen a manuscript of a theory proposed by Italian physicist Elio Conte (b. 1946). Conte had suggested that neutrons were created by the capture of an electron from the innermost electron shell, called the K-shell, by a proton. However, according to Larsen, Conte did not identify how electron capture, which takes place entirely inside a nucleus, could be triggered in LENRs. (Conte, 1999)

Larsen also knew that Tadahiko Mizuno, in his 1997 book, had suggested an $e + p \longrightarrow n + v$ reaction to explain the results of his experiments. (Mizuno, 1997) In 1998, Yasuhiro Iwamura had proposed the same reaction to explain his results. "D+D fusion cannot explain the phenomena," Iwamura wrote. "We may at least say that D+D fusion cannot play an important role." (Iwamura, 1998)

"Intuitively," Larsen said, "Mizuno and Iwamura knew that it had to be some kind of $e + p \rightarrow n + v$ reaction, that it was not a K-shell electron capture but, rather, that it was taking place outside a nucleus. There was no sensible way to trigger K-shell capture from outside the nucleus."

Larsen knew that LENRs took place on the surfaces of the metal cathodes — within the first micron — rather than within the bulk of the material. This significant fact had been reported in 1989 by the Naval Research Laboratory in Washington, D.C. (Chapter 11), in 1992 by a Russian group (Chapter 8), by John Dash (Chapter 13) and by other researchers. Larsen didn't know about the early research, but he knew that Miley, as well as Iwamura, had performed depth-profiling analyses of their LENR transmutation samples. When Miley and Iwamura went more than 0.15 microns below the surface, they saw no transmutations.

The researchers who ignored the heavy-element transmutations and isotopic shifts missed critical clues revealed by these surface effects.

"Virtually everybody in the field that I spoke with in the late 1990s, including Fleischmann, was saying it was a bulk effect," Larsen said. "The exception was the people who had done the transmutation experiments, for example, Miley and the Japanese researchers. As these researchers looked deeper into the material, they found fewer and fewer transmutation products, which meant it had to be a surface effect."

By 2003, Larsen was confident about the following information: 1) the various metals used in LENR experiments all formed hydrides (or deuterides) — compounds of the metals that absorb many times their own volume of hydrogen or deuterium; 2) free protons, outside of nuclei, were reacting with electrons; 3) the reactions were taking place on the metallic film surfaces of the cathodes and not the interior and 4) the laser-triggering was doing something important. Larsen began studying metallurgy and surface physics. The implication became clear: Whatever was going on had to do with the phenomenon known as plasmonics, and Larsen suspected that surface plasmon electrons held an important clue to the LENR mechanism. Although surface plasmon polaritons and surface plasmon electrons are involved in the reactions, only the latter term is necessary to describe these reactions.

Nicholas A. Melosh, an associate professor of materials science and engineering at Stanford University, wrote, "Plasmonics is the study of the interaction between the electromagnetic field and free electrons in a metal. Free electrons in the metal can be excited by the electric component of light to have collective oscillations." The study of plasmonics has broad applications, including in the fields of physics, material science and biology.

Specifically, Larsen determined that the effect involved a type of wave known as a surface plasmon. Surface plasmons occur when light is directed and bound to the surface of the metal, coupling with electrons. Electrons that compose surface plasmons oscillate coherently together, are quantum-mechanically entangled, and behave as a collective many-body system.

A key to exciting, or activating, the surface plasmons with light, for example by a laser, is the matching of the frequency and polarization of the light emitted by the laser to the natural frequency of the oscillation of the surface plasmon electrons. Letts and Cravens had found that the laser effect worked particularly well with gold-coated cathodes.

"That clinched it for me," Larsen said. "I knew that surface plasmons were key to the $e + p \rightarrow n + v$ reaction in LENRs because gold loves to form surface plasmons. This helped me to identify which electrons were reacting with the protons (or deuterons) to make the neutrons."

Islands of Electrons and Protons

By this time, Larsen had found the scientific literature that gave him additional insights into what was happening on the surfaces of the metals in LENRs. Larsen explained:

> Ever since the mid-1990s, C. Aris Chatzidimitriou-Dreismann, at the Technical University of Berlin, had studied the behavior of protons and deuterons on the surfaces of metal hydrides, and he had looked at them with deep inelastic neutron and electron scattering. A loaded hydride will create patches of protons or deuterons on the surface, like little islands. He discovered that these many-

body patches oscillate collectively and that the protons or deuterons that comprise them are quantum mechanically entangled with each other.

When the LENR experimentalists were talking about fully loading a metal hydride, you get a kind of breathing mode in which protons or deuterons start leaking back out onto the surface of a metal that is absorbing hydrogen.

Here's an analogy: Imagine that you take a bone-dry sponge and slowly drip water onto it. Eventually, it will become fully saturated, and you'll begin to see small droplets of water appearing on the surface. When you fully load a metallic hydride lattice with protons or deuterons and you continue to load even more, protons will begin to move out onto the surface. An elastic film of electrons already covers the surface layer of the metal hydride. You finally have a "charge-neutral" surface filled with islands of electrons and protons waiting for the right type of energy input that will trigger a reaction.

The Power of Many-Body Collective Effects

According to Widom and Larsen, the concept of many-body collective effects is fundamental to their theory. An example of such a collective effect is a concentrating solar power system. Typically installed in a barren desert area, thousands of mirrors are arranged on the ground to reflect light collected from the sun to a central tower, where the light is converted to heat, which is converted to electricity.

The intensity of light reflected from each mirror is harmless by itself, but as each mirror contributes to the more concentrated electromagnetic irradiation at the tower, the energy density becomes high enough to be deadly.

Here's why the concept is important. Making a neutron by way of the $e + p \longrightarrow n + v$ neutronization process in the cores of collapsing stars requires a lot of energy and a very high density of matter. Specifically, the reaction needs 0.78 MeV of energy. That amount of energy among free particles does not exist at room temperature, as Larsen explained to me in 2016:

> Normally, the only place where an $e + p \rightarrow n + v$ reaction takes place outside of a nucleus is in the core of a dying star as it's going supernova. Dying stars have tremendous pressure, density, and temperature. But you have none of that in condensed matter. Instead, what you have is many-body collective effects with huge electric fields. You can pull a little bit of energy from a whole lot of particles and concentrate that energy on a fewer number of particles.
>
> So many physicists were incredulous about our work because most of those people know about supernovas, and they said, "How can you possibly get a supernova r-process in condensed matter?" The reason they didn't understand it is that they didn't understand many-body collective effects.

It was obvious to Larsen that the reaction was not precisely $e + p \rightarrow n + v$, a two-body reaction between an electron and a proton. He concluded that some sort of many-body collective effect was intrinsic to the reaction process. This concept was not evident to a group of six scientists in Italy who published an erroneous critique of the Widom-Larsen theory. (Ciuchi et al., 2012; Widom et al., 2012)

"The only way to create an electron with an energy of 0.78 MeV in condensed matter without extremely high temperatures," Larsen said, "is to take some energy from many less-energetic electrons, feed it into a subset of those electrons, which will then pump some of them to 0.78 MeV or higher. That's what collective many-body effects do. That's what surface plasmons do; they can concentrate energy."

In my discussions with Larsen, I suggested that the reactants and products in his and Widom's theory could be more clearly expressed as $e_n + p_n \rightarrow n + v$, indicating that a number of electrons and a number of protons work collectively to create a neutron and a neutrino. A more-detailed, four-step mechanism is shown later in this chapter.

In electrolytic LENR systems, the electrical current flowing through the cathode creates a high electric field across the cathode surface. This provides sufficient input energy needed for protons or deuterons to react with surface plasmon electrons and thus produce neutrons. At LENR-active sites on cathodes, the electrical-field energy concentrates further in small spots, which leads to even higher nuclear-strength local electric fields at those locations. Patches of protons or deuterons at these sites oscillate collectively and, in light of the breakdown of the Born-Oppenheimer approximation (see below), the patches begin to electromagnetically couple loosely with nearby surface plasmon electrons.

As energy is injected into what will become LENR-active sites, the value of the local electric field increases the effective masses (the energy) of a smaller quantity of electrons at the reaction site to the required 0.78 MeV.

Larsen uses the infrared video taken by the SPAWAR researchers (Chapter 20) to demonstrate the reality of tiny, short lived, infrared heat-producing LENR-active sites.

In the video, visible hot spots that appear and disappear in split seconds are, in fact, groups of even smaller 100-300 micron patches that compose the LENR-active sites. Those patches are born and die, Larsen said, all within a lifespan of 200-400 nanoseconds.

Century-Old Computational Shortcut Fails

Sometimes, mathematical calculations are so complex that a shortcut to approximate a rigorous solution is not only practical but also necessary. Calculations required for the Widom-Larsen theory are time-consuming and can be extremely difficult. Therefore, scientists wanting to check the Widom-Larsen calculations might be tempted to use shortcuts. Brevity won't work in this case; the calculations need to be done in their entirety.

A well-known shortcut, proposed a century ago, is known as the Born-Oppenheimer approximation, named after physicists Max Born (1882-1970) and J. Robert Oppenheimer (1904-1967). Since 1927, the Born-Oppenheimer approximation has been an accepted mathematical shortcut that allows scientists to quickly perform basic quantum mechanics calculations involving simple molecules, but it is known to break down (that is, not be accurate) under certain conditions. Larsen explained how he learned about the limitations of this shortcut and why it did not apply to LENRs:

> I was trying to learn everything I could about what happens on the surfaces of metallic hydrides in the presence of high electric fields. I found that, in the late 1970s, some researchers found that the Born-Oppenheimer approximation doesn't apply to all systems at all times. Then I ran across a paper published in *Nature* that experimentally verified that the Born-Oppenheimer approximation on a surface does indeed break down — that is, it doesn't work.
>
> I knew that the failure of this shortcut was important and that it consequently dictated the kinds of calculations that would be necessary, but I wanted to talk with someone who specialized in the topic. So I found an expert on the breakdown of Born-Oppenheimer, John Tully (b. 1942), professor of physics and chemistry at Yale. He taught me that, in those regions, the Born-Oppenheimer approximation has got to break down.
>
> Typical practitioners in physics in the late 1990s assumed that the Born-Oppenheimer approximation worked in all situations, and they would use that century-old assumption in calculations involving LENRs.
>
> But for the previous 10 or 20 years, the surface chemists and physicists knew there were exceptions. The breakdown of Born-Oppenheimer wasn't more widely accepted until sometime between 1999 and 2000. The idea that the Born-Oppenheimer approximation didn't always apply was controversial at the time, but now it's an accepted part of physics and chemistry.

But Where Are the Gamma Rays?

Once Larsen had conceived of the mechanism to create ultra-low-momentum neutrons (ULMNs), the next step in the reaction process was relatively straightforward. The ULMNs are captured by a nearby nucleus. But the neutron-capture process is almost always associated with gamma-ray emission. In transition metals such as palladium, gamma-ray fluxes arising from neutron capture have energies in the MeV range, which is deadly.

But no dangerous gamma rays had ever been observed in LENRs. Larsen knew something had to be locally converting the gamma rays into something less harmful, such as infrared radiation. He was fairly certain it had something to do with the surface plasmon electrons, but he didn't yet understand exactly what that mechanism was. Chapter 27 discusses the gamma-suppression mechanism in detail.

Allan Widom: Key Collaborator

By early 2004, Larsen had most of the general concepts put together, but he needed an academic collaborator who was well-published and who had the physics and calculation skills to help him complete the development of the theory. Larsen hadn't done these kinds of calculations for many years. After an extensive search, Larsen said, he found Allan Widom, a professor of condensed-matter physics at Northeastern University:

> Around March or April 2004, I began to look for a theoretical physicist with strong experience in many-body collective effects, quantum electrodynamics and condensed-matter physics. I knew that these were the required disciplines, but they are a rare combination. Eventually, I found Allan and hired him in September 2004. Coincidentally, Allan was a close personal friend of Giuliano Preparata, a well-known LENR theorist. Allan understood all the required disciplines. On top of that, he had an

excellent reputation as a theoretical physicist and a good publication record.

Widom had had no involvement or interest in LENR at the time. He was skeptical but was willing to look at the experimental evidence and consider my theoretical concepts. Together, we reviewed hundreds of papers dating back many decades, all the way back to 1922; it took us about six months. We found that most of the important work had happened outside the U.S. The researchers overseas generally had much more open minds. Together, we worked out the remaining details of the physics and mathematics of our theory of LENRs.

One of the key aspects that Widom and Larsen figured out together was why deadly gamma rays from the neutron-capture processes were not detected. A simplified explanation of the gamma-conversion mechanism is that, in LENR systems, the gamma radiation is locally converted to infrared radiation at the LENR-active sites, which are roughly two-dimensional and are generally less than 100 microns across. LENR-active sites are situated on the surface of metal hydrides and do not extend more than a short distance into the bulk metal.

Widom and Larsen's gamma-conversion mechanism explains why heavy external shielding — for example, blocks of concrete or lead — has not been required for conducting safe LENR experiments and why LENR researchers have not died from gamma irradiation.

Widom and Larsen also worked out the mathematics and physics details of the mechanism for creating nuclear-strength local electric fields, the reason that the neutrons were ultra-low-momentum, and the details of the many-body collective electroweak process that creates those neutrons.

The Widom-Larsen Ultra-Low-Momentum Neutron

Textbooks generally recognize that neutrons can have a wide variety of energies, ranging from nano-electron-volts to mega-electron-volts.

Their energies, or momentum, are also often described as having certain temperatures. Until the Widom-Larsen theory, nobody had proposed the existence of neutrons with lower energy than ultracold neutrons, 0.0000003 eV. The Widom-Larsen theory proposes and names an ultra-low-momentum neutron (ULMN) that is even colder: a neutron with kinetic energies that are effectively zero, on the order of 10^{-12} eV or less – that is, 0.000000000001 eV.

The primary significance of the low energies of ULMNs is that, unlike higher-energy neutrons, ULMNs generally do not go flying off, scattering in all directions, and producing biologically hazardous neutrons. Instead, they are captured, or absorbed, by many of the thousands of nearby atoms, which leads to elemental transmutations and isotopic shifts in nearby nuclei.

Almost all ULMNs are absorbed locally, within a few microns, by the materials in the environment in which they were created. In rare instances, a small percentage of ULMNs are not captured by nearby nuclei. In those cases, interactions with materials elevates small numbers of neutrons up to thermal energies where they can potentially be detected. In fact, very small bursts of low energy neutrons have been observed in some experiments. The tiny percentage of LENR neutrons that are never captured simply decay benignly into a proton, an electron, and an electron antineutrino within several minutes.

Because there is an inverse function between momentum and wavelength, ULMNs, which are created ultra-cold, have extremely large quantum mechanical wavelengths. This results in extraordinarily large nuclear absorption cross-sections compared to more typical neutrons. This large cross-section provides two beneficial effects: 1) high reaction rates and 2) the mean free path for a neutron is reduced to atomic distances, minimizing the chances of it escaping the metal surface.

This means that ULMNs are difficult to detect directly outside of LENR cells. For the same reason, significant activation products are unlikely to be observed if tests are performed to look for the effect of neutrons on targets located outside of electrolytic cells. These two unusual attributes of ULM neutrons impeded laboratory measurements and experimental investigation of these phenomena, and this helps explain why conceptual understanding took so long.

Throughout the history of this research, there have been countless reports of brief, low-flux emissions of neutrons detected from experiments using electronic detectors. These devices detect neutron emissions in real time, and such measurements are compared to measurements from other detectors, placed some distance away from the device under test, to check for background levels. The magnitude of neutrons detected in this manner has always been close to background levels. Large emissions of dangerous energetic neutrons were not seen in LENRs.

However, a non-electronic type of detector, called a solid-state nuclear track detector, informally known as a CR-39 detector, has been used with success for this purpose. The reason is that these detectors are constantly integrating detectors that count results over time. They are placed inside or outside an experimental cell and can register cumulative neutron emissions over days or weeks.

Four-Step Widom-Larsen Process

According to the Widom-Larsen theory, LENRs are not predominantly strong-interaction fusion or fission reactions. Instead, LENRs are driven mainly by the weak interaction. The broad outline of the Widom-Larsen process can be shown in the $e_n + p_n \longrightarrow n + v$ equation, indicating that a number of electrons and a number of protons create a neutron and a neutrino. More specifically, the process is broken down into four basic steps, all of which are based on textbook physics.

In Step 1, below, a high electromagnetic field concentrates energy from a large number of surface plasmon electrons to a much smaller number of heavy-mass surface plasmon electrons. This is an electromagnetic interaction.

Step 2a, below, an electroweak interaction, depicts a reaction with hydrogen. Step 2b, above, depicts the same reaction using deuterium. The heavy surface plasmon electron reacts with a proton or a deuteron and creates one or two ultra-low-momentum neutrons (ULMNs), respectively, plus a neutrino.

In Step 3, below, the ULMN is captured by a nearby target atom on the surface of the reaction site. This is a strong-force interaction. This neutron-capture process increases the target atom by one neutron and, if stable, creates a new heavier isotope of the same element, and the reaction stops. If the target atom is unstable, the reaction continues to Step 4.

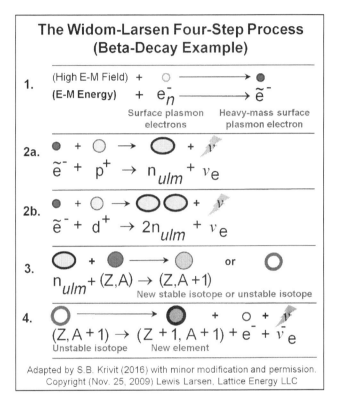

The Widom-Larsen Four-Step Process (Beta-Decay Example)

Adapted by S.B. Krivit (2016) with minor modification and permission. Copyright (Nov. 25, 2009) Lewis Larsen, Lattice Energy LLC

In Step 4, above, an unstable nucleus undergoes a weak interaction called beta decay. (A neutron within an unstable nucleus decays into a proton, an energetic electron and an antineutrino.) The energetic electron released in a beta decay exits the nucleus as a beta particle. Because the number of protons in that nucleus has increased by one, the atomic number has increased, creating a different element and transmutation product. If more ULMNs are available, the transmutation network may keep going, repeating Step 3 and possibly Step 4.

Once neutrons are created in the system, a variety of nucleosynthetic reactions may occur. The reactions can be short and simple or longer and more complex.

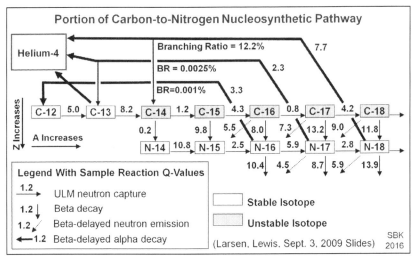

Nucleosynthetic reaction-example of carbon-12 transmutation to helium-4

Energy Release Mechanisms

In the diagram above, based on Larsen's Sept. 3, 2009, slides, the first reaction represents a carbon-12 atom capturing a neutron and going to carbon-13. This releases 5.0 MeV of energy through the neutron-capture process, creating gamma rays that are converted into infrared heat at the material surface by nearby heavy electrons.

Given a source of ULMNs, the reactions can continue to carbon-18, nitrogen-18, and beyond, creating helium-4 as a byproduct from beta-delayed alpha decays of neutron-rich isotopes of nitrogen.

In addition to being released through neutron-capture processes, energy may be released through other decay processes and emissions. Defined values of energy release are associated with each process and each set of particles involved in the reaction network.

In experiments with light hydrogen, 0.78 MeV of energy is required to create each ultra-low-momentum neutron. In the case of deuterium, there is an energy requirement of 0.39 MeV to create each ULMN. The calculations shown below are based on well-known, conventional nuclear physics processes.

As shown below, a LENR reaction in which carbon-12 is present as a seed element may produce 22 MeV of energy on its way to nitrogen-15. In another example, if lithium-6 is present as a seed element, it may produce 26.9 MeV of energy on its way to helium-4. (Widom et al., 2006)

Sample Energy Calculations in LENRs Based on the C-12 to N-15 via C-15 reaction pathway as shown in Lewis Larsen's Sept. 3, 2009, slides, page 11						
Source Element	Particle or Projectile	Target Element	Process	Q-Value in MeV		
				Release	Cost	Net
C-12	neutron	C-13	Neutron capture	5.0	0.39	
C-13	neutron	C-14	Neutron capture	8.2	0.39	
C-14	neutron	C-15	Neutron capture	1.2	0.39	
C-15	beta	N-15	Beta Decay	9.8		
			Estimated neutrino loss from beta decay	1.00		
			Total Energy Release for Pathway	22.0		

Stable Isotopes Produced in LENRs

In 1996, Thomas Passell (Chapter 18) was perplexed by experimental data reported by scientists in the LENR field. "The most troubling observation," Passell wrote, "is the almost complete absence of radioactivity in palladium and nickel cathodes which have apparently produced excess heat. That is, 'Why should nuclear reactions producing only stable isotopes be the heat producers?'" (Passell, 1996)

Because the nuclear reaction most familiar to the public is nuclear fission, the process used in nuclear power plants, people naturally expect nuclear reactions to produce dangerous, medium-lived unstable isotopes.

Commercial nuclear fission reactors produce radioactive products, such as cesium-137 and strontium-90, which have half-lives of 30 and 28 years, respectively. If these radioisotopes escape containment, they can enter the ecosystem and cause harmful biological contamination.

In contrast to nuclear fission, LENRs do not emit deadly fluxes of high-energy neutron and gamma radiation. Moreover, their decay processes quickly lead to stable isotopes, so they do not generally produce significant amounts of hazardous medium-lived nuclear wastes.

The reaction paths in LENRs can be traced in a visual chart called the "valley of stability," a nuclear physics tool that provides a conceptual view of the elements and isotopes. A chart of known nuclides plots the atomic number of each nuclide on the y-axis, its number of neutrons on the x-axis, and the stability (binding energy) of the nucleus, represented by the depth, on the z-axis. The stable isotopes for each element are located in the center of the valley. The stability of a nuclide is determined primarily by its ratio of neutrons to protons. When a nuclide obtains an additional neutron, its position moves away from the center of the valley toward the right. In some cases, the addition (capture) of one neutron changes a stable isotope to an unstable isotope. In other cases, the isotope becomes unstable only after several successive neutron captures. The interactive "Live Chart of Nuclides" at www-nds.iaea.org is a useful tool for seeing this behavior.

I discussed Passell's question with Larsen, who provided helpful information. "There are unstable products produced in LENR transmutation networks," Larsen said. "However, these intermediate products are not observed because, being typically very neutron-rich, their half-lives are quite short, normally ranging from tens of minutes down to milliseconds. Instead, they transmute in a very rapid, sequential series of beta decays that I call a 'beta-decay cascade.' The end-products of these fast cascades are stable isotopes, which can be observed at the end of an experiment. This is why LENR systems do not typically produce large quantities of medium-lived radioactive isotopes with problematic half-lives, as commercial fission reactors do; importantly, there are no known nuclear waste disposal issues with LENR systems."

In LENRs, when a stable nuclide captures an ULMN, the unstable radionuclides tend to have short half-lives of seconds, minutes, hours or days. For example, if palladium-110 captures between one and four successive neutrons, there are 16 possible unstable decay products before the beta-decay cascade terminates at a stable nuclide. With three exceptions among the 16, all of those unstable radionuclides have very short half-lives: seconds to hours. The three exceptions have exceedingly long half-lives, so long that they are considered virtually stable: cadmium (Cd-113), with a half-life of $8x10^{15}$ years; Cd-116 with $3x10^{19}$ years; and indium (In-115,) with $4x10^{14}$ years. By comparison, the age of the universe is estimated to be $13.5x10^{9}$ years.

Neutron capture on nickel-64 can lead to a strontium-90 nucleus through LENRs, but that transmutation would require 16 successive neutron captures, then 10 successive beta decays. However, there are innumerable other, more benign, pathways that can occur when nickel-64 captures neutrons.

"Statistically speaking," Larsen wrote, "the LENR networks are vastly more likely to produce stable end-products; therefore, inconsequential amounts of unstable isotopes with intermediate lifetimes are produced. They may well be present in very minute quantities, but their relative abundances would be so small that they would not be detectable, which is exactly what is shown by a large body of experimental observations in LENRs."

No Chain Reactions

Moreover, LENRs don't involve chain reactions. In nuclear fission, a single neutron capture by a fissionable atom triggers the emission of multiple energetic neutrons, driving an accelerating chain reaction. This is the principle behind both the atomic bomb and a nuclear reactor. In nuclear fission, radioactive starting fuel materials are required and are intrinsic to the reactions.

LENRs, on the other hand, do not require and, in general, do not use radioactive fuels. In LENRs, unless fissionable isotopes are present, the transmutation network's reactions normally don't produce more neutrons than they consume.

According to Larsen, LENR fuels can be any comparatively inexpensive element that captures neutrons. Some elements, such as lithium, make better LENR fuel targets than others because neutron captures on those particular fuels, and the subsequent nuclear decays, release a greater quantity of nuclear binding energy, much of which becomes usable heat.

LENRs stop producing ULMNs very soon after the input energy stops, which terminates the electroweak reactions. The input energy is provided either by an external source and/or from locally generated infrared photon radiation (heat) that is reflected back into reaction sites. Continued production of neutrons in LENR-active sites absolutely requires input energy from one or more sources.

"This Will Change Everything"

During a phone call with Larsen in January 2012, I asked him whether he remembered a defining moment in the development of his and Widom's theory. Yes, Larsen said, there was:

> It was sometime in February or March of 2005. Allan and I had finally gotten the math to work for the collective electromagnetic coupling between the surface plasmon electrons and the micron-scale, many-body patches of protons or deuterons found on fully loaded metallic hydride surfaces. We had been struggling for several weeks to see how to get the local electric fields high enough to make neutrons using the weak interaction. It was around 2 in the morning Chicago time. We were doing all this work back and forth via e-mail while simultaneously chatting on the telephone. When it was clear that we were right and that we definitely had it nailed down rigorously, that's when Allan said, "Lew, do you realize what we've done? This will change everything."

Larsen explained to me that "this will change everything" meant the ideas that he and Widom developed are applicable to a host of scientific disciplines, not only to LENR research but also to astrophysics, geophysics and isotope geochemistry:

In the field of astrophysics, scientists have believed that nucleosynthesis takes place only in the cores of dying stars. But for many years, they have seen isotopic and elemental anomalies in meteorites, dust grains, and atmospheres of stars with unusually high magnetic fields. Astrophysics has assumed that these anomalies were created strictly inside stars, but our theory shows that nucleosynthesis can take place in many environments outside of stellar cores. There have also been unsolved mysteries in geochemistry, where isotopic anomalies have also been observed. Our theory can explain these anomalies, as well.

On May 2, 2005, Larsen and Widom went public, uploading the pre-print of their first paper to the arXiv server. The paper published on March 9, 2006, in *European Physical Journal C – Particles and Fields*. (Widom et al., 2006)

Extending the Theory

Widom and Larsen wrote three more papers together that provided additional details of the theory (Widom et al., 2005, 2006, 2007). In 2007, Larsen engaged physicist Yogendra N. Srivastava (b. 1941) for his expertise in collective magnetic phenomena and Standard Model high-energy particle physics. Together, they extended the theory beyond condensed matter into magnetically dominated types of reactions, such as, for example, what occurs in exploding-wire phenomena. (Appendix A) The three authors wrote four more papers, one of which published in *Pramana Journal of Physics*, a refereed publication of the Indian Academy of Sciences, and provides a summary of the theory. (Widom et al., 2007, 2008, 2012; Srivastava, et al., 2010) According to Larsen, the *Pramana* paper represents the most informative peer-reviewed publication of the theory as of 2015.

During the writing of this book, I made many attempts, by e-mail and telephone, to contact Widom and learn more about his role in the theory. He has not answered my phone calls or returned messages or e-mails. He doesn't appear to have spoken with any other journalist, either. However, Widom has spoken at science conferences and has continued to publish interesting papers on a variety of physics topics.

On Oct. 30, 2008, by mutual agreement, Larsen terminated Lattice Energy's business and technical collaboration agreements with Widom and Srivastava, and the two academics pursued their own independent attempts to develop the research commercially. Larsen then developed and extended the body of theory to other areas, including carbon fullerenes, aromatic rings, and ultra-high-temperature superconductors, as well as pursuing detailed, proprietary device engineering.

Larsen explained why they were not able to get all of their papers published in peer-reviewed journals. "Ironically," Larsen said, "many reviewers take a quick look at our work and think we are proposing 'cold fusion.' Of course, nothing could be further from the truth."

CHAPTER 27

Solving the Dead Graduate Student Problem

Physicsts assumed, in 1989, that Martin Fleischmann and Stanley Pons were seriously mistaken in their fusion claim for three fundamental reasons: 1) There was no conceivable way to overcome the Coulomb barrier at room temperature; 2) there was no evidence of strong neutron emissions from their experiments; and 3) there was no evidence of strong gamma-ray emissions. This chapter discusses and explains the missing gamma-ray emissions that had such an important role in the initial dismissal of this new science.

Apart from the ultra-low-momentum neutrons, the aspects of the Widom-Larsen theory discussed so far involve standard physics and chemistry that were simply not recognized as elements of LENRs until Lewis Larsen and Allan Widom put all the pieces together.

Any type of fusion or neutron-capture reaction that produces significant heat should emit enough gamma radiation to kill anyone in its vicinity. But why isn't appreciable gamma radiation detected in LENRs? Why haven't experimenters been harmed?

Throughout the history of this controversy, critics have wondered, if it's really a nuclear reaction, then where are the gamma rays? In other words, where were the dead bodies of graduate students who had performed the experiments? No researchers in the field reported observing any MeV-energy gamma rays, only occasional low-energy gammas. And nobody had reported dead graduate students.

Were gamma rays being converted into something else? The idea that gamma rays could be converted was not new. For example, a process called pair production converts the energy of an incident gamma photon into an electron-positron pair. But the idea that gamma rays could be converted to infrared radiation was new.

Widom and Larsen's U.S. Patent #7,893,414, "Apparatus and Method for Absorption of Incident Gamma Radiation and Its Conversion to Outgoing Radiation at Less Penetrating, Lower Energies and Frequencies" is informative. (Larsen, 2011) As the text of the patent, written by Larsen, explains, gamma photons are known to interact with normal matter through three main processes. The first is the photoelectric effect, which applies to gamma photon energies less than or equal to 0.5 MeV. In this process, an incident gamma-ray photon strikes and is fully absorbed by an atomic electron.

In the second process, Compton scattering, an incident gamma-ray photon having energies between 0.5 MeV and about 5.0 MeV strikes an atomic electron, ionizes it and, after several other steps, results in a lower-energy gamma photon.

In the third process, pair production, an incident gamma-ray photon with energies greater than or equal to 5.0 MeV penetrates the outer cloud of atomic electrons and encounters the Coulomb field of an atomic nucleus. Through a series of steps, the gamma-ray is converted into a electron-positron pair.

In LENRs, the expected gamma rays from neutron-capture processes are, in fact, produced when ULMNs are locally — that is, at the reaction site — absorbed by nuclei in LENR systems.

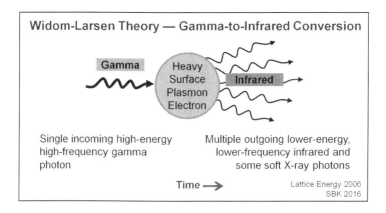

Widom-Larsen Theory — Gamma-to-Infrared Conversion

Gamma | Heavy Surface Plasmon Electron | Infrared

Single incoming high-energy high-frequency gamma photon

Multiple outgoing lower-energy, lower-frequency infrared and some soft X-ray photons

Time ➞

Lattice Energy 2006
SBK 2016

For example, neutron capture on a proton produces a deuteron plus a 2.24 MeV gamma ray. Neutron capture on deuterium produces a 6.250 MeV gamma ray. However, as Larsen wrote, LENRs can quickly and efficiently absorb hard gamma rays (having energies greater than 1 MeV) and convert that energy into other soft radiation: in this case, infrared radiation.

The information about the gamma conversion process is not discussed in the Widom-Larsen scientific papers; rather, Larsen and Widom disclosed the key details of the gamma conversion in U.S. Patent #7,893,414, which issued on Feb. 22, 2011. Larsen also provided additional details in a slide presentation he published on March 22, 2013. Two paragraphs in his slides contain the essential details. They are excerpted below and edited for clarity:

> When an ultra-low-momentum neutron captures onto an atom located inside the entangled three-dimensional quantum-mechanical structure of a LENR-active patch, there is normally a prompt gamma photon emission by that atom. The key point to remember is that the DeBroglie wave functions of the entangled, mass-renormalized heavy electrons are also three-dimensional, not two-dimensional.
>
> Because the neutron-capture gamma-photon emission occurs within the structure of a LENR-active patch, there are always heavy electrons available nearby to absorb such gamma emissions and convert them directly into infrared photons. Therefore, it doesn't matter where a gamma emission occurs inside a given patch; it will always get converted to infrared, which is exactly what has been observed experimentally. Large fluxes of hard gammas will not be emitted from such a patch, no matter which direction, on any axis, they are measured from.

Therefore, according to Larsen, the patches, or reaction sites, are also locally converting the gamma rays. This, in effect, is the built-in gamma conversion mechanism for LENRs and explains why emissions of significant amounts of deadly high-energy gamma rays (above ~1 MeV) have never been observed in LENRs. It also explains why heavy shielding is not required to conduct safe LENR experiments and why LENR researchers have not died from radiation exposure. In Larsen's slides, he used data measured by the Francesco Piantelli group to illustrate the gamma suppression (Focardi, 1998).

One of the Piantelli group's key experiments ran for many weeks and produced significant excess heat. According to the Widom-Larsen theory, gamma emissions are suppressed through their local conversion directly into infrared radiation (that is, heat) by heavy-mass surface plasmon electrons. According to the theory, gamma radiation starting at about 0.75 MeV, and going up to roughly 11 MeV, is converted.

In the first figure below, the black curve shows the background gamma counts, and the red curve shows the measured gamma counts from one of the Piantelli group's nickel-hydrogen LENR experiments. The second figure below shows the difference between the measured and the background spectrum, revealing a 660 keV gamma-ray peak produced by the experiment and no other peaks beyond 1000 keV. (Campari, 2004)

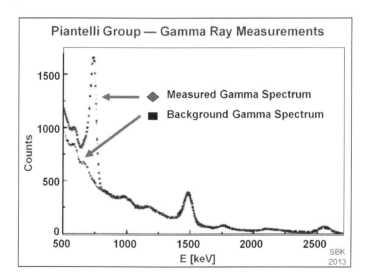

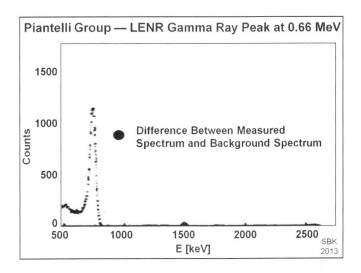

Another paper by the Piantelli group reported additional data from the same or a similar experiment. The first figure below shows gamma emission acquisitions during the first five days of the experiment. The difference between the measured spectrum and the background spectrum reveals two gamma peaks produced by the experiment, one around 660 keV and another just below 1,500 keV. After the experiment was running for five days, as shown in the second figure below, the first peak dropped from 300 counts to 50, and the second peak disappeared. (Focardi, 2004)

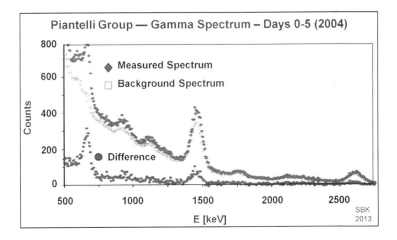

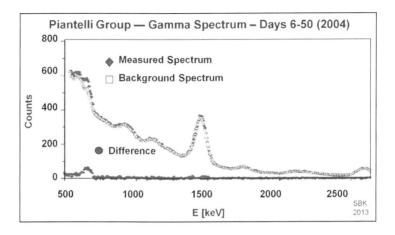

Once the experiment has been running for a few days, no gamma radiation is seen beyond the small peak at 0.66 MeV. This is particularly significant because neutron-captures on nickel isotopes normally produce gamma emissions in the MeV range.

"This is a direct experimental observation of the suppression of gammas via conversion to infrared by heavy electrons per our theory," Larsen said. "After 600 keV, the gamma emissions drop off drastically, right on the edge of the cutoff region, around 500 to 1000 keV, that we predict. All gamma emissions with higher energy than this region drop to background and stay there through the duration of the experiment. Our theory predicts that the suppression also extends up to around 11 MeV. This works well for LENRs because very few neutron-capture reactions are more energetic than 11 MeV."

The Piantelli group wrote that, after Day 50, the entire gamma spectrum went completely to background and stayed there while the experiment ran for another 35 days, during which it produced 25 MJ of heat. (Campari, 2004) "After this time period," the Piantelli group wrote, "the spectrum went abruptly to the background one. Later on, the cell produced excess power, a maximum of 25 Watts ... for about 35 days, after which the cell was shut down, in order to repeat the experiment."

In the history of the field, intermittent, low levels of gamma radiation have been reported often. Until now, there was never any logical explanation.

"This is an example of an experiment that produced a huge amount of heat," Larsen said. "Initially, you don't have a large population of heavy electrons. But once you start making them, which are required to start making the heat, the gamma is cut off; you get the heat but no gammas. Some people think the gamma conversion mechanism is something that covers the whole surface. This isn't necessary, and this isn't what happens. All you need to do is have the mechanism at the LENR-active sites. It's intrinsic to the system; the LENR-active sites have their own built-in gamma conversion mechanism."

Some people have suggested that testing the gamma suppression mechanism might be as easy as directing a gamma beam toward a LENR cell. This is unlikely to work. The suppression mechanism is active specifically at the LENR-active sites, which, until more advanced engineering is developed, occur randomly on the surfaces of metal hydrides. Moreover, as the SPAWAR video shows, each of these sites heats up quickly, then dies in milliseconds.

The gamma conversion-to-infrared also helps to explain the LENR experiments that have shown the ability to be self-sustaining for brief periods.

"The reason it works is that the gammas are converted into infrared, and the walls re-radiate it back into the active area of the experiment." Larsen said. "If they didn't convert, the gammas would go through the stainless steel walls, escape, and cool the system."

According to Larsen, the Piantelli experiment provides an easy way to understand the natural ability of LENRs to self-sustain. For the system to start, it needs the addition of enough energy to create the first ultra-low-momentum neutrons. The Piantelli design does this with an electric heater. Once the system gets going and the gamma radiation converts to infrared, that heat reflects from the stainless steel walls and feeds the production of the neutrons. Once the cycle starts, because the net energy gain is positive, the system doesn't need the electric heater, and it runs until the fuel is consumed.

Prior Patents?

Widom and Larsen's U.S. Patent #7,893,414 is unusual in that it cited no prior patents that are directly related. The patent is also remarkable in that the patent examiner issued no objections to the application.

"There was limited back-and-forth dialogue," Larsen wrote, "because the U.S. examiners found it almost impossible to find any prior art to cite against us. The European Patent Office examiners had the same vexing (for them) problem. In the end, they found a few things to cite but nothing that was able to limit our claims. This absence of prior art is extremely rare in the world of patents these days. In the lexicon of patent lawyers, this is called a fundamental patent because it represents the first stake in the ground in a totally new area of applied technology."

Many theoreticians had tried to explain the LENR phenomena but were unable to envision a complete set of pieces to the puzzle. If anyone before Widom and Larsen imagined such a gamma-this conversion process, they didn't have all the required pieces to explain it, or they didn't have the courage to publish such a bold idea.

Broad recognition of the Widom-Larsen theory will illuminate the results of prior experimental nuclear research in which gamma radiation was expected but not observed and, until now, was an unsolved mystery.

Key Concepts of Widom-Larsen Theory

Fully exploring the physics of the Widom-Larsen theory requires expertise in either particle physics or astrophysics, condensed matter physics and nuclear physics. However, the general concepts of the theory can be understood by nonscientists. There are six key concepts of the Widom-Larsen explanation of the LENR mechanism:

LENRs are primarily a surface rather than bulk effect.
The Born-Oppenheimer approximation breaks down.
Collective, many-body effects are essential in making the
heavy electrons.

Quantum electrodynamics (a quantum field theory) is
intrinsic to the mechanism.
Ultra-low-momentum neutrons are created.
Subsequent neutron-capture processes create the isotopic
and elemental changes.
Gamma radiation is converted to infrared radiation locally.

Larsen created a diagram that shows an overall perspective on LENR
reactions. It shows how LENRs begin in the chemical realm and end in
the nuclear realm. Below is a simplified version of his diagram.

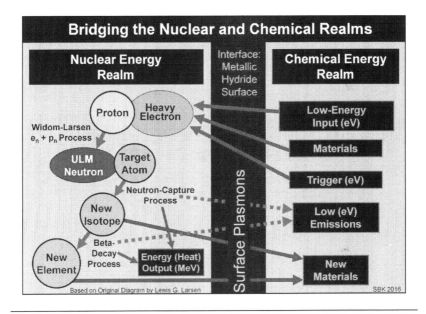

Improbable, Even to Einstein

In 2006, Larsen was browsing through the $2.99 discount section at
Border's bookstore, and a 1997 book by Ernest Sternglass, *Before the Big
Bang — The Origin of the Universe and the Nature of Matter*, grabbed his
attention. In the book, a memoir, Sternglass reviews modern particle
physics and cosmology while recounting his personal experiences with
some of the most renowned physicists in the 20th century. Larsen read
the book that weekend and was astounded.

The story begins in 1947, when 24-year-old Ernest Sternglass (1923-2015), after completing his undergraduate degree in electrical engineering at Cornell University, was working at the Naval Ordnance Laboratory in Maryland. (Sternglass, 1997)

At the lab, he had begun to study secondary electron emission, the ejection of electrons from solids when they are bombarded by fast electrons. Sternglass had been asked by the Navy to look into the phenomenon to examine the possibility of developing a new television technique for night vision by detecting normally invisible infrared radiation emitted from objects. His research was later used in the video camera on Apollo 11 to transmit the live images of the first astronauts walking on the moon.

As he began examining the topic, he learned that it was loosely related to the photoelectric effect for which Albert Einstein (1879-1955) had won a Nobel Prize. Sternglass wrote to Einstein and explained his ideas. Three days later, he received a letter from Einstein asking him to come visit. "I was all shaken up," Sternglass wrote. "After all, I was just a young kid who hadn't even finished graduate school." (Freeman, 1981)

After they talked for an engaging five hours on Einstein's back porch, the Nobel Laureate encouraged the young Sternglass to continue pursuing his ideas. But to Sternglass' astonishment, Einstein urged him to remain at his present job doing applied physics rather than going to graduate school. "Don't go back to school. They will try to crush every bit of originality out of you," Einstein said. "Be careful. There will be enormous pressures to conform." (Freeman, 1981, 57)

Sternglass did not follow Einstein's advice and did go back to Cornell. Sternglass' original hope was to do his thesis work on the theory of secondary electron emission under Hans Bethe (1906-2005) who, later in his life, was awarded the Nobel Prize in physics for his work on the theory of nucleosynthesis in stars. However, Bethe discouraged Sternglass, telling him that he himself had done earlier work on secondary emission and concluded that a theory about such a complex phenomenon could not readily be achieved. Sternglass remembered Einstein's encouragement and found someone else.

Philip Morrison (1915-2005), who was most well-known for his work on the development of the atomic bomb as part of the Manhattan

Project, agreed to serve as Sternglass' principal advisor in theoretical physics on his thesis committee.

Sternglass had a radical idea: An electromagnetic model that suggested neutrons could be "formed from protons and electrons at very low energies, far below the energy predicted by the existing theory." The $e + p \rightarrow n + v$ reaction (in the case of free protons rather than protons inside nuclei) was believed to occur only at the extreme temperatures available in the cores of collapsing stars. Sternglass had read about experiments performed in the 1920s in which researchers had passed thousands of volts of electricity through hydrogen-filled tubes. The tubes were similar to modern fluorescent bulbs used for lighting, although the newer devices use only a hundred volts.

In the 1910s and 1920s scientists experimented with such tubes and found evidence of atomic transmutations. The experimental data made no sense to any of these scientists, and by the 1930s, the research was largely abandoned. Vol. 3, *Lost History*, reviews this forgotten work in detail.

Sternglass does not seem to have known about the experiments from the 1910s and 1920s, but he was aware of earlier theoretical work published by Charles Galton Darwin (1887-1962), who described collective many-body excitations of electrons (Darwin, 1920). According to Darwin's calculations, Sternglass wrote, neutrons might be formed by capturing an electron, even at low energies. (Sternglass, 1997, 85)

Therefore, in 1951, Sternglass began experiments to test his bold idea. He had seen one of these old tubes in the basement laboratory of Lyman Parratt, a senior member of the Physics Department at Cornell, and asked Parratt to build one for him. Sternglass knew that neutrons would be extremely difficult to detect so, instead, he devised a system to look for secondary evidence, short-lived beta radiation from neutron captures on target materials such as silver and indium foils.

His advisers allowed him to pursue the experiment even though he knew that, according to existing theory, "there was no chance that such an experiment could possibly succeed." (Sternglass, 1997, 85) Sternglass knew that the $e + p$ reaction would require 0.78 MeV of energy and that Parratt's tube could provide only about 0.035 MeV, a difference of 22 times the amount of energy input. As an alternative, he knew that, if the experiment failed to produce neutrons, it wouldn't endanger his prospects for receiving a doctoral degree because he could fall back on his theory of secondary electron emission.

Around June 1951, Sternglass was ready to begin. Within two months, he obtained successful experimental evidence for the formation of neutrons from protons and electrons in a high-voltage hydrogen discharge. He measured a background neutron count of 15 counts per minute. With the experiment running, he observed neutron counts of 6-7 per minute above background — that is, about 40 percent above background. He also saw each of the expected decay rates for silver and iridium. Using a standard neutron source, he calibrated that the rate of neutron formation was about 10-20 neutrons per second. (Sternglass, 1951)

Sternglass performed a variety of checks to eliminate alternate explanations. These included exposing the foils without turning the discharge on, replacing electrodes with freshly machined parts, and calculating the possibility of a deuteron-deuteron reaction. Nothing could explain the data, and many of his colleagues didn't believe the results. He wrote to Einstein on Aug. 26, 1951, and told him about the results and his dilemma.

"So far," Sternglass wrote, "no one in the [Physics] Department here has been able to suggest any nuclear reaction that could be made to account for the activity at such a low energy." (Sternglass, 1951)

Einstein replied with encouragement on Aug. 30, 1951. "In order to form a neutron," Einstein wrote, "an electron is needed that has passed through 7.8×10^5 volts in order to provide the required additional energy. I can hardly imagine that electrons of such high voltage are formed in your tubes. Perhaps reactions occur in which multiple electrons simultaneously transfer energy to one proton. According to quantum theory, this is somewhat conceivable, although not probable." (Trost, 2013) Thus, in 1951, Einstein had intuited a key concept of many-body collective effects that Widom and Larsen, without any knowledge of Einstein and Sternglass' communications, later used in their LENR theory.

Sternglass was well aware of the significance of his findings. "One would have a ridiculously simple neutron formation process, which might even be used in atomic energy applications, for instance in reducing the size of a reactor!" he wrote. (Sternglass, July 7-8, 1951)

Although Sternglass wrote in his 1997 book that he was not discouraged in his 1951 research, his memory at age 74 seems inconsistent with his own notes written half a century earlier in his lab books. His notes indicate that several professors were encouraging and agreed that neutrons were being produced, but, in the absence of any sensible theory, they thought the neutrons came from a room-temperature D+D fusion reaction. (Sternglass, July 7-8, 1951)

His lab notes indicate that he felt discouraged, however, because he could not increase the neutron counts when using hydrogen. (Sternglass, Nov. 16, 1951)

In 1953, Edward Trounson, a colleague of Sternglass' at the Naval Ordnance Laboratory, independently confirmed his 1951 results and tried deuterium. Sternglass described the thrilling news in his lab book:

> Ed finally [got] his equipment running last week, and he found the same activity induced in his iridium foils that I did. Thus, it looks as if neutrons are really formed in a hydrogen discharge. ... Ed said that he made five runs with hydrogen at about 50 ma and 36 kV, which gave close to one count per minute above a background of 16 cpm. He also made one run with pure deuterium, which gave 900 cpm.
>
> This seems to show quite conclusively that the D+D reaction cannot account for the phenomenon since the neutron production should go up as the square of the concentration. As soon as he checks this for another concentration, we will decide on publication of our results. (Sternglass, Jan. 12-13, 1953)

His confidence was renewed, as he wrote the following day: "Thus, it is clear that, for *whatever* reason, I am getting neutrons; it is entirely at variance with contemporary ideas on nuclear phenomena." (Sternglass, Jan. 12-13, 1953)

I did not find further updates in Sternglass' notes from Trounson; presumably, the subsequent checks failed. In 1960, Sternglass tried a variation of the experiment suggested by Einstein but got null results. (Sternglass, 1997, 87)

In 2006, when Larsen stumbled on Sternglass' book, he called the author on the telephone, and, over the course of a few months, the two had several stimulating phone calls. Larsen explained to Sternglass the deeper significance of Sternglass' 1951 theoretical and experimental work, and on Nov. 25, 2011, Larsen published his summary of the story in a document online.

A decade after Larsen and Widom published their theory, several scientists contacted Larsen and told him that they, too, had seen experimental results that, to them, had seemed inexplicable. Some of these results included unexplained neutron fluxes in deuterium plasma-discharge and radio-frequency-excitation experiments and unexplained isotopic shifts in battery materials. In all such cases, they told Larsen, they never published their anomalous results because they thought doing so would be too controversial and risky for their careers. Sternglass was also at a loss.

"No one could find any explanation other than that suggested by Einstein," he wrote. (Sternglass, 1997, 86)

Either Einstein learned about the importance of collective effects from Darwin, or he made a tremendous intuitive leap. Sternglass' earlier experiments supported his theoretical prediction that protons could react directly with electrons at surprising low energies to make neutrons.

However, Sternglass was burdened with self-doubt, worried about reproducibility, and influenced by the preconceptions of top physicists at Cornell, who thought that such an idea was improbable. With the exception of his 1997 book, Sternglass never formally published the results of his successful experiments or his theoretical ideas in a scientific journal.

CHAPTER 28

Doomsday for Cold Fusion

The response to the Widom-Larsen theory of LENRs ranged from deafening silence to predictions of doomsday. Larsen anticipated some of the reaction and planned accordingly.

Apart from George Miley, with whom Larsen had contracted to perform LENR research, Larsen had kept secret the fact that he was developing a theory. He told Edmund Storms, an employee of his company at the time, that he had hired Widom to work with him on a theory only the day before the Widom-Larsen theory paper appeared on the arXiv preprint server on May 2, 2005.

For the previous eight years, Larsen had enjoyed friendly relationships with many LENR researchers and learned a tremendous amount from them about their experimental research. The camaraderie came to an abrupt halt when scientists read the conclusion of Larsen and Widom's first paper. After displaying several nuclear reaction equations besides fusion that could lead to helium-4, Widom and Larsen wrote that, just because researchers had observed helium-4, that data didn't necessarily constitute evidence for deuterium-deuterium (D+D) "cold fusion." Weak interactions and neutron-capture processes were superior to the room-temperature fusion idea, Widom and Larsen wrote, because their model was not impeded by the Coulomb barrier. When the preprint came online, Larsen found that his relationship with the researchers had changed overnight.

Larsen was not completely surprised by the negative response from the scientists who believed in fusion, but he did not anticipate the fierceness of their reaction. "I remember what they were saying," Larsen told me. "They said that you couldn't get excess heat with light water or transmutations and that Miley and the Japanese people who were reporting that data were crazy because there was no deuterium in the system, and therefore the D+D fusion reaction couldn't explain the normal hydrogen results."

At the time, Larsen had been talking with all the well-known American researchers, such as Storms, Michael McKubre, and Peter Hagelstein. He was in communication with all of them until May 2005, when his and Widom's paper was published. After that, many of them stopped talking with him, didn't return his phone calls and didn't respond to his e-mails.

I knew nothing about the social dynamics at the time. Nearly a year later, when the first Widom-Larsen theory paper published on March 9, 2006, in the *European Physical Journal C - Particles and Fields*, I began to take a greater interest in the theory, thanks to David Nagel. Eventually, I began having conversations with Larsen.

Only One Theory

One of the things Larsen told me about was a government meeting to which he was invited to discuss his and Widom's theory. As discussed in Chapter 25, beginning in July 2001, Larsen and Widom participated in numerous invitation-only federal government meetings and briefings to discuss their theory. On Dec. 12 and 13, 2006, the Defense Threat Reduction Agency (DTRA) held a High-Energy Science and Technology Assessment meeting at its headquarters auditorium at Fort Belvoir, Virginia.

The first day was dedicated to unclassified work, but the second day's discussion was at the Secret level. Aside from LENRs, three intriguing topics were discussed: nuclear isomers, antimatter and fourth-generation nuclear weapons. These three topics were well-established as legitimate areas of scientific research but, like LENRs, were constrained by significant engineering-related and technological challenges. Years later, after filing a FOIA request, I obtained the final report of the meeting.

About 100 people participated in the workshop, according to Larsen.

Many of them were military, and some top Department of Energy (DOE) officials were there, as well. In 2006, when I first heard about this, I was surprised. It was one of the first indicators that a shift was occurring and that there was new acceptance of the idea of LENRs by the U.S. government scientific community.

Five people presented as part of the LENR panel. The first two were Nagel and Michael Melich (both of whom had primarily managed LENR projects rather than directly performed research). The third was a close colleague of Nagel and Melich, Mitchell Swartz, an experimentalist who worked in his own lab in Massachusetts. Larsen and Widom spoke there, as well; they were the only LENR theorists who had been invited to speak on the panel.

Yeong Kim (b. ~1934), a "cold fusion" theorist and professor of physics at Purdue University, was in the audience, but he was not invited to speak. The final report of the meeting said that Kim had spoken at the meeting; he did not. Moreover, the agenda, as listed in the report, omitted Widom's name as one of the speakers.

The report's authors also wrote that "the Widom construct appears promising but lacks robust experimental verification and rigorous peer review," despite the fact that the Widom-Larsen theory had been published in a prominent peer-reviewed journal nine months earlier.

In 2012, I contacted George Ullrich, one of the two authors on the report, to inquire about the errors. He explained that the speaker participation errors were introduced by his co-author Rich Sutton and that Ullrich had caught the errors after the draft was submitted in 2007. (The copy of the report I received in 2011 did not include any such corrections.) Ullrich also explained his statement about the Widom-Larsen theory to me.

"It was my opinion at the time," Ullrich wrote, "that the theory lacked rigorous review. I certainly did not intend to disparage the *European Physical Journal C - Particles and Fields*, but given the controversy and heated debate regarding LENR at the time, I did not consider one refereed journal article to be a rigorous validation of theory. This meeting occurred five years ago. Since then, a number of things have fallen into place, and I think Widom-Larsen seems to be standing the test of time."

Hell Freezes Over

The most significant part of the meeting, however, was the luncheon talk given by the archenemy of test-tube fusion, Robert Park, former spokesman for the American Physical Society. As mentioned in Chapter 11, Park was well-known for his caustic public comments about the subject and its practitioners.

The history of the 1989 conflict is filled with emotional and unjustified attacks by dozens of well-known characters in the drama. However, after 17 years of continued progress in experimental research and the publication of a potentially viable theory, none of the field's previous adversaries had admitted that they might have been wrong – until this workshop. On Dec. 12, 2006, hell froze over, and, at least behind closed doors, Park effectively conceded.

As Larsen remembers, Park said, "Low-energy nuclear reactions are real phenomena, though poorly understood. There's probably something real there, but it's not well-understood yet." It was a far cry from the way Park described it in his book. (Park, 2000; Krivit, 2008)

I asked Larsen whether he was surprised that Park did not attack his and Widom's theory. "No," Larsen said, "because I sent the paper to him in advance, and he didn't respond with any technical objections. In fact, I sent it to many of the previous critics of the field and personally invited them to send me their critique. I also sent it to John Huizenga, the chairman of the 1989 DOE cold fusion review panel, Richard Garwin, who was on that panel, and other previous critics."

"I Didn't Say It Was Wrong"

As fate would have it, I ran into Garwin a few months later, on Feb. 15, 2007, at the American Association for the Advancement of Science meeting in San Francisco. We had communicated by e-mail and phone many times by then. Despite making a very public bet against "cold fusion" in *Nature* magazine on April 20, 1989, he continued to be interested in the topic. I would often present research papers to him for comment, and Garwin always offered his critique. Almost certainly, he knew that the scientific question of "cold fusion" — or whatever it actually was — had never been resolved.

The meeting hall was filled with several thousand attendees, who watched a slide presentation that honored a dozen of America's greatest physicists, some still alive. Garwin's photo was among these, and I wondered if he was in the room. Moments later, to my great surprise, Garwin got up from a seat in front of me and headed toward the escalator. I recognized his face and gave chase.

I caught up with him on the escalator, introduced myself, and immediately asked whether he had seen the Widom-Larsen paper. He said he had. I asked him for his comments. In a grumbling-complaining tone, he said that Widom and Larsen had failed to explain the gamma conversion mechanism of the theory. We reached the bottom of the escalator, and he respectfully departed my company.

The next day, I spoke to Larsen and told him that Garwin seemed to think that the gamma conversion mechanism had an error. The next thing I knew, Larsen sent an e-mail to Garwin. "Dear Richard," Larsen wrote, "could you please elaborate in writing on the specifics as to exactly where you think that we are in error? If the above attribution [by Krivit] is accurate, we are very curious about how you came to such a conclusion. By the way, have you read all four of our papers? We look forward to hearing from you further."

"I didn't say it was wrong," Garwin replied to Larsen and me. "I said that I had not received a reply to my question/suggestion about using such a material as a shield against high-energy gamma or X-rays."

"You are correct," Larsen replied. "We have not responded to your prior questions about the LENR gamma shielding application. We did not answer those particular questions because of underlying intellectual property issues. Nonetheless, we thank you again for your interest in our work." Widom and Larsen's patent US 7,893,414, which revealed the details of the gamma shielding, was pending at the time.

Garwin sent me a copy of Larsen's response and prefaced it with one sentence: "Now you have a story." Indeed, it was the first time I had spoken with Garwin about any LENR experiment or theory in which he didn't suggest that it was wrong. Nevertheless, Garwin couldn't bring himself to say anything positive or affirmative about the research or the

Widom-Larsen theory.

Garwin's 1989 article in *Nature* had concluded with a comment that somebody was going to end up eating their hat. The D+D "cold fusion" idea didn't succeed, but the validity of excess heat and a variety of nuclear phenomena were well-confirmed by 2007. Certainly, Garwin did not expect that he might eat even part of his own hat.

The Changing Tide

Public confirmation of Park's acceptance of LENRs came the following month, in a March 22, 2007, article in *Chemistry World* by science reporter Richard Van Noorden. He wrote that Park was still not without criticisms, but he conceded that "there are some curious reports — not cold fusion, but people may be seeing some unexpected low-energy nuclear reactions."

Of course any low-energy nuclear reaction is unexpected, to say the least, according to current textbook science. Park had clearly made the distinction between "cold fusion" and LENRs. No major recent experimental breakthroughs had occurred. The only logical explanation for Park's acceptance was the Widom-Larsen theory.

The concession by Park and the lack of critique by Garwin marked a significant turning point in the field. Although experimental evidence, according to the scientific method, should take precedence over theory, the merits of the Widom-Larsen theory clearly had changed the tide and was fostering acceptance from prominent members of the wider scientific community. Politically, it seemed to be an unprecedented breakthrough.

I began paying serious attention to the Widom-Larsen theory at that time and started discussing it with everyone I knew in the field. What happened next was surprising: Not a single LENR researcher acknowledged Park's concession. Park had just given them a golden endorsement. The researchers should have been trumpeting his concession all over the Internet.

Certainly, most LENR researchers who knew Nagel, Melich, Swartz and Kim had to have heard from them that Widom and Larsen had been selected to present their theory at the DTRA workshop. It was a major step forward for the entire field. But silence. Nothing but silence. Why? In time, I began to understand. The ideology of the researchers who had been fighting for recognition since 1989 was inseparable from the D+D "cold fusion" idea. Park had said it was real but not fusion. So did the Widom-Larsen theory.

I was stunned to see the apparent rejection of the Widom-Larsen theory by well-recognized LENR researchers. Granted, the dozen or so theorists who were active at the time weren't likely to be thrilled at possibly losing the race. But I expected the experimentalists to at least appreciate the value and benefits of a potentially viable theory, certainly one that was gaining respect outside the field. I was never more wrong.

Absolute Conviction

In March 2007, I was finishing my LENR chapters for the 2009 Elsevier *Encyclopedia of Electrochemical Power Sources*. I had sent a draft of the theory section to Storms and asked for his critique, and I told Storms that Garwin had not identified anything wrong with the Widom-Larsen theory. Thus began a multiyear period in which Storms and other scientists, both overtly and behind the scenes, campaigned against its recognition.

"The Widom-Larsen theory is wrong for many reasons," Storms wrote. "I would not pin any advantage to cold fusion on it." Of course the theory offered no advantage to the fusion idea. Widom and Larsen's non-fusion idea was far more sensible.

At the time, I had perceived Storms to be one of the top authorities on LENRs. He had an encyclopedic knowledge of the subject matter, although most of his reviews of the topic were self-published. Early on, Storms had served, in a way, as my mentor. He had been the first scientist I had interviewed, and he had always been eager and willing to answer my questions. Nevertheless, his absolute conviction was a red flag for me, and I was unwilling to accept his point of view. "I need to explicitly know what is wrong with the Widom-Larsen theory," I wrote back to him, much as I had done with Garwin.

Storms wrote back with a longer answer, but I was unable to make any sense out of it at the time. Later, I realized that Storms had found nothing wrong with the theory. Instead, he disputed the theory because he didn't understand how collective effects could increase the mass of electrons, didn't understand that ultra-low-momentum neutrons were unlikely to activate materials beyond the LENR reaction surface, and didn't understand why the nucleosynthetic reaction networks tended to terminate in stable isotopes. I certainly didn't understand any of that, either, in 2007; I needed more time and many conversations with Larsen to get a grasp of the concepts.

Storms' absolute conviction that the Widom-Larsen theory was wrong and that I should ignore it put me in a difficult position. Storms' opinion stood in stark contrast to Nagel's opinion. Nagel had suggested to me that the Widom-Larsen theory was the first viable model to explain the diverse set of observed LENR phenomena. Storms' reaction also conflicted directly with that of Garwin, who had never hesitated to offer his criticisms to me of anything related to LENRs. I had to seek other opinions.

I sent a question about the Widom-Larsen theory to all researchers in the LENR field by way of the Condensed Matter Nuclear Science (CMNS) e-mail list on Feb. 21, 2007. Storms was the first to reply to me and the rest of the list. He asserted that the Widom-Larsen theory did not explain any of the experimental data in the field, and he concluded with an aggressive statement. "When people examine this theory carefully and see its obvious flaws," Storms wrote, "the field is going to look silly in supporting it."

I was intrigued by the intensity of his response. It reminded me of the comments by Nathan Lewis, a chemist at Caltech who had argued so vehemently against Fleischmann and Pons' claims in 1989 (Vol. 2, *Fusion Fiasco*).

Rather than ignore the Widom-Larsen theory, as Storms wanted me to do, I mentioned it during my talk at the American Physical Society meeting on March 5, 2007. During the question-and-answer session, Hagelstein, who had abandoned his own neutron-based theory years earlier, spoke politely but critically of the theory. When theorist Scott

Chubb spoke about the Widom-Larsen theory during the APS session, he made personal, defamatory and false statements.

Four Critiques

On March 16, 2007, I published four comments about the Widom-Larsen theory. The first Widom-Larsen paper had published 12 months earlier. Nobody had published a peer-reviewed critique of the Widom-Larsen theory in any journal; nor had anyone uploaded a critique to the arXiv preprint server.

The first non-peer-reviewed critique was from Akito Takahashi, a theorist and a professor of physics at Osaka University. Like Storms, he wrote that the Widom-Larsen theory had "many fundamental problems." Yet the full page of items he called problems appeared instead to be things he did not understand, or they were irrelevant or nit-picks.

The second non-peer-reviewed critique was from Chubb. He, too, did not understand how collective effects applied to the theory. There wasn't much substance to his review, which was peppered with emotive words like "bizarre," "crazy," "misguided," and "unethical."

The third critique was from David Rees, a particle physicist at SPAWAR. Rees had no involvement in the field. He had been asked by the SPAWAR LENR researchers to evaluate the Widom-Larsen theory. Unlike Storms (a radiochemist), Chubb (a condensed-matter physicist), and Takahashi (a nuclear engineer), Rees had no difficulty understanding how collective effects led to the creation of heavy electrons.

"The bottom line," Rees wrote, "is that the techniques and calculations are straightforward and (as far as I can tell) correct. There could be some assumptions that need to be dealt with in more detail (could spell trouble). My 'feel' is that they are OK. The paper looks good to me. It's a clever idea, I think (clever ideas are always obvious in hindsight)."

While at the APS meeting, I met professor Robert Deck of the University of Toledo, a nonlinear optics physicist. He had no involvement in the field, and I asked whether he'd be willing to provide a critique for me. Like Rees, he understood how Widom and Larsen supported their explanation of the creation of heavy-mass surface plasmon electrons. "The argument in support of this explanation is given elegantly," Deck wrote.

The trend was developing very clearly for me; qualified uninvolved analysts were comfortable with the theory, describing it as clever and elegant. People inside the field, particularly scientists who were wedded to the fusion idea, were describing it as fatally flawed and worse. I therefore decided not to follow Storms' advice to ignore the theory.

Show Me the Data

On Nov. 1, 2007, I sent a set of questions to members of the LENR scientific community, including questions about the Widom-Larsen theory as well as questions about the purported 24 MeV ratio.

Again, Storms was the first to reply. "Because the Widom-Larsen theory violates basic concepts and ignores accepted data," Storms wrote, "many of the detailed questions you ask are irrelevant. Nevertheless, perhaps the answers you obtain will finally eliminate this distraction."

For a while, I tried to reason with Storms. However, when he wrote that "people who do not believe cold fusion is possible are irrelevant," I began to suspect that further dialogue with him would be unproductive.

From this conflict with Storms, I began to ask the crucial question: "What is the experimental evidence that shows that 'cold fusion' is actually fusion?" Storms insisted that "at least six experiments show a relationship between helium-4 and heat, giving a ratio of 24 MeV."

Storms and other scientists had asserted that the proof of "cold fusion" was the fact that, in the experiments, for every 24 MeV of heat released in the experiment, a helium-4 atom was produced. They believed that this supposed ratio was proof of "cold fusion" because it resembled the third branch of D+D thermonuclear fusion, in which one of every million fusion reactions produces helium-4 and a hazardous 24 MeV gamma-ray. Over the next few months, I began looking closely at their claim.

Concurrently, Larsen helped me understand the problem of energy balance — that is, the effects of the production of other nuclear products and reactions in LENR systems, in addition to the helium-4. The experiments produced not only helium-4 but also a wide variety of effects and products. This made the heat/helium-4 ratio argument moot. If helium-4 wasn't the only significant product of the reactions, apportioning all the heat only to producing helium-4 was meaningless.

Throwing Down the Gauntlet

By the end of 2007, I had received a number of letters about the Widom-Larsen theory from scientists with a vested interest in the fusion idea. Their letters convinced me that the Widom-Larsen theory merited a feature article in *New Energy Times*. The criticisms in those letters were no more insightful than those from Takahashi, Scott Chubb and Storms: They wrote that they did not understand parts of the theory. However, they wrote in a way that suggested that the theory was wrong rather than that they just didn't understand it. There were also more hostile responses.

In a letter to *New Energy Times*, theorists Krityunjai Prasad Sinha and Andrew Meulenberg, colleagues and frequent co-authors of Hagelstein, predicted doomsday for the LENR field.

"We see a disaster if the [Widom-Larsen model] were to be published and acclaimed by [the members of the field]," Sinha and Meulenberg wrote. "It would certainly confirm most physicists' view of the field. Mostly, those [people] looking for flaws would read it. They would easily find and advertise them. ... It looks like a snow job. And that makes us wonder why/how it was done."

Two weeks after I published my Jan. 11, 2008, "Widom-Larsen Not-Fusion Theory" article, Hagelstein and his colleague Irfan Chaudhary submitted a critique to the arXiv server. (Hagelstein, 2008) Twenty-two months had passed since the journal publication of the first Widom-Larsen theory paper. During that time, no physicists in the wider scientific community had criticized their published theoretical work.

A few weeks later, Widom and his colleagues submitted a paper to arXiv and pointed out Hagelstein and Chaudhary's errors. (Widom et al., Feb. 5, 2008) According to Larsen, Hagelstein and Chaudhary did not respond further.

The fusion believers could not beat Widom-Larsen using scientific arguments. They used other strategies, instead. Nearly always, but usually in the background, a shadowy figure named Michael Melich was involved.

Michael Melich sketched by photographer D. Tran during the International Conference on Cold Fusion, August 2008, Washington, D.C.

CHAPTER 29

Futile Resistance

Despite the concerns articulated by the fusion believers that the Widom-Larsen theory would mean disaster for the field, the broader scientific community was showing signs of accepting the reality of LENRs. In response, the believers took steps to deny certain experimental data and escalated their resistance to the new theory.

Michael McKubre, Edmund Storms, Peter Hagelstein and Talbot Chubb began saying publicly that no neutrons whatsoever were produced in LENRs. In fact, low levels of neutron emissions, as well as bursts, had been detected all around the world in 1989, including at the most prominent U.S., Italian and Indian government laboratories. (Vol. 2, *Fusion Fiasco*) They continued to be observed in later years.

In October 2007, at the Eighth International Workshop on Anomalies in Hydrogen/Deuterium-Loaded Metals, in Catania, Italy, Francis Tanzella reported the results of palladium-deuterium co-deposition experiments he performed at SRI International. Tanzella had replicated the U.S. Navy SPAWAR palladium-deuterium electrolytic co-deposition experiment. (Tanzella, 2007)

In at least three experiments, Tanzella observed neutron counts above background with a boron trifluoride neutron detector that, according to McKubre, had worked "reliably and in which they developed trust over many years of observation."

In one of these experiments, Tanzella measured a sustained (14-hour) burst of neutrons 14 times greater than background. Along with the burst, Tanzella observed a distinct drop in cell voltage potential, suggesting that anomalous heat was being produced.

In January 2008, Michael McKubre and I were in India; we had been invited on a small tour to speak about LENRs. On Jan. 7, 2008, McKubre, in a presentation to the division heads of the Bhabha Atomic Research Centre in Trombay, India, said there were no neutrons in LENRs. Two days later, I spoke at the Indian National Institute for Advanced Studies, and I displayed Tanzella's graph with the prominent neutron signal.

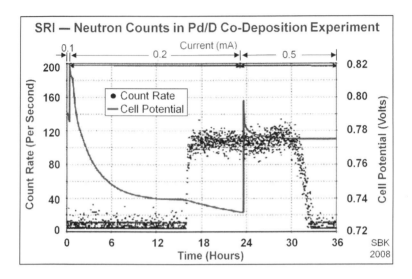

After the meeting, McKubre pulled me aside and said, "I don't want you showing Fran's graph with the neutron signal anymore. It's bullshit. I don't believe it."

In follow-up e-mails with McKubre in which I cited his statement of disbelief, I asked him whether he had found any flaws in the data Tanzella had reported. He hadn't, and he apologized. "Sorry if I seemed heavy-handed," McKubre wrote. Tanzella did not submit his paper to be included in the conference proceedings.

Meanwhile, Italian researcher Domenico Cirillo was designing an experiment that used solid-state nuclear track detectors, also known as CR-39 plastic detectors, enclosed in hermetically sealed polystyrene cylinders filled with boron. Boron is a neutron absorber, and when boron captures neutrons, it emits alpha particles, which then create visible tracks on the plastic detector. In 2011, he reported the detection of significant fluxes of neutrons. His paper, which included Widom as a co-author, was published online in November 2011. (Cirillo, 2011)

In response, in 2013, an eight-person team at the Italian National Agency for New Technologies, Energy and the Environment (ENEA), in Frascati, Italy, led by LENR researcher Vittorio Violante, claimed an attempted replication of Cirillo's experiment, but the team failed to detect similar fluxes of neutrons. (Faccini, 2013; 2014) The team was part of a consortium organized by Michael McKubre.

The experiment was sponsored by Giancarlo Ruocco, at that time the vice-rector of the University of Rome. Ruocco had attacked "cold fusion" theories and experimental results for years. He had broadly criticized LENRs in popular Italian magazines and e-mail discussion lists. In 2012, Ruocco was also a co-author on a paper that tried but failed to discredit Widom-Larsen's calculated rates of neutron production in LENR electrochemical cells. (Ciuchi, 2012; Widom, 2012)

For the experiment, Ruocco invited three scientists from the University of Rome (Riccardo Faccini, Alessandro Pilloni, and Antonio D. Polosa) to assist with data analysis and theoretical interpretation.

Violante wrote that his group's failure to detect fluxes of neutrons established an upper bound for the neutrons Cirillo could have observed in his experiments. This was not valid because Violante's experiment was not a replication of Cirillo's. Violante had placed his detector in nearly the worst possible location to detect neutron fluxes.

In fact, the placement of Violante's boron-CR-39 neutron detector would have been better-suited to detect background neutrons because it was about 2.5 times farther away from the source of the LENR-produced neutrons, and its broadest face was oblique to, rather than perpendicular to, the long axis of the cathode. Violante also surrounded the cathode with a quartz cylinder; Cirillo's design had no such component. Three fusion neutronics experts from ENEA assisted Violante; they must have known that light water is an excellent neutron moderator and that increasing the distance from the cathode to the detector would have significantly decreased the fluxes of low-energy neutrons that could be observed. Furthermore, Violante and his colleagues did not find any specific error in the Cirillo experiment. (Widom, 2013)

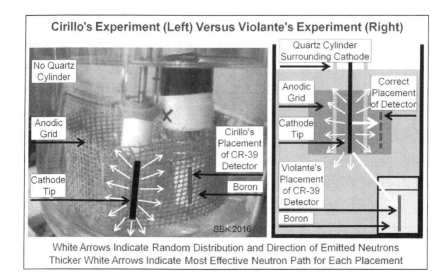

Cirillo's Experiment (Left) Versus Violante's Experiment (Right)

White Arrows Indicate Random Distribution and Direction of Emitted Neutrons
Thicker White Arrows Indicate Most Effective Neutron Path for Each Placement

A Bad Feeling

In early 2008, the preparations for the Fourteenth International Conference on Cold Fusion (ICCF-14), in Washington, D.C., were well under way. David Nagel, the chairman, had invited Michael Melich to be the technical chairman. By 2008, certain researchers were taking overt measures to cast doubt on experimental data that conflicted with the benchtop fusion idea, data that instead would support the Widom-Larsen theory.

In addition to the detection of low-level neutron emissions, the SPAWAR group had detected emission of energetic (MeV-level) alpha particles. These results conflicted directly with the $d + d \longrightarrow Helium\text{-}4 + 23.8\ MeV\ (heat)$ idea, which, according to the equation, leaves no room for other energetic reaction products. (Szpak, 2007) Hagelstein then wrote a theory paper in which he asserted that LENRs produced a maximum of only 20.2 keV alpha particles. (Hagelstein, 2010)

On April 18, 2008, Nagel and Melich posted the topics that would be covered at the ICCF-14 meeting. SPAWAR researcher Pamela Mosier-Boss sent an e-mail to me that day. "I have a very bad feeling about ICCF-14," she wrote. "Melich has commandeered the conference and is working on taking over ICCF-15. And it looks like he's trying to take out anyone who he feels is in his way."

What she meant was that Melich had placed himself in a strategic position that allowed him to take control of the scientific agenda for the current ICCF meeting and that he was working to form alliances in Italy with Vittorio Violante to have some control of the following year's ICCF meeting.

Transmutation and Nuclear Particles Marginalized

In April 2008, I looked at Nagel and Melich's outline for the conference: For the first time in many years at ICCF meetings, the topic of LENR transmutations was gone. After I brought this to the attention of the LENR community, George Miley threatened to hold a separate conference on LENR transmutations. In a concession to Miley, Melich and Nagel offered him a separate room, after the conclusion of the main conference midday on Friday, in which he could conduct an independent session on LENR transmutations.

Eventually, Melich and Nagel also allowed an hour and a half during the main program for five transmutation-related papers. One paper during the session was by Yasuhiro Iwamura on his gas-permeation experiment. Two groups (one of them, Toyota Central Laboratories) reported tentative confirmations of Iwamura's experiment at Mitsubishi. The fourth paper was from John Dash. Nagel and Melich had an incentive to accept his paper: Dash's philanthropic sponsor, "Sam," had also co-sponsored the ICCF-14 conference, in the amount of $50,000.

Eleven days before the start of the conference, Melich and Nagel added another paper to the transmutation session. Melich had invited David Allan Kidwell, a research scientist and spectroscopist at the Naval Research Laboratory (NRL), to present a paper. Kidwell suggested that all scientists in the LENR field who had reported LENR transmutations were either incompetent or fooling themselves. (Chapter 31)

Half of the 150 conference participants attended Miley's transmutation workshop on Friday afternoon. The marginalized session was not part of the official program; its papers were not published in the conference proceedings. Melich and Nagel had succeeded in preventing that research from becoming part of the scientific record.

Melich had an uncanny ability to direct, control, and influence scientists in the LENR field. One of the ways in which he did this was to use his position as a well-connected government employee to elicit research funds from Defense Advanced Research Projects Agency (DARPA) and the Defense Threat Reduction Agency (DTRA) and to direct that money to LENR researchers and projects of his choosing.

His other primary strategy was to tell people that he was specifically tasked by the U.S. government to learn about LENRs and to advise the federal government on the research. Many people believed him, particularly those who hoped the federal government was secretly supportive of "cold fusion." However, his very overt efforts to promote the dubious fusion idea, and his ownership stake in ENECO, strongly indicate that his dominant motives were vindicating his belief in the discredited idea and/or seeking personal gain rather than to provide a benefit for the country.

In late summer 2007, Frank Gordon, at SPAWAR, who knew that DTRA was interested in LENRs, contacted George Peter Nanos, a senior DTRA officer, through a mutual acquaintance. Nanos is a former director of the Los Alamos National Laboratory and a retired vice admiral in the U.S. Navy. Gordon sent Nanos information about his, Mosier-Boss, and Larry Forsley's experimental results, and Nanos invited them to the DTRA office in Washington, D.C., so Nanos could learn more, as Gordon explained:

> Larry and Pam were the primary presenters. There were three or four other people there whom Nanos had invited, including Nagel. Melich arrived after our presentations had started. In retrospect, it was apparent that Melich had advance knowledge of the meeting and had pre-briefed Nanos, convincing him that heat was the only valid metric for nuclear reactions. As the meeting was breaking up, I was having a one-on-one discussion with Nanos about the evidence for nuclear particles that we had detected.
>
> Melich, who was several feet away talking to others, raised his voice, jumped into the middle of our conversation

and said that our results were speculative and unconfirmed. Melich said this as he quickly moved to where I was standing with Nanos. I reminded Melich that our results had been analyzed by experts and that our results had been published in refereed peer-reviewed journals.

Melich's current thrust that heat is the only valid metric that will convince anyone that nuclear reactions are occurring is puzzling. Heat results have been used on many occasions to try to convince skeptics that nuclear reactions are occurring. For the most part, those arguments have been unsuccessful. It seems to me that direct measurements of nuclear particles using instrumentation that the nuclear industry has used for years are much more compelling. I can't understand why Melich is pushing heat production and spending DTRA's money on such experiments.

Further evidence of Melich's personal interest in the field comes from the fact that he was a shareholder in and technical advisor for a privately held company called ENECO, which had acquired the license rights to the University of Utah patent applications for Martin Fleischmann and Stanley Pons' research. Melich's ownership in ENECO was revealed in court documents after the company filed for bankruptcy in 2008. Melich and Nagel were listed as technical advisors (while Melich was a government employee) on the ENECO Web site. The ENECO Web site said, "Dr. Melich advises ENECO on government research programs as well as other institutional or commercial opportunities related to the development of the company's energy conversion."

Also while Melich was a government employee, he worked with PHLburg Technologies Inc., a company that specializes in exporting scientific personnel and technology from Russia to the U.S. The company's website said, "Dr. Melich is a senior scientific advisor to the Pentagon, the U.S. Naval Research Laboratory, and other world-renowned research laboratories, private business, and is a university professor. Dr. Melich is PHLburg's chief scientist."

Melich used his authority in other ways at the ICCF-14 meeting to promote the Fleischmann-Pons fusion concept. He paid Melvin Miles to write a paper for the conference on Fleischmann and Pons' heat measurements. "Melich more than paid my way to the conference," Miles said. "$4,000 plus hotel and conference registration fee waived."

In contrast, Irina Savvatimova, a Russian researcher who had reported significant heavy-element LENR transmutation results (Chapter 8), was unable to obtain a U.S. visa in time to come to the conference because Melich told her she should not apply too early to the State Department.

Tadahiko Mizuno, another researcher who had published significant heavy-element LENR transmutation results, did make it to the conference, but, as he told me, the organizers made it very difficult for him. Melich and Nagel published the ICCF-14 conference schedule only five days in advance of the conference. Without knowing which day he would speak, he was forced to make his reservations to cover the whole conference.

In conjunction with the conference, Melich ramped up his long-running effort to name the field in honor of the would-be fusion forefathers. He asked researchers to use the term "Fleischmann-Pons Effect" instead of "cold fusion," which he said was better than LENRs because FPE was "clear and encompassing." As the proceedings reveal, researchers eager to speak at ICCF-14 knew how to name their abstracts so as to increase their chances of acceptance.

A Not-So-Friendly Warning

In 2008, *New Energy Times* did not support the agenda of the fusion faction but instead began the year with the feature article "The Widom-Larsen Not-Fusion Theory." Scientists had insisted to me that a) the proof of "cold fusion" rested in the fact that helium-4 was the only energetic product from LENRs, b) for each helium-4 atom produced, 24 MeV of heat was released, and c) because this 24 MeV ratio distantly resembled part of the well-known D+D thermonuclear fusion reaction, the LENR reaction was, by inference, a fusion process.

On June 19, 2008, Storms sent me an e-mail containing a warning.

"You need to be more careful in how you reveal the truth about the field," Storms wrote. "Eventually, the field will be big enough and so well-accepted that a little plainly spoken truth would not cause you any problem." Apparently, my inquiries were beginning to ruffle feathers.

On July 10, 2008, I published an editorial criticizing Nagel and Melich for saying on the ICCF-14 Web site that the 24 MeV ratio was an established fact and that it provided the foundation of evidence for the D+D "cold fusion" idea. In fact, in the history of the field, I had found only three sets of experiments that reported the ratio between excess heat and helium-4, and the values varied greatly from 22 to 103. The only experiments that came close to 24 MeV, and which researchers in the field said held up to scrutiny, were the 1994 experiments performed by Michael McKubre's group at SRI International. The proof, as I wrote, was underwhelming, and I said that the evidence for "cold fusion" was a myth. This infuriated fusion-believing researchers and their supporters.

I published one other article on July 10, 2008, that was surely unappreciated by the deuterium-deuterium fusion proponents. The title was "Deuterium and Palladium Not Required." It was my report of the Francesco Piantelli group's dramatic results in a nickel and normal-hydrogen gas system. As discussed in Chapter 13, in 1994, the Piantelli group had obtained record levels of excess heat: tens of Watts rather than the typical tenths of Watts in palladium-deuterium electrolytic systems.

Neither Piantelli nor his colleague Sergio Focardi regularly participated in ICCF meetings. In 2000, at ICCF-8, Focardi had presented the group's results for the first time at an ICCF meeting. As discussed in Chapter 18, their group had even more astounding results to report. But the disbelief among fusion proponents that normal hydrogen worked as a reactant in LENRs had grown strongly by then, and the Piantelli group's results were ignored in 2000. Until July 2008, one month before the start of ICCF-14, the Piantelli group's research had been largely forgotten.

My article on their research forced two issues: 1) it questioned the foundation by the D+D "cold fusion" proponents that the underlying reactions were caused by D+D fusion and 2) it questioned the logic of using expensive materials, such as heavy water and palladium.

On Aug. 11, 2008, the first day of ICCF-14, Sam, my sponsor, told me that Nagel and Melich had complained to him that I was writing critical news stories about how they were handling the conference. "Could you please back off," Sam asked me, "just until the conference is over?" I agreed. It was the first time that Sam had tried to intercede in my journalism.

Two days later, Leona Neighbor, a woman I had known for a few years who had identified herself simply as a fan of the research, approached me. She said she had been asked to deliver a message to me on behalf of some of the American researchers.

"Some important people in the field want you to stop making trouble," Neighbor said. "They want you to keep your opinions to yourself and just report the facts. They want you to know that, if you continue digging, your reputation might be harmed and that they might go to the people who fund you and try to get your funding terminated. Also, if you continue on the path you are on, it might become harder for you to do your job, harder to get interviews and sources to talk to you."

I made it clear that I was not going to compromise my reporting in response to their threats. In a later phone call, Neighbor named McKubre and identified Storms and Melich, not by name but by their unique personalities in this small research community.

A Spooky Scientist

Melich's responsibilities as a government scientist appeared to have little to do with LENRs or energy research. For many years, Melich's activity as a research professor with the Naval Postgraduate School (NPS) in Monterey, California, involved the study of phased-array antenna systems for communications and radar. Melich began working at NPS in 1984 after leaving NRL. He lives in Niceville, Florida, with his wife, Marianne Macy, author of the 1986 book *Working Sex: An Odyssey Into Our Cultural Underworld.*

Melich's personal interest in energy followed in the footsteps of his father, Mitchell, who in the 1950s was a central figure in the Utah uranium boom. The older Melich was a Salt Lake City attorney who had run for governor and had served as Solicitor at the Department of the Interior under President Nixon. When the state of Utah was eager to profit from Fleischmann and Pons' research in early 1989, Mitchell Melich proposed that the state authorize $5 million to induce Fleischmann and Pons to remain in and work for the state. (Vol. 2, *Fusion Fiasco*)

In the early 1990s, Michael Melich wrote several papers defending Fleischmann and Pons' heat measurements and their fusion idea. (Melich, 1992; Hansen, 1993; Melich, 1993) By 2005, according to an NPS report, most of Melich's official research activity still involved radar and antenna systems.

However, Melich became involved in two multi-year programs, sponsored by NRL, the Chief of Naval Operations and DARPA, called "Unconventional Weapons of Mass Destruction: Detailed Investigation of Novel Nuclear Physics and Its Implications." According to the NPS report, the program involved research and analysis required to develop doctrine, tactics, techniques, procedures and maritime and joint operational concepts. Melich served as liaison for the researchers on the project. He appears to have used funding from this program for NRL LENR research that he directed. (NPS, 2006)

By 2009, according to another NPS report, the only research projects listed for Melich were in LENRs. All of those projects were designed to validate excess heat or, in collaboration with Hagelstein, to search for theoretical understanding of what he called the "Fleischmann-Pons Effect." (NPS, 2010)

Tall and lanky, with piercing eyes, Melich is known for speaking in intimidating, commanding tones and for his cold, emotionless facial expressions. Melich played up his position as a government employee in his interactions with LENR researchers, as shown in the following letter sent to me by a researcher who asked to not be identified:

> Mike McKubre introduced us to Mike Melich. Knowing he worked for the Navy and was "involved" in cold fusion, Melich wanted to meet us and learn more about [company name]. I picked Melich up from the train. He arrived in a

trench coat, which prompted me to tease him about looking like a spy. As we spoke, he was very evasive about who he specifically worked for other than the U.S. government and the Navy.

I questioned why he had homes in multiple locations in the U.S., and his answer was evasive. Ultimately, there was a clear insinuation that we were being "watched" and that he was here for that reason. It was better to cooperate than make the process difficult. Given that I have nothing to hide, I welcomed Melich into our home.

Every time he was here, he took many photos — family gatherings, meals, etc. It was actually McKubre that pointed out to me that Melich was taking pictures. I finally told Melich that I didn't want any more photos taken and was uncomfortable with it. I asked the purpose, and he joked with me that it was part of intelligence gathering. He never said he worked for Navy intelligence, but it had become a "fact" through innuendo.

This sounds so circumstantial, but there is no denying that I have never doubted that Melich worked for the U.S. government and Navy intelligence. It would actually shock me to hear that this isn't so. I've even been led to believe that phones were tapped and emails read. Only after Scott Chubb said in Catania that DOD was paying Melich to follow me that it all sounded ridiculous and I began asking questions.

Of course, no real intelligence officer would do such things. Nor would he or she do anything to be noticed, such as promoting the idea of "cold fusion" and defending Fleischmann and Pons or, as Melich did later, promoting Andrea Rossi, a convicted Italian fraudster.

It Doesn't Look Like Fusion

On Aug. 20, 2008, a week after ICCF-14, I gave my presentation summarizing my perspective on the subject at the American Chemical

Society national meeting in Philadelphia, Pennsylvania.

Two years earlier, after learning about the general concept of the Widom and Larsen theory, I had made the decision to switch terminology. I stopped using the term "cold fusion" and referred to the research as LENRs. I had spent the next two years examining the experimental distinctions between "cold fusion" and LENRs. By August 2008, the distinctions were clear to me, but I had not publicly expressed a definitive conclusion. I wanted to see what arguments and disagreements the fusion-believing researchers would have.

A colleague of mine, Mike Carrell, had seen a preview of my slides and had advised me. "Steven, you have made an excellent point, in effect, that the emperor has no clothes." Carrell wrote. "I recommend that you pick words carefully and use them gently but precisely, for the issue is emotionally charged by the sincere efforts of good people. You can quote Fleischmann and Pons' words 'hitherto unknown nuclear process.'"

As I explained at ACS, the experimental data revealed that LENRs did not look anything like nuclear fusion. The experimental data disproved the test-tube fusion idea. The product types, pairings, ratios, and associated energies from thermonuclear fusion reactions looked nothing like those of LENRs. In total, I identified eight major categories of inconsistencies between deuterium-deuterium fusion and LENRs:

1. Missing or suppressed gamma radiation
2. Wrong neutron-to-tritium ratio
3. Wrong helium-4-to-neutron ratio
4. Missing first branch of D+D fusion
5. Missing second branch of D+D fusion
6. Poor support for the 24 MeV heat/helium-4 ratio
7. The existence of heavy-element transmutations
8. Normal water and normal hydrogen experiments

Two "cold fusion" theorists, Akito Takahashi and Xing Zhong Li, attended my entire presentation. McKubre was supposed to give a talk there, but he pulled out a week before the conference. Hagelstein showed up late and missed my talk. The room was filled with other people who were knowledgeable about the research. Takahashi and Li had never been hesitant about correcting or critiquing me in the past.

As a nonscientist, I was aware of the risk I was taking. I was prepared for someone to reveal something obvious and embarrassing that I had overlooked. In fact, this was a primary reason for my talk; I wanted to see how scientists in the field who believed in room-temperature fusion would respond to these direct experimental contradictions to their idea.

Takahashi and Li had nothing to say after my presentation. In fact, nobody had a single critical question or argument about my presentation. There was no rebuttal to my distinction then or since.

Besides Widom and Larsen, only two proponents of neutron-based theories were active in the field in 2008. One was Japanese scientist Hideo Kozima, who had respectfully acknowledged Widom and Larsen's method of producing neutrons.

The other was American theorist John Crocker Fisher (b. 1919). During his long career at General Electric (GE), he managed a group of physicists from whom he learned quantum mechanics. He also researched the magnetic and electrical properties of metals. At GE, he worked with Ivar Giaever, a Norwegian-American physicist who shared the Nobel Prize in physics in 1973 with Leo Esaki and Brian Josephson "for their discoveries regarding tunneling phenomena in solids." Esaki credits Fisher, his mentor at GE, for teaching him about the principle of tunneling. (Giaever, 2007) After writing a book about energy, Fisher retired from GE in 1980 and the following year was one of eight people elected to the National Academy of Engineering. (Fisher, 1974)

Although Fisher's theory didn't articulate a mechanism that could produce neutrons, he was convinced that neutrons were the answer. An independent thinker, he didn't care for the fixation on fusion so prevalent in the field, including by Brian Josephson, as Fisher said in an e-mail to the CMNS list in July 12, 2008.

"In my opinion," he wrote, "our field of science has been crippled by wide acceptance of the belief that deuterium fusion of some sort is responsible for energy generation and by [our] rejection of alternative mechanisms. Progress is stunted when we reject a mechanism, because we then fail to undertake the experiments it suggests."

It wasn't a message that "cold fusion" believers wanted to hear.

Cold Fusion at All Costs

Scientists who believed in room-temperature fusion didn't want to listen to the advice offered by theorist John Fisher. They didn't appreciate the insight offered by the Widom-Larsen theory. Instead, they went to great lengths — including redefining terminology and disavowing some of their own data — to preserve the idea of D+D "cold fusion."

Although leaders of the field agreed in 2002 to change the name of its conference series from the International Conference on Cold Fusion to the International Conference on Condensed Matter Nuclear Science, not until 2007 did ICCF-13 conference chairman Yuri Bazhutov use the new name for ICCF-13. The acronym, ICCMNS, however, never replaced ICCF.

David Nagel, the organizer of ICCF-14, couldn't settle on which name to use for his conference so he used both names to officially identify the August 2008 ICCF-14 conference.

For many years, critics of the subject had accused its scientists of being believers, a derogatory term for scientists. Some critics suggested that these scientists behaved like members of a religion. Thus, I was surprised at ICCF-14 when Peter Hagelstein's slide presentation said that he "wanted to believe" that the purported ratio of 24 MeV of heat produced for every helium-4 atom existed.

By August 2008, I had begun to suspect it didn't exist. In October 2008, I surveyed LENR scientists to ask what experimental evidence supported the 24 MeV ratio. As I learned, Hagelstein's belief was typical. "The proof is the 24 MeV! McKubre nailed it," Scott Chubb wrote.

Yet one more essential fact remained: Other energetic products besides helium-4 were observed in the experiments. The existence of other energetic products belied the idea that the reactions could be D+D fusion. Larsen had told me this in 2006, but I needed until 2010 to figure out how to explain it clearly to my readers.

The fusion proponents could not avoid the fact that just one solitary experiment, by Michael McKubre's group, provided thin evidence for their belief.

Growing Desperation

The stubborn and increasingly illogical insistence by the scientists and their supporters that the processes underlying LENRs were D+D fusion can be ascribed to their growing need for professional vindication, both for themselves and for Fleischmann and Pons. Despite the availability of a non-fusion theory that could explain most of the reliable experimental data, scientists like Edmund Storms and Melvin Miles hounded editors at *Nature* and *Chemical and Engineering News* to recognize what they believed was evidence of "cold fusion."

In February 2009, after an editor at *Nature* rejected Storms' manuscript, "Judging the Validity of the Fleischmann-Pons Effect," Storms told his colleagues that he sent it to the journal *Physical Review B*. "We will keep submitting it to various major journals until it is accepted," Storms wrote. "At least they will know that we are still alive."

Storms was 78. He and his colleagues knew that the time left in their careers was limited and that each year that went by decreased the likelihood that they would be recognized in their lifetimes for their participation and achievements.

Vindication for "cold fusion" was unlikely, but recognition for LENRs was. However, such recognition was a double-edged sword because it would attract many newcomers. These newcomers and younger scientists, many of them working in established laboratories, would then start uncovering the remaining mysteries of LENRs. It seemed unfair, but this new generation likely would reap the benefits of battles fought by the earlier generation of LENR scientists.

Park Concedes Again

On March 22-26, 2009, in Salt Lake City, Utah, the American Chemical Society hosted a third LENR symposium. The symposium was organized by Jan Marwan, with my assistance. For the third time in four years, Robert Park, the former archenemy of the field, acknowledged that LENRs were real. His grudging public admission appeared on March 27, 2009, in his weekly newsletter.

"The American Chemical Society was meeting in Salt Lake City this week, and there were many papers on cold fusion, or, as their authors prefer, LENR (low-energy nuclear reactions)," Park wrote. "These people, at least some of them, look in even greater detail where others have not bothered to look. They say they find great mysteries, and perhaps they do. Is it important? I doubt it. But I think it's science."

Neutron Capture Is Not the New Cold Fusion

In early 2009, as a last-ditch effort to salvage even the appearance of fusion, some scientists began trying to convince each other and members of the public that the neutron-capture process could be considered to be a form of fusion.

Neutron capture, according to the *Oxford Dictionary of Physics*, involves a single particle, such as a neutron, with no electric charge, entering a nucleus. Nuclear fusion, according to the Oxford dictionary, involves two nuclei having like charges that overcome electromagnetic forces, such as the Coulomb barrier.

Some of the e-mail messages exchanged among scientists on the CMNS discussion list at that time revealed their desperation to retain the term "fusion." For them, the word "fusion" had deep-seated, personal meaning. McKubre's May 13, 2009, e-mail provides an example.

"I submit that John Fisher's neutron-addition reactions also qualify as fusion," McKubre wrote. "We are certainly not bound to use the jargon of particle physics and should not! We are free to use the English language without apology or deference, so long as we use it correctly. ...

The fusion 'purists,' with their corrupted definition, want ownership of the products and process. ... But for definition, this is not important. Is the final state the result of 'joining two or more things together'? If yes, then I submit it is fusion."

For McKubre, neutron capture was the new fusion. In response to McKubre's e-mail, Fisher diplomatically declined McKubre's new use of nuclear terminology, explaining the well-established definition of neutron capture by the physics community. McKubre responded indignantly.

"[I] don't see why we should be frightened off by physicists' jargon usage of the word fusion," McKubre wrote. "The high-energy and particle physicists don't own the word in our context. In the future, we will."

They All Said It Was Wrong!

As the concept of LENRs gained credibility in various parts of the federal government, the Defense Intelligence Agency (DIA) took note. In August 2009, Beverly Barnhart, an analyst with the DIA, conducted a LENR workshop at the U.S. Navy's SPAWAR San Diego laboratory so her agency could learn more.

On Nov. 13, 2009, she released a Technology Forecast report based on that workshop to some of the participants in the workshop. I received a copy of the DIA report from one of them. In addition to Barnhart, contributing authors were McKubre, Pamela Mosier-Boss, Patrick McDaniel and Larry Forsley (colleagues of Mosier-Boss), and Louis DeChiaro (a physicist at the Naval Surface Warfare Center, in Dahlgren, Virginia, and a colleague of Peter Hagelstein's).

The report said that some researchers thought that LENRs could be explained by D+D fusion, that other researchers thought LENRs could be explained by chemical reactions, and that some people thought it was explained by a combined fusion-fission reaction. Essentially, the authors considered every possibility except the Widom-Larsen theory, which they did not mention.

I was perplexed by this significant omission from the government report. I called Barnhart and spoke with her on the phone for two hours. She explained her entire process to me and how the whole idea of a workshop and report came about. I asked her about her understanding of the proposed LENR theories, which speakers had been invited, and how the report was written.

Unaware at the time of the significant contributions of her co-authors to the report, I told Barnhart that I was concerned about her omission of the Widom-Larsen theory. I asked her how and why she had excluded their work.

"As I understand," Barnhart said, "it's not even Larsen's theory. He's trying to raise money so Widom can start a company, and people have looked at it, and they say it's wrong. And Larsen and Widom don't even like each other anymore. How could there be anything to Widom-Larsen, when everybody — I mean everybody I spoke to — told me that it was wrong?"

I was astounded at the false information she had been given by the fusion promoters. Moreover, the information she described sounded as if it had come from a newspaper tabloid.

I told Barnhart that she had been duped, and I asked her why she never contacted Larsen or Widom to seek the facts for herself. It never occurred to her to check, she told me, because she was so convinced that their theory was worthless.

Violante Denies His Own Data

In October 2009, Vittorio Violante tried to cast doubt on his own earlier LENR results, which showed evidence of significant isotopic shifts in silver and copper in light-water electrolysis. His data contradicted the D+D "cold fusion" idea.

Back in 2002, at the ICCF-9 meeting in Tsinghua, China, Violante reported the results of two nickel-hydrogen thin-film experiments that showed significant isotopic shifts, 29% and 47%, in the silver-107/silver-109 ratio. His group, which included McKubre as a co-author, also observed a significant increase in the tritium level in one run, about 10 times the background.

In 2002, Lewis Larsen wasn't alone in thinking about plasmonics. Violante knew, and wrote, that the isotopic shifts indicated that "nuclear processes different from $d + d \rightarrow 4He + heat$" took place and that they were directly tied to surface-plasmon electrons. (Violante, 2002) This contradicted McKubre's claim, two years earlier, at the ICCF-8 conference, that he had measured data confirming the $d + d \rightarrow Helium\text{-}4 + 23.8\,MeV\,(heat)$ idea.

At the ICCF-10 meeting in Cambridge, Massachusetts, in 2003, Violante continued with his nickel-hydrogen thin-film system, but now he was using a low-power laser to excite surface plasmons and trigger the effects, which Dennis Letts and Dennis Cravens revealed in 2000, as discussed in Chapter 19. Violante reported significant shifts in the ratio of each of the copper isotopes measured on a nickel film, and he wrote that he had reproduced his ICCF-9 data. (Violante, 2003, 405; Violante, 2003, 421)

Violante presented another paper at ICCF-10, in which he proposed theoretical ideas based on surface plasmons. (Violante, 2003, 667)

In the fall of 2004, at the ICCF-11 meeting in Marseille, France, Violante continued talking about surface plasmons and how they could produce a local electromagnetic field enhancement. (Castagna, 2004) Larsen was in the audience, listening intently.

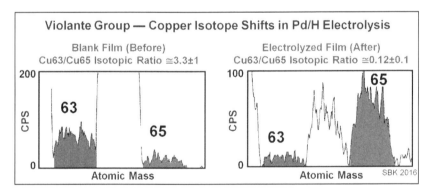

"All of a sudden," Larsen said, "I heard Violante start talking about surface plasmons again. Violante said that when he inverted the polarization of the laser, the heating effect stopped. That further confirmed for me that the $e_n + p_n \rightarrow n + \nu$ reaction utilized surface-plasmon electrons, but I thought I was going to throw up. I was afraid Violante was going to conceptually connect surface plasmons to the $e_n + p_n \longrightarrow n + \nu$ reaction and that he would figure out the mechanism responsible for producing the neutrons. I thought he was going to beat us to it and publish the details of the mechanism before we did. But he didn't connect the dots."

Sometime between ICCF-10 and ICCF-11, Kenneth Grabowski and Graham Hubler, both NRL scientists, and Michael Melich, a retired NRL scientist, began to help Violante, according to Violante's ICCF-11 slides. At that point, Violante began a strange, gradual, and convoluted process of trying to discredit some of his own data — specifically, the data that contradicted the D+D "cold fusion" idea.

In their November 2004 slide presentation at ICCF-11, Violante's group began to suggest that the copper shifts were artifacts. (Apicella, 2004)

In late 2005 or early 2006, more than a year after ICCF-11, Violante submitted for publication in the ICCF-12 proceedings four papers that had been presented at ICCF-11. Violante did not explain why he failed to submit the papers in time for publication of the ICCF-11 proceedings.

In one paper, Violante's group created a complex scenario to suggest that their copper isotope shifts might have been in error, but they reported no direct evidence that invalidated their 2003 data. Instead, the paper contained only speculations and the suggestion that the older experiment was invalid. (Apicella, 2006)

The group continued its attempts to discredit the data at ICCF-15 in 2009. (Apicella, 2009) Yet none of their speculations could explain the difference between their results with the electrolyzed film and their reference film. (Violante, 2003, 405)

It was not generally known that Violante had tried to disavow his own isotopic shift data. I learned about this only when Mahadeva Srinivasan, whom I had invited to contribute a paper on LENR transmutations for the John Wiley & Sons *Nuclear Energy Encyclopedia,* brought it to my attention. Srinivasan had shown a draft to Violante, who then told Srinivasan that he had retracted his isotopic shift data.

On hearing this, I asked Violante where and how he had retracted that data. He failed to direct me to any formal retraction. Instead, he directed me to his group's papers, as discussed above.

A Good Team Player

Back in October 2009, Violante and his group had given another presentation at ICCF-15, in Rome, at which Violante served as conference chairman. The paper reported only excess heat, no heavy-element transmutations, and no isotopic shifts. In his paper, "Evolution and Progress in Material Science for Studying the Fleischmann and Pons Effect," Violante wrote that LENR phenomena occurred only with deuterium, not with hydrogen (Violante, 2009). He now shared the same philosophy, research priority, and terminology as Melich.

Violante was a co-author of another paper presented at ICCF-15, an ambiguous paper that reported no positive results. Instead, it implied that anomalous elements found in Violante's 2002 experiment were not heavy-element transmutations but pre-existing impurities. (Rosada, 2009) It was reminiscent, as discussed in Chapter 11, of NRL electrochemist Debra R. Rolison's strained 1991 attempt to disown her transmutation results.

By 2009, Violante had made a complete reversal from his 2002 conclusion, and he was now fully aligned with the agenda of the American LENR researchers to promote the $d + d \longrightarrow Helium\text{-}4 + 23.8$ $MeV\ (heat)$ idea. As McKubre revealed in his ICCF-15 slides, Violante's lab had received U.S. government funding, thanks to a subcontract with SRI International from DARPA contract HR0011-05-C-0089. (McKubre, 2009) The American fusion believers had neatly taken care of the problem caused by Violante's data that contradicted the idea of "cold fusion." And, along the way, Violante had missed his opportunity to identify a key physical mechanism explaining the phenomena.

CHAPTER 31

U.S. Assault on Japanese Science

M ichael Melich, with the help of David Kidwell, orchestrated a coordinated attack against the work of LENR researcher Yasuhiro Iwamura at the International Conference on Condensed Matter Nuclear Science (ICCF-15) meeting in October 2009. At the time, Iwamura was working at Mitsubishi Heavy Industries, in Yokohama, Japan. He had begun his LENR research in 1993.

By 1998, in addition to observing excess heat, Iwamura had observed isotopic shifts and a variety of heavy-element transmutation pairs. In these pairs, a starting element gradually decreased at the same time that a product element gradually increased. His results were published in the *Japanese Journal of Applied Physics* in 2002. (Iwamura, 2002)

In one set of the Mitsubishi deuterium gas-permeation experiments, Iwamura and his colleagues observed the gradual decrease of the element cesium (atomic number 55) and the gradual increase of the element praseodymium (atomic number 59). Mitsubishi scientists realized that these LENR experiments offered a potential way to convert one element into another, for example transmuting radioactive elements into benign, stable elements. As a major Japanese manufacturer of nuclear fission reactors that produce radioactive cesium-137 as a common byproduct, Mitsubishi would benefit greatly from finding such a way to make radioactive isotopes harmless.

The results of the Mitsubishi experiments were independently confirmed at Osaka University. (Higashiyama, 2003) In 2002 and 2003, physicist Francesco Celani (b. 1951) and his colleagues at the Italian

National Institute of Nuclear Physics in Frascati reported replication of the older Mitsubishi experimental design using the electrolytic method.

Before the experiment, Celani's group used inductively-coupled mass spectrometry to identify all elements in the cell, including the electrolyte. Afterward, they found that some strontium had transmuted to molybdenum and that the isotopic composition of the molybdenum was different from the one found in nature. (Celani, 2002, 2003)

Weak Interaction

Unlike many researchers in the late 1990s who believed that LENRs were based on a fusion process, Iwamura thought otherwise. To him, a weak-interaction, neutron-based theory with deuterium most likely explained the phenomena. He published his idea, represented by the reaction $e + d \rightarrow 2n + v$, in *Fusion Technology* in 1990.

Iwamura focused on deuterium. He never claimed to observe any positive results with normal hydrogen; nor has he disputed such results from other researchers. For many years, he has maintained a cordial professional relationship with Akito Takahashi (b. 1940), a senior figure in the Japanese LENR community. Takahashi has been proposing a theory based on the D+D "cold fusion" idea since April 1989. According to Takahashi's theory, deuterium, but not hydrogen, works as a reactant in LENRs. (Takahashi, 1989)

At the ICCF-10 meeting in 2003, American scientist Talbot Chubb, despite being a fusion promoter, praised the Iwamura results, as discussed in Chapter 20. "The results presented at this meeting seem destined to affect the course of solid-state and nuclear science," Chubb wrote. "Probably the most important of the results were those concerned with a unique form of nuclear transmutation reported a year ago by Iwamura."

Melich's Strategy

Although independent replication of experimental claims is the bedrock of validation in the scientific community, for at least a century the concept has sometimes been misused by competing scientists to instill public doubts about a competitor's results.

In such cases, competitors report that their replication attempt failed to yield positive results, claiming that they performed an exact (or better) replication of their opponent's experiment.

Rarely do the opponents then suggest that they may have failed to mirror the materials, the procedures and the analytical criteria exactly. Instead, they assert that the originator's results are erroneous, the result of a mistake or an assumed but unproven artifact.

In response to the non-fusion implications of the Iwamura experiment, Michael Melich devised a federally funded plan that tried to falsely discredit the Mitsubishi results. Melich obtained funding from the Defense Threat Reduction Agency (DTRA) and the Defense Advanced Research Projects Agency (DARPA) and arranged for the money to go to the Naval Research Laboratory (NRL). At NRL, he collaborated with Graham Hubler. The pair hatched a plan to replicate the Mitsubishi experiment, and they identified NRL scientists to work on the project.

The key participant in the project was David Kidwell, who was new to the LENR field. Between 2002 and 2005, collaborating with Mitsubishi, Kidwell and other NRL researchers tried to replicate the Mitsubishi results, but NRL was unsuccessful each time.

Those NRL replication attempts couldn't have worked, according to what Iwamura told me. "NRL didn't do a precise replication," Iwamura said. "They didn't have suitable equipment." In my conversations with Iwamura, he seemed unaware of the organizational politics, motives, and biases of the U.S. scientists who advocated the fusion concept.

Kidwell's first public appearance in the LENR community was in August 2008, when he presented a paper at ICCF-14. He did not present successful results of his own but focused on criticizing the work of other LENR researchers, particularly the Japanese researchers who had reported evidence of isotopic shifts and heavy-element transmutations.

In 2008, a nondisclosure agreement between NRL and Mitsubishi legally constrained Kidwell in what he could say publicly. Instead, he tried to discredit another set of isotopic shifts measured in samples from the Arata-Zhang LENR experiments.

Yoshiaki Arata (b. 1924), a retired professor of physics at Osaka University, is a respected physicist in Japan who, in 2006, received a prestigious award, the Order of Culture. He was the first person in Japan to perform a thermonuclear fusion experiment demonstrating large numbers of D+D fusion reactions. His associate is professor Yue Chang Zhang, of Shanghai Jiao Tong University, in Shanghai, China. Thomas Passell, a former program manager for the Electric Power Research Institute, reported the isotopic shifts from an Arata-Zhang LENR experiment in 2003 at ICCF-10.

"Probably the greatest revelation in this work," Passell wrote, "is the possibility that trace elements may be significant participants in nuclear reactions in solids such as palladium; [therefore], focusing entirely on [the idea of] deuterium-deuterium fusion is not necessarily the only path forward in understanding these phenomena." (Passell, 2003)

Passell tried diplomatically to encourage his peers who believed in the fusion idea to consider an alternative explanation. The data he reported, and his conclusion, became a headache for them.

At ICCF-14, Kidwell laid the groundwork for his future criticisms of the Mitsubishi research. He said that researchers who were observing heavy-element transmutations in LENR experiments should beware of making such claims. "If you have what you think you're making all over your room, do you really have it?" Kidwell said. "Or are you just fooling yourself from some random event?"

Kidwell presented himself as an expert on spectroscopy specifically for LENR research and told everyone in the room how he thought they should perform their analyses. "Without all these precautions," Kidwell said, "I wouldn't go talk to my colleagues. You're not going to rewrite 100 years of chemistry, you're not going to pass go and you're not going to win a Nobel Prize! With these cautions, you might go talk to your colleagues and say, 'Something unusual occurred.' But without them, I would just kind of be embarrassed."

I was perplexed about why someone with no experience in the field would attend an ICCF meeting with such a hostile attitude and without positive results of his own to report. After his presentation, I asked him why he presented his paper. "Melich asked me," Kidwell said. John Dash, who, during the question-and-answer session, had challenged Kidwell's methodology, explained the problems to me in greater detail in an e-mail:

> Kidwell seemed totally ignorant of the large body of knowledge that suggests that the action in LENR experiments happens on the cathode surfaces rather than in the bulk. Kidwell uses inductively-coupled mass spectrometry to perform elemental trace analysis. The problem with his use of this technique is that he dissolves away the outermost layers of the sample and throws that away in his belief that everything on the outer layers is possible contamination.
>
> It's a shame because Kidwell has facilities that most of us can only dream of. That's been going on for years at NRL, beginning with Debra Rolison. She failed to recognize the relevance of what some people have called "hot spots." She failed to do precise mapping of the isotopic anomalies and, instead, averaged the signal over the entire sample. This resulted in [her] interpretation that there was nothing there. She set us back 10 years. She had it, but she blew it, too.

Lucky Tweezers or Lucky Swabs?

By 2009, at ICCF-15, in Rome, Italy, the nondisclosure agreement had expired, and Kidwell was now free to talk about the NRL-Mitsubishi collaboration, and to attempt to discredit the Mitsubishi heavy-element transmutation results. He said that the praseodymium measured in the Mitsubishi LENR experiments, and the apparent transmutation of cesium to praseodymium, was the result of pre-existing praseodymium contamination in the Mitsubishi laboratory.

Kidwell also suggested that a former Mitsubishi employee used "lucky tweezers" to spike the experiment with praseodymium. Kidwell's statement implied that the Mitsubishi researcher, placed the rare element praseodymium in the multilayer substrate composing the experiment either by fraud or by incompetence.

However, Mitsubishi researchers said they never had used praseodymium in that cleanroom laboratory. Mitsubishi found praseodymium in the lab only after Kidwell, who visited the Mitsubishi lab as a guest, used swabs that he brought to perform a surprise environmental survey after the other NRL guests had gone back to the U.S. Funding for this portion of Kidwell's work came from DARPA.

In his rebuttal, Iwamura explained that, because of multiple aspects of the experimental protocol, Kidwell's claimed scenario was virtually impossible as an explanation for the reported transmutation data.

Patent Competition: Mitsubishi Versus U.S. Navy

Iwamura had filed for a U.S. patent on his LENR method on Oct. 19, 2001. Eight years later, in 2009, it was still stuck in the patent office. A week before the ICCF-15 meeting in 2009, Kidwell applied for the first of two of his own LENR-based patents. Funding for this part of Kidwell's work came from the DTRA. There were strong similarities between the ideas discussed in the patent applications of Iwamura and Kidwell. Each method for triggering LENRs used nanometer-sized particles placed on a metal-oxide support mechanism, subjected to pressurized deuterium gas.

The biggest difference in experimental data used to support claims in the respective patent applications is that Iwamura analyzed only for transmutation products whereas Kidwell analyzed only for excess heat. Kidwell's claims were consistent with Melich's agenda to promote the fusion idea and discredit the neutron-based idea. Kidwell's patent application failed to cite Iwamura's application, which had publicly disclosed Iwamura's earlier ideas.

As discussed in the next chapter, Iwamura endured an unsuccessful 10-year ordeal with the U.S. Patent and Trademark Office.

"Over-the-Top Brutal"

In July 2013, at the ICCF-18 meeting in Missouri, Kidwell continued battering the Mitsubishi LENR claims. He had been invited by the conference organizers to deliver the keynote presentation. A businessman who attended ICCF-18 sent me the following report.

"Kidwell was over-the-top brutal at ICCF-18 in his keynote speech regarding Mitsubishi," he wrote. "Kidwell directly attacked Iwamura and, per several ICCF veterans, added nothing new to his contamination argument. Iwamura was very upset and confronted Kidwell during the question-and-answer session. I personally witnessed both Kidwell and his NRL colleague David Knies making attacking arguments during the poster sessions. For someone like me who doesn't have a dog in the fight, their behavior seemed mission-oriented and goes way beyond an argument about science."

It was mission-oriented, actually. The businessman did not know the history of the fusion believers' battle against non-fusion experimental results, such as isotopic shifts, heavy-element transmutations and excess heat with normal hydrogen.

Although many attendees of ICCF-18 were shocked at Kidwell's remarks, the organizers of the meeting certainly knew what to expect from Kidwell when they invited him to deliver the keynote presentation. The main organizer of the meeting was Robert V. Duncan, at the time vice-chancellor for research and a professor of physics at the University of Missouri. Duncan was responsible for the management of the school's major research facilities, including the Missouri University Research Reactor, the most powerful university research reactor in the United States.

Three months after the ICCF-18 meeting, Duncan announced that he would be leaving the University of Missouri for Texas Tech University, in Lubbock, Texas, a university that had no nuclear research programs and no nuclear physics or nuclear engineering department.

Four years earlier, in April 2009, Duncan had appeared on the high-profile CBS television news program *60 Minutes*, which featured LENR research by a private Israeli company, Energetics Technologies. *60 Minutes* reported that DARPA had performed its own analysis of LENRs and that DARPA had confirmed that LENRs produce excess heat.

In his interview, Duncan called LENRs "cold fusion," failing to make a scientific distinction between the fusion idea and LENRs, which does not presume or assert the idea of nuclear fusion. I had numerous interactions with Duncan, and he, like everyone in the field, was clearly aware of the Widom-Larsen theory by 2009. Regardless, he asserted that nobody understood the theoretical mechanism.

After his television appearance, Duncan and his wife, Annette Sobel, took paid consulting appointments with Star Scientific Ltd. of Australia, a company promoting muon-catalyzed fusion as a practical energy production technology, according to a former investor in the company. The company used Duncan's endorsement to convince investors to help fund the company despite the fact that muon-catalyzed fusion had long been deemed incapable of producing net energy.

The second organizer of the ICCF-18 meeting in 2013 was Graham Hubler. The evidence that *60 Minutes* provided to support its claim that DARPA had performed its own analysis of LENRs was a document written by Hubler. The document was not written on official letterhead, had no official government cover page and instead looked like a personal document and opinion written by Hubler.

Moreover, it wasn't an analysis of work performed by DARPA. It was just Hubler's informal analysis of work performed by SRI International that was funded by DARPA and managed by Hubler. The SRI experiments performed under the DARPA contract were replications of experiments performed at Energetics Technologies, a subcontractor to SRI on the DARPA contract and the same company that was featured on the *60 Minutes* program.

After Hubler retired from NRL in 2012, he went to work at the University of Missouri in the Department of Physics and Astronomy.

The third organizer was Yeong Kim at Purdue University. Kim had been trying since 1990 to develop a theory that would explain LENRs as D+D "cold fusion." (Kim, 1990)

After ICCF-18, Kidwell asked Tatsumi Hioki, the lead researcher at the Toyota LENR group, whether he could come to Toyota to look at their apparatus. Hioki refused. The Toyota researchers had apparently recognized that Kidwell's interests were not purely scientific.

CHAPTER 32

Innovation Stagnation

Mitsubishi's Yasuhiro Iwamura went through an unsuccessful 10-year ordeal in an attempt to patent Mitsubishi's LENR technology with the U.S. Patent and Trademark Office (USPTO). For many years, patent applications in the subject were separated from other applications for special handling and were almost always rejected.

As early as June 5, 1989, the USPTO had advised its examiners not to let "cold fusion" patent applications pass through the pre-examination screening process. In 1994, the USPTO established a secret policy. The Sensitive Application Warning System (SAWS) was established to weed out applications for a mixed bag of controversial or potentially embarrassing patent subject matter, as well as a mixture of nonsensical topics, such as antigravity devices, free-energy claims, and other matters that violate the general laws of physics. "Cold fusion" was listed among these.

The SAWS policy remained secret until a memo describing SAWS was leaked by an employee of the USPTO and it received press coverage. Intellectual-property experts have criticized the notion of secret rules, without any notice or opportunity for comment by the public, that strangled certain types of subject matter without providing adequate disclosure and due process to inventors and applicants. The patent office publishes a very detailed guidebook called the Manual of Patent Examining Procedure, on which the public and practitioners rely, but the USPTO failed to mention SAWS in that document.

After the SAWS program was revealed, and in response to a public uproar, the USPTO said it conducted a review of the program. On

March 2, 2015, the USPTO retired the program with the explanation that "only a small number of applications examined over the last twenty years were ever referred to the SAWS program ... [and it] has only been marginally utilized and provides minimal benefit."

Nevertheless, patent examiners were well aware that the subject matter identified by SAWS seemed to include LENR-related applications, and that awareness likely had a significant impact on patentability over a period of several years, even if only a handful of applications were formally designated under the program.

The USPTO allowed and issued some patents for LENR-related inventions before 2015, but only a handful mention "cold fusion" in the context of energy production in the claims or the disclosure.

For example, George Miley filed an application on Feb. 26, 2001, and it issued as patent number US 7,244,887 on July 17, 2007. Lewis Larsen filed an application on March 26, 2002, and it issued as US 6,921,469 on July 26, 2005. Kidwell filed two applications in 2009, and they issued as patent numbers US 9,182,365 and US 9,192,918 in November 2015. Kidwell did not even use the term LENR in his patent applications; nor did he make any suggestion that the excess heat he claimed was from a nuclear reaction.

Iwamura did not use the term "cold fusion" in his application; nevertheless, patent examiner Jack W. Keith failed to make the distinction between LENRs and "cold fusion." According to the patent file wrapper paperwork, Keith believed Iwamura's claim was based on "cold fusion," and because he believed that all such claims were "allegations that border on the incredible," he believed they didn't work. In patent lingo, the term for such devices is "inoperable." If the patent examiner determines that a device is inoperable, it cannot be patented because it fails the fundamental test of providing utility.

"Because the allegations would not be readily accepted by a substantial portion of the scientific community," Keith wrote, "sufficient substantiating evidence of operability must be submitted by applicant."

Keith's approach and response to Iwamura's application were typical of USPTO examiners reviewing LENR-related patent applications, and the primary cause was the secret SAWS policy. The secrecy of the

SAWS rules impeded LENR patents and the development and protection of intellectual property in the U.S.

Mitsubishi's USPTO Nightmare

What happened to Iwamura provides a useful account of what happened to many other applicants for two decades. In his initial rejection on Oct. 28, 2003, Keith cited a litany of 1989 news stories reporting, in his words, "negative results from ... scientists skilled in the art."

"The examiner," Keith wrote, "has presented evidence showing that, in such cold fusion systems, the claims of transmutation, excess heat as well as of other nuclear reaction products, are not reproducible or even obtainable. It consequently must follow that the claims of excess heat are not reproducible or even obtainable with applicant's invention."

Keith cited the 1989 Department of Energy (DOE) "cold fusion" review panel, as well as informal speculations about errors discussed in Internet newsgroups, in e-mails and on personal Web sites. The examiner also cited the books by Frank Close (1991), John Huizenga (1992), and Gary Taubes (1993) that depicted the new science as an erroneous body of experimental and theoretical research. In hindsight, it is clear that Keith did not expect to issue a patent to Iwamura.

After providing these citations for the general subject matter, which, according to Keith, identified numerous "sources of error in cold fusion systems," Keith provided two references specific to the Iwamura application that, he wrote, "further attested" to experimental error or misinterpretation of experimental data.

One reference was a message posted to an online news group, sci.physics.fusion, on July 10, 2002, from Kirk Shanahan (b. 1955). Shanahan is a chemist who works at the U.S. DOE's Savannah River National Laboratory, in South Carolina, and who had been an aggressive opponent of LENRs for many years. Shanahan and I have communicated on the topic of LENRs on several occasions. He has often responded to published journal papers with critique that is of questionable validity.

Shanahan expressed two criticisms in his sci.physics.fusion message. First, he speculated that Iwamura did not adequately check for sources of possible contamination. Iwamura never observed transmutations in control experiments and he also observed concurrent reductions in the starting elements, so Shanahan's complaints were largely irrelevant. Shanahan's second complaint was that the Iwamura experiment had not been independently replicated. This was an invalid criticism; the Iwamura paper had just published that year.

The other reference cited by patent examiner Keith was a July 22, 1998, message posted to the sci.physics.fusion news group, this one by Rich Murray (b. 1942). Murray had been a reader of mine for many years, and we had exchanged some e-mails, though I never knew much about him. I called him for the first time on May 20, 2016, and we had a delightful conversation. Although he didn't have enough information in 1998 to accept Iwamura's transmutation claim, he was quite surprised when he learned that the patent office had used his posting to reject Iwamura's patent application.

"I'm not scientifically qualified," Murray said. "I would say I'm more of a spiritual adventurer. My training was in psychology, and I've spent the last two decades as a home hospice caregiver."

Keith asserted that, because of the many reported failures to replicate Fleischmann and Pons in 1989, Iwamura's device could not be considered real unless it was replicated at a credible independent laboratory. The examiner could not prove that Iwamura's device didn't work; nor could he prove that the science was wrong.

Thus began a frustrating and wasteful exercise by both the examiner and Iwamura. The examiner made demands that might have satisfied his judgment that Iwamura's device was operable. But once Iwamura satisfied one demand, the examiner presented another, and so on. This type of response is often referred to as "moving the goalposts."

The Nightmare Continues

On July 7, 2004, examiner Keith again rejected Iwamura's claim and demanded that his experiment be replicated by an "independent unbiased source."

"Reproducibility," Keith wrote, "must go beyond one's own lab." Accordingly, on January 7, 2005, Iwamura informed the examiner that researchers at Osaka University had replicated his experiment. Even though the experiment took place in another laboratory, the examiner denied the claim because one of the Mitsubishi researchers helped the Osaka researchers.

In response, on April 4, 2006, Iwamura advised the examiner that, on May 20, 2005, Mitsubishi had begun a cooperative research agreement with NRL in an attempted replication effort. Kidwell had failed at his initial attempt sometime in the fall of 2005, but neither he nor Iwamura had given up on the collaboration.

On July 10, 2006, three months after Iwamura advised the patent office of the NRL replication attempt, Kidwell, Grabowski and one other person from NRL visited the Mitsubishi lab and observed experiments. That's when Kidwell performed his surprise environmental survey and asserted that the Mitsubishi cleanroom was contaminated with praseodymium.

It is unknown whether the information about Kidwell's contamination claim reached the patent office, but the examiner allowed the application to remain open, and he gave Iwamura more time to produce other replications that, ostensibly, might change the examiner's mind.

A Second Examiner Joins the Fray

By 2008, another examiner, Daniel Lawson Greene, was assigned to the application, replacing examiner Keith. On April 21, 2008, Greene issued a final rejection, again based primarily on Greene's belief that Iwamura was trying to patent "cold fusion." But Iwamura's patent application made no such assertion or implication. To the contrary, Iwamura had been proposing a weak-interaction-based neutron theory since 1998 to explain his experiments. The patent record indicates that the examiner didn't seem to understand the distinction. The examiner also didn't seem to know about the three replications that had been performed at Japanese universities.

"Applicant has been given more than two years to supply the office with evidence in support of the invention," Greene wrote. "No such clear and convincing evidence has been provided. This supports the examiner's contentions that applicant's invention is drawn to the theory of 'cold fusion,' as there is no ability to reproduce [it]."

The hoped-for NRL replication didn't pan out. However, Iwamura told the patent office on Sept. 22, 2008, that the experiment had been replicated by researchers at Iwate University and at Kobe University, without any direct assistance from Mitsubishi researchers. The examiner still refused to award a patent to Mitsubishi.

A Third and Final Examiner

On Dec. 12, 2008, prosecution of Mitsubishi's patent application was again reassigned, this time to examiner Johannes P. Mondt. Mondt applied the same reasoning as his predecessors: He assumed that the application had been claiming tabletop fusion; therefore, the invention was summarily dismissed as inoperative. Mitsubishi allowed that application to expire.

On June 12, 2009, Mitsubishi resubmitted the patent application and began the whole process again. On Oct. 9, 2009, Mondt gave the same reason for rejection.

"There is no reputable evidence of record to support any allegations or claims that the invention is capable of operating as indicated in the specification," Mondt wrote.

Iwamura and a patent lawyer went to meet with Mondt on July 9, 2010, but that didn't help. Iwamura told me that Mondt asked irrelevant questions. Iwamura tried his best to explain the status of the research to the examiner. Ron Rudder, an attorney from the law firm representing Mitsubishi Heavy Industries, also tried to help explain the research, but their efforts were in vain. Mondt just didn't seem to understand the science, Iwamura wrote to me in an e-mail.

The application sat on Mondt's desk for another year. On Oct. 10, 2011, Mondt used a journal comment written by Shanahan as a reason to reject Mitsubishi's patent application. The reason Mondt cited Shanahan's journal comment was that, in it, Shanahan had cited Kidwell, who had asserted that Iwamura's reported transmutation products were the result of contamination. Kidwell's suspicious 2009 claim of contamination, which he made at ICCF-15, had scored a direct hit against Mitsubishi's U.S. patenting efforts.

Mondt did not cite Kidwell. Instead, Mondt cited the Shanahan journal comment because, although Kidwell gave an oral presentation at the ICCF-15 conference, his presentation was never published as a paper in the proceedings. Understandably, the patent examiner found it preferable to cite a published paper.

Kidwell's NRL colleague, Kenneth Grabowski, also gave an oral presentation on the Mitsubishi-NRL collaboration. That presentation, too, was not published as a paper in the proceedings.

According to Iwamura, Kidwell has never published a critique of the Mitsubishi cesium-to-praseodymium LENR transmutation results in any peer-reviewed journal or distributed any related preprint. Iwamura told me that, in fact, no one has ever published any critical peer-reviewed comment on either his group's 2002 paper or the Toyota group's 2013 paper published in the *Japanese Journal of Applied Physics*.

I wrote a brief response to Shanahan on the *New Energy Times* Web site on July 30, 2010. Ten researchers, including Iwamura, joined forces and wrote their own response to Shanahan. That paper published on Aug 6. 2010, but Mondt ignored both responses.

Iwamura's patent was not issued in the U.S. He did, however, obtain a LENR patent in Japan, 04346838 (P2001-201875), and two related Japanese patents, 0434726 (P2005-142985) and 04347262 (P2005-142986), all issued on July 24, 2009. He was also awarded a European LENR patent, EP1202290B1, on April 12, 2013.

In October, 2013, in the *Japanese Journal of Applied Physics,* researchers at Toyota Central Research and Development Laboratories reported that they, too, had replicated the Mitsubishi findings: an increase in the amount of praseodymium in a cesium-ion-implanted Pd/CaO support mechanism subjected to pressurized deuterium gas. (Hioki, 2013)

In March 2015, the Condensed Matter Nuclear Reaction Division of the Research Center for Electron Photon Science at Tohoku University,

Japan, was established, and Iwamura was asked to head the LENR research group there.

The most telling and perhaps accurate statement made by Mondt about Iwamura's application was this: "Applicant's finding, if true, would represent a serious revamping of standard nuclear physics and, as such, would represent an extraordinary achievement."

Indeed, overwhelming evidence does indicate that Iwamura's transmutation results are correct and that they represent an extraordinary milestone in nuclear science.

CHAPTER 33

I Killed Cold Fusion

Yasuhiro Iwamura's transmutation results were bad news for the "cold fusion" believers. Things were about to get worse. I soon found out and revealed that the supposed best evidence for room-temperature fusion had been fabricated. As a result, the last vestiges of support for the fatally flawed idea disappeared.

The scientists who believed in "cold fusion" had assured me that a sizable experimental database proved that 24 MeV of excess heat was produced for every helium-4 atom produced in palladium-deuterium experiments. In the fall of 2008, I searched for experiments that provided evidence of this 24 MeV/helium-4 ratio. I found only one experiment that members of the field accepted as convincing; it was performed by Michael McKubre's group at SRI International.

McKubre's 24 MeV value still didn't prove the existence of a D+D fusion reaction because many other energetic reaction products had been measured in LENR systems and the fusion idea didn't encompass other reaction products. However, the scientists who believed in room-temperature fusion ignored this fact and insisted that McKubre's experiment proved that it was real.

As the editor-in-chief of the 2011 John Wiley & Sons *Nuclear Energy Encyclopedia*, I invited several researchers to contribute chapters to the LENR section of the book. I had asked Pamela Mosier-Boss, at SPAWAR, for a chapter on nuclear products observed in LENRs because she had direct experience with a wide variety of related research. On the other hand, she also believed that the underlying process responsible for LENRs was fusion. I therefore asked her to limit her manuscript to experimental data rather than theory.

Mosier-Boss submitted an excellent draft to me in December 2010. However, she also wrote that McKubre's 2000 paper, presented at the Eighth International Conference on Cold Fusion (ICCF-8), and his and Hagelstein's 2004 paper presented to the Department of Energy (DOE) reported experimental evidence for the $d + d \rightarrow$ *Helium-4 + 23.8 MeV (heat)* idea. (McKubre, 2000; Hagelstein, 2004) I read the papers she referenced for McKubre's data, but both papers were sparse on details for the 24 MeV/helium-4 claim.

I asked Mosier-Boss to help me find the data that supported her assertion that McKubre had found the elusive 24 MeV (or, more precisely, 23.8 MeV). She couldn't find it. I asked her to check with SRI chemist Francis Tanzella, who had worked with McKubre on the experiments. Neither of them could answer my questions about the experiment, so I searched for it myself.

McKubre's 2000 paper listed reference #1 as TR-107843-V1, a 1998 Electric Power Research Institute (EPRI) technical report. I had tried to obtain a copy of this report several years earlier, but for nonmembers, EPRI charged $20,000. Now, fortuitously, the 379-page report was available free online, and it provided voluminous, detailed information about the heat-and-helium-4 experiment performed in 1994, identified as "M4."

I spent several weeks studying the data and experiment. The authors had explained the entire experimental process clearly, stated their assumptions, and described the variety of events that took place during the 1,700-hour experiment. However, the data was spread across dozens of pages, and no single graph gave a complete picture of the experiment.

The report provided measurements for helium concentration, current density, deuterium loading, excess heat, and excess energy. I began by making single-page graphs for each type of measurement. I then layered them together so I could see all of the data and events on a single graph. The graphs are available on the *New Energy Times* Web site.

When I was done, I understood the larger picture as well as the details of the experiment. The researchers who wrote the 1998 EPRI report had been meticulous. They showed all their data and clearly explained what they did. But the helium measurements in the 1998 EPRI report didn't match the helium measurements for the same experiment, as shown in the 2000 McKubre ICCF-8 paper.

I first contacted Tanzella, who had been unable to answer my and Mosier-Boss' initial questions. Tanzella had been part of the original experimental team for M4, was an author of the 1998 EPRI report and was an author of the 2000 paper. I told him that I had carefully studied M4 from the EPRI report but that I still had some questions. He agreed to meet with me at SRI headquarters, a short drive away.

Over the course of a two-and-a-half-hour meeting with him, I reviewed my understanding of the experiment. Tanzella was cordial, fully cooperative and open with me. He concurred with my analysis of every page of the M4 data in the EPRI report. I also identified two minor errors in the report, and again, he concurred with me.

A key assertion McKubre had made in his 2000 paper was that, during the electrolysis experiment, helium-4 dissolved into and desorbed out of palladium. During our meeting, Tanzella told me that helium did not dissolve into metal. I had already researched the behavior of helium in metals (Appendix D) and knew that Tanzella was correct.

When we got toward the end of my questions, I asked Tanzella to look at the pages of the 1998 EPRI report that discussed the helium measurements. I pointed out the discrepancies between the 1998 report of the helium measurements and the 2000 and 2004 publication of the same helium measurements. He read the relevant pages in front of me, and he could not explain the discrepancies; he was perplexed. Like me, he was at a loss for a scientific explanation for the changes. He seemed unaware of the changes that were made and said that I would have to ask McKubre because he had written the papers.

On Jan. 22, 2010, I sent a formal news inquiry to McKubre, and to Ellie Javadi and Lindsay Sheppard at the SRI public affairs office. None of them responded.

Unscientific Changes

In my investigation, I found that, between 2000 and 2007, McKubre gradually changed, added and deleted data points and values for experiment M4. McKubre made a total of 11 changes without any scientific explanation, most changes without notification.

As a result of his changes, McKubre made his data more closely match the D+D "cold fusion" hypothesis. In some cases, where McKubre did not invent values, he shifted the theoretical baseline to effectively bring his measurements closer to the value predicted by his fusion hypothesis. His 11 changes and the complete investigation record are available on the *New Energy Times* Web site.

During the March 21, 2010, press conference at the American Chemical Society meeting in San Francisco, California, I had the opportunity to ask McKubre about M4 in person. In his response to and denial of these allegations, he said that, sometime after 1998, he had found a single error in the 1998 report and that he had sent that error to EPRI, the sponsor of the research.

The day after the press conference, experiment M4 was expunged history. McKubre had mentioned experiment M4 in nearly every presentation in every conference presentation since 2000. McKubre's proof of electrolytic fusion disappeared silently at 4:29 p.m. on March 22, 2010, when he concluded his technical presentation at the meeting. For the first time in a decade, without any notice, let alone formal retraction, McKubre had failed to mention his $d + d \longrightarrow Helium\text{-}4 + 23.8$ $MeV\,(heat)$ claim, let alone say that experiment M4 provided its proof.

A few days later, when I contacted Brian Schimmoller, a senior communications manager at EPRI, he responded that the company had no record of receiving any such correction. "After checking," Schimmoller wrote, "there is no record in our system of any corrections or errata published for those reports, and the retired project manager tells us that he's not aware of any corrections or errata, either."

That project manager was Thomas Passell. Schimmoller also contacted Albert Machiels, the other manager on that project. Machiels was also not aware of any corrections or errata. McKubre continued to repeat the "correction" story to his colleagues, for example in a Dec. 10, 2011, e-mail to his colleagues.

"Neither I nor anyone else at SRI 'fabricated' data associated with M4 (or anything else that I am aware of)," McKubre wrote. "We had occasion to re-analyze those data, found an error in the EPRI report (a private document at that point), and communicated that promptly to the only person who was aware or cared (in the mid-to-late '90s) — the EPRI program manager. Later published reports in the open literature are, I believe, correct, and I have had no reason to doubt or refine them in the past dozen years."

From 2000 through his 2009 ICCF-15 presentation, all published reports by McKubre, including the paper he presented to DOE in 2004, contained the same uncorrected and unjustified changes, additions, and deletions.

Even if there was an error and McKubre had reported it to EPRI, that leaves 10 other manipulations for which McKubre never offered a defense. McKubre failed to identify a single factual or technical error in my investigation. In private e-mails to researchers in the field, McKubre denied that he had done anything wrong, and he made disparaging personal comments about me.

Distinction Between LENRs and Cold Fusion

Since 1989, scientists who did not accept the idea of room-temperature fusion had struggled to offer a scientific explanation for their views. It was easy to dismiss the controversial idea *a priori*, based on theoretical expectations. However, no critics had gotten close enough to the topic to learn exactly where the experimental flaws lay.

In the 2009 CBS *60 Minutes* program on the research, Richard Garwin was introduced as "one of the most respected physicists in the world." Garwin, for example, told CBS, "I think probably [McKubre] measures the input power wrong." He also told CBS that, to be scientifically credible, tabletop fusion devices had to be able to heat a cup of water for tea and do so reliably, 100 percent of the time. None of his points was defensible as a definitive scientific critique.

Instead, in August 2008, at the American Chemical Society national meeting, I had presented eight specific experimental discrepancies with LENR data that contradicted the D+D "cold fusion" idea. In October 2008, I had revealed that only McKubre's experiment had come close to the 24 MeV ratio.

Now, in January 2010, my M4 investigation had revealed the flaws in the final piece of the experimental data on which the believers had rested their case. On March 20, 2010, I published my formal rejection of the "cold fusion" idea and continued, as I had been since 2008, to distinguish between it and LENRs.

All Gone — Ashes

A thoughtful and poetic response to my investigation came from Lewis Larsen:

> For the majority of passive, honest experimentalists in the field of LENRs who never really questioned the doctrine laid down by the "cold fusion" subgroup for 20 years (or directly participated in any of their shenanigans), reading your investigation must have been a soul-searing experience. A deafening silence has prevailed within the field since it published.
>
> After believing fervently in their shared D+D fusion dream, toiling ceaselessly day and night in their labs and in their minds for all those years, enduring all the painful ridicule and scorn dished out by mainstream science and the media, quietly collecting and disseminating their good experimental data at conferences of the faithful, and keeping the flame alive — it is suddenly all gone with a stroke of your journalist's pen.
>
> Their familiar, long-standing champions and heroes (Fleischmann, McKubre and Hagelstein) and the dream of a simple, easy-to-understand D+D reaction are gone, now to be replaced by a complex, alien, abstract collection of Widom-Larsen concepts comprising arcane mathematics, physics, dynamic nucleosynthetic networks, and quantum chemistry that they simply do not understand.
>
> Overnight, they lost their anchors, their pantheon of heroes, and a cherished conceptual paradigm that they had

fought for and worked on all those years (in many cases with little or no funding except what they scraped out of their own pockets) in the hope of the field's eventual redemption, maybe some real money, and a shining place in the sun of science, and maybe a few of them even getting Nobel prizes. All gone. Ashes. Bitter feelings of despair and emptiness.

It is always hard to put oneself in another man's shoes, but I can imagine that their sense of loss and confusion at this moment could be profound.

McKubre's Hidden Helium

In the same 2000 paper in which McKubre first publicly reported experiment M4, he also publicly reported another helium-producing LENR experiment. He displayed helium measurements obtained from his group's replication of Leslie Case's 1998 deuterium-gas experiment. The paper contained a graph similar to the one below, with two major exceptions. McKubre and Peter Hagelstein also displayed the graph in their paper presented to DOE reviewers in 2004. (Hagelstein, 2004)

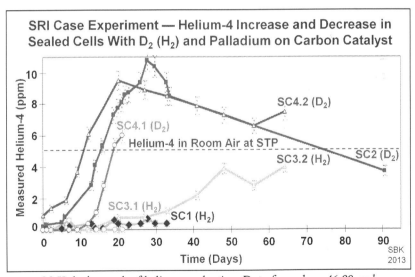

McKubre's graph of helium production. Data from days 46-90 and identification of hydrogen isotopes were never displayed by McKubre.

The first difference is that McKubre displayed the results only from Days 0-45. He did not inform readers that any of the experiments ran past Day 45. The second difference is that, although McKubre provided the labels for each run — for example, SC4.1 — he failed to identify which runs used deuterium gas and which ones used hydrogen gas. I know of no document in which McKubre has ever publicly distinguished the deuterium from the hydrogen runs.

In the text of his paper, McKubre wrote that the experiments "show no increase of helium-4 over long periods of time (including all cells operated with H_2 instead of D_2)." This is not true. I learned the truth by examining the data embedded in McKubre's slides.

In 2004, while I was writing *The Rebirth of Cold Fusion*, I told McKubre that I wanted to publish his helium results in my book but that I needed to adjust the aspect ratio of the image so it would fit. He sent me his PowerPoint slide with the graph. A few years later, I became curious about the poorly identified curves, and to my great surprise, I found that the raw experimental data were embedded in the slide. With a few clicks, I was able to modify the graph to see the data out to Day 90. The embedded data also clearly showed which runs had been performed with deuterium gas and which ones had been performed with normal hydrogen gas. I learned not only that the deuterium runs — the three curves that rise steeply — produced helium-4 but that one of the normal hydrogen gas runs (SC3.2) also produced helium-4. McKubre had concealed the data because it invalidated the D+D fusion idea.

Italians' Inconceivable Helium-4

Martin Fleischmann and an Italian group also observed the production of helium-4 in normal-hydrogen electrolysis experiments with palladium cathodes but didn't report it. In 2002, physicists Antonella De Ninno (b. 1961) and Antonio Frattolillo (b. 1958), at the Italian National Agency for New Technologies, Energy and the Environment in Frascati (ENEA), performed electrolysis experiments expecting that only the deuterium experiments would produce helium-4. They were assisted by Italian theoretical physicist Emilio Del Giudice.

De Ninno reported the results at the Twelfth International Conference on Emerging Nuclear Energy Systems, in Brussels, Belgium, in August 2005. At the conference, I presented a paper in which I gave a review of the research. At the time, I also helped De Ninno edit her group's paper for English grammar. I did not notice the light-water helium-4 at the time.

In May 2016, Larsen brought that light-water helium-4 result to my attention. When I discussed the results with De Ninno, however, she said that she and her co-authors determined that the helium-4 in the light-water experiments was a pre-existing contaminant. The group ran 11 experiments with palladium cathodes: six runs with heavy water and five control experiments with light water. The paper displays data for only three of these experiments.

In run #1 with heavy water, which obtained the required minimum loading of deuterium into the cathode, the group measured at least 1,000 times more total helium-4 atoms than in run #2, which also used heavy water but did not obtain the minimum loading ratio. In run #3 with light water, which did obtain the required minimum loading of hydrogen into the cathode, the group also measured at least 1,000 times more total helium-4 atoms than in run #2.

The group performed helium-4 measurements on light-water experiments and used those results to calibrate their system, based on their assumption that light water could not any produce helium-4. When, as described in their paper, they found unexpected high values of helium-4 atoms in the headspace of the control experiments, they initially assumed that the helium-4 came from atmospheric contamination, and they removed that helium by flushing the headspace with additional nitrogen. When they continued to measure additional helium-4 as the light-water electrolysis continued, they attributed that helium-4 found in the headspace to desorption from the light-water electrolyte.

De Ninno told me that the group believed that the light water, but not the heavy water, was contaminated with helium-4 absorbed from the atmosphere. According to the group, the heavy-water experiments were not contaminated with helium-4 because the heavy water was shipped and stored under argon. Additionally, when the researchers poured the heavy water out of the bottle, they always replaced the empty space with argon. The light water, De Ninno said, was never protected with argon and thus was contaminated.

I spoke with chemists Tanzella, Melvin Miles, and Mosier-Boss about how they stored heavy water for use in their LENR experiments. None of them stored it under argon. Nevertheless, they only measured helium-4 in some of their heavy-water experiments. Furthermore, the helium-4 measured by Miles and at SRI was always observed with the production of excess heat. Experiments without excess heat never produced helium-4. These factors contradict De Ninno's idea of contamination.

And, according to Lenntech, a Netherlands company involved in water treatment and air purification systems, helium, a noble gas, does not dissolve in water and is not a water contaminant. The company's Web site says that concentrations in seawater are no higher than 4-7 parts per trillion. The helium-4 levels observed in the Italian experiments were far larger.

Widom-Larsen Theory Explains Helium-4

The production of helium-4 in the Italian experiments can be explained by neutron capture on lithium, as described in equation #30 of Widom and Larsen's first paper. (Widom, 2006) The production of helium-4 in the SRI experiment can be explained by neutron capture on carbon (see diagram in Chapter 26).

The Widom-Larsen theory explains helium-4 production when either deuterium or hydrogen is available as a reactant. Furthermore, equation #32 in the 2006 Widom-Larsen paper shows how helium-4 can be not only be a product but also can be consumed as a reactant. This can explain the three runs measured at SRI that showed the decrease of helium-4 in the leak-proof cells.

Data that contradicted the SRI and Italian researchers' fusion hypothesis were staring the researchers in the face, yet their belief was so strong that they could not or would not recognize it.

CHAPTER 34

Their Own Worst Enemies

Years passed, and the believers saw their hopes for vindication fade. Despite having significant, credible experimental data, they became their own worst enemies.

The revelation in January 2010 that Michael McKubre had manipulated experimental results to manufacture evidence of fusion was undoubtedly discouraging to people in the LENR field, particularly those who were attached to the idea of room-temperature fusion.

This discouragement was evident in Scott Chubb's somber editorial in *Infinite Energy* magazine and McKubre's March 2010 presentation at the American Chemical Society (ACS) meeting, both dwelling on the 1989 conflict. (Chubb, 2010)

Doors Closing on "Cold Fusion"

In 2008 and 2009, I worked with Jan Marwan (b. 1972), a German electrochemist, to produce and edit two technical sourcebooks on LENRs that were published by the American Chemical Society and Oxford University Press. (Marwan, 2008; Marwan, 2009)

Marwan, 35, was a member of the second generation of scientists to enter the field. He had fresh ideas, and he did not hold any grudges from the 1989 conflict. In 2007, Marwan had organized the first LENR symposium at an ACS meeting in 18 years. The ACS publishing division thought the topic was worthy of attention and invited him to develop the sourcebooks.

Because he was new to the topic, Marwan asked me to help him co-edit and organize the 2009 symposium. Marwan and I worked well together: I knew the breadth of the field and the players, he knew electrochemistry, and we enjoyed a wonderful camaraderie. Michael Bernstein, Michael Woods, and Mark Sampson, in the ACS office of public affairs, offered to organize a press conference in conjunction with our 2009 session, and I advised them to identify the research primarily as LENRs, rather than "cold fusion," which they did.

At the time, Marwan was independent from the American scientists. All that ended when the second book, which included a chapter on the Widom-Larsen theory, published in December 2009. The scientists, rather than continuing to ignore Marwan, began to befriend him.

Aligned with his new allies, he began using the ACS platform to promote "cold fusion." In December 2009, without telling me, Marwan, the main organizer for the ACS LENR symposiums since 2007, deleted my abstract for the March 2010 symposium from the ACS computer system. When I learned about this and asked him for an explanation, he told me that there wasn't enough space in the program for my talk.

ACS also hosted a press conference in conjunction with Marwan's 2010 symposium and identified the subject as "cold fusion," rather than LENRs. I was the only journalist in the room, and I wondered whether it would be the world's last "cold fusion" press conference. Katharine Sanderson, writing for *Nature*, viewed the press conference remotely. In her *Nature* article, Sanderson cited my *New Energy Times* article "On the Reality of LENR and the Mythology of Cold Fusion." (Sanderson, 2010)

Apart from Sanderson's article and Ira Flatow's interview of McKubre on National Public Radio's *Science Friday*, there was no other news media attention. (Flatow, 2010) However, the ACS was berated by the American Physical Society (APS). James Riordon, the APS media relations, wrote a commentary titled "Chemists Taken In by Cold Fusion ... AGAIN!"

Marwan submitted to the ACS a proposal for a third LENR sourcebook. Many of the papers in his proposed book were about room-temperature fusion theories. ACS rejected it. He sent the proposal to John Wiley & Sons. Wiley rejected it. Marwan sent the proposal to the American Institute of Physics. AIP rejected it. Eventually, Marwan published the papers online in the fusion believers' *Journal of Condensed Matter Nuclear Science*. (Biberian, 2011)

Marwan was able to organize another ACS LENR symposium; it took place in March 2011 in Anaheim, California. It, like the 2010 ACS symposium, promoted the D+D "cold fusion" idea. The following year, ACS rejected Marwan and his colleagues' request to host another LENR symposium; his meeting series was finished.

LENR Products: Much More Than Helium-4

On July 7, 2008, Lewis Larsen had explained to me why the purported 24 MeV ratio between excess heat and helium-4 was irrelevant, even if it was measured accurately: There were many products other than helium-4. On July 30, 2010, I figured out how to convey this concept effectively.

I began with a diagram (shown below) that depicted the well-understood D+D fusion reaction. The only reactant, deuterium, is shown on the left. The reaction products — helium-3, neutrons, tritium, protons, helium-4 and gamma radiation — are shown on the right. The diagram also depicts the product pairing, the energies associated with each of the products and, on the far right of each pair, the probability of each branch occurring during fusion reactions. Two of the branches occur with slightly less than a 50% probability; the third branch occurs only very rarely.

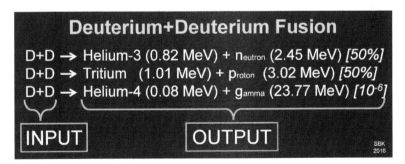

In the hypothetical "cold fusion" idea, *d + d —> Helium-4 + 23.8 MeV (heat)*, shown below, everything but helium-4 and 24 MeV of heat was eliminated. According to the concept, helium-3, tritium, neutrons, and protons were not produced at appreciable rates in LENRs. Normal product pairing (helium-3 and neutrons; tritium and protons), and associated energies with each product, according to the concept, also did not occur.

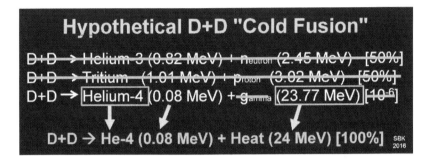

D+D "cold fusion" proponents ignored the fact that there were many discrepancies between the products of LENRs and the products of D+D fusion. In addition, they ignored at least five major categories of LENR output products, many of which were associated with MeV-level energies.

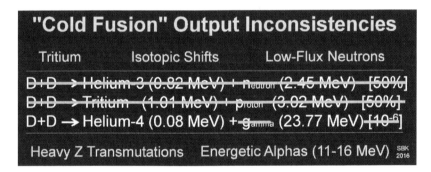

Also, D+D fusion, by definition, requires deuterium. LENRs do not, and therefore D+D fusion cannot be a necessary part of the nuclear reaction process. Furthermore, other reactants that are used in LENR systems appear to have a contributing role in the processes and products.

On Sept. 22, 2011, two months after I published these three diagrams on *New Energy Times*, Physicist Joseph Zawodny, at NASA's Langley Research Center, displayed the third diagram at a LENR workshop held at the NASA Glenn Research Center. He included it in his slide "You Can't Ignore Data," although he forgot to attribute the source.

On June 11, 2010, Edmund Storms submitted a manuscript, "Status of Cold Fusion (2010)" to the journal *Naturwissenschaften*. On Aug. 4, 2010, five days after I published the diagrams, Storms submitted a revised manuscript. (Storms, 2010) I didn't notice what Storms had done in his paper until I read it for the first time in March 2013.

First, he wrote that, except for helium-4, all other products in LENRs were a "side issue" and that their associated energies were trivial.

Second, he gave examples of control experiments in which researchers found no excess heat with normal hydrogen — and he omitted normal hydrogen experiments that did produce excess heat.

These errors required correction. However, Sven Thatje, the editor-in-chief of *Naturwissenschaften*, had, around June 2011, asked Storms to serve as the topic editor for all manuscripts submitted to the journal on the topic of LENRs. I contacted Thatje in 2013 and asked whether he would receive my comment and send it to peer review even though the Storms' paper was by then three years old. He agreed.

I received the first set of reviews on my submission a month later. Reviewer #3 wrote a single sentence, suggesting that it should be published as is: "Krivit's discussion and comments on helium-4 and light hydrogen are reasonable based on the referred experimental papers."

Reviewer #2 wrote a lot about me personally, criticized the Widom-Larsen theory even though I did not mention the theory, and wrote a lot about supposed evidence for the *d + d —> 4He +24 MeV (heat)* idea. Eventually, however, the reviewer argued directly with my comment.

Reviewer #1 wrote that I didn't provide evidence to support my claims of Storms' errors, and the reviewer made personal comments about me. Both reviewers recommended rejecting my manuscript. "My suggestion," Reviewer #2 wrote, "is that you simply reject the submission. Nothing will be lost to science." Thatje's cover letter to me that accompanied the reviews revealed his surprise:

> In the few years as editor-in-chief of *Naturwissenschaften*,
> I have come across scientific dispute, rivalries, emotionally
> fuelled "discussions" etc. However, the level of anger within

this community by far exceeds what I have experienced so far. Energy production and storage is likely the most important issue mankind is facing. Ignoring any scientific evidence, even if perceived as little evident as possible, should not be ignored. It would be good if both parties would come together and pulled the same wagon, wouldn't it?

Thatje, of course, had no idea about the extent of hidden factionalism and academic politics. I sent him my responses. A month later, I received a second round of comments from the reviewers. I pointed out to Thatje several factual inconsistencies in the comments from the reviewers. I also brought to his attention one reviewer's remark, as I wrote, that was "inappropriate and constitutes an abuse of the peer-review process." Ten days later, Thatje overrode reviewers #1 and #2 and accepted my comment for publication. (Krivit, 2013)

Tea and Cookies With Garwin

On Aug. 9, 2010, I sent a copy of my "Cold Fusion Is Neither" report to Richard Garwin and asked for his comments, questions and critique. "I will read this with great interest," Garwin replied. "I am moving in mid-September from my house of 55 years, so I will probably not get to this until October." Garwin lived in Westchester County, New York, I was heading there to visit family for the Thanksgiving holiday in November. I asked whether he would welcome a visit from me, and he agreed. I told him that I wanted to hear his perspective and reflections and offer mine to him.

On Nov. 30, 2010, I got off the Metro-North train in Scarsdale, New York, walked a few blocks and arrived in his new apartment at 3 p.m., the appointed time. Garwin, 82, greeted me at the door gruffly: "Give me your coat." He introduced me to his wife, Lois, and he offered me jasmine tea and cookies she had just baked. We discussed the LENR topics of excess heat, helium-4, isotopic shifts, neutron emissions, heavy-element transmutations, the field's history, and muon-catalyzed fusion. We had a delightful, engaging discussion; however, space constraints here permit only a brief summary of our conversation.

He first criticized something I wrote about him that I had taken out of context. He read aloud my quote of Robert Park's acknowledgment that there was something real about LENRs, but Garwin gave no indication that he was prepared to do the same.

Garwin handed me a list of nine items he wanted to discuss about my "Cold Fusion Is Neither" report. We began with McKubre's heat measurements and came to an agreement about our interpretations. Next, we discussed the matter of the isotopic shifts reported by Thomas Passell. (Chapter 18) Garwin disputed my statement that there was "no conventional explanation" for the data. I asked him what conventional explanation he could suggest. He stared silently at the data for several minutes.

"I can explain it," he finally said. I asked him how. He said that, even though the measured data was beyond the uncertainty range of the measurements, the researchers could have incorrectly measured their range of uncertainty. I chose not to argue with Garwin and, instead, felt disappointed at his imaginative denial of the experimental data.

But he made no blanket statement to broadly dismiss the research. Instead, he left room for the possibility that LENRs were a valid field of scientific research. "If it were true," Garwin said, "it would be *really very interesting*. And when something that you don't believe turns out to be true, well, you never know what the consequences would be."

After a couple of hours, Garwin and I were done with our respective questions. I thanked him for his time, interest, and critical questions. Garwin took my coat out of the closet and handed it back to me, subtly revealing a kind, grandfatherly side that I hadn't seen before.

"Have another cookie," Garwin said. "Have one for the road."

Blinded by Fusion

In 1999, David L. Goodstein, the vice provost of the California Institute of Technology (Caltech), sent a letter to Eugene Mallove. Mallove had provided Goodstein a copy of his new video *Cold Fusion:*

Fire From Water, along with a cover letter. "I'm afraid both the letter and the video tended to confirm my sense that the cold fusion community is its own worst enemy," Goodstein wrote.

A decade earlier, Goodstein had watched his Caltech colleagues Nathan Lewis and Steven Koonin recklessly and viciously, respectively, attack Martin Fleischmann and Stanley Pons at the March 1989 APS meeting in Baltimore, Maryland. (Vol. 2, *Fusion Fiasco*). From time to time, Goodstein wrote critically — but respectfully — about the subject. The responses he received from test-tube fusion enthusiasts, Goodstein wrote, had been "mostly vitriolic abuse."

I, too, have seen many examples, public and private, of such hostilities from tabletop-fusion-believing scientists and their supporters. One stands out. In March 2006, at another APS meeting in Baltimore, Koonin was a keynote speaker at an evening lecture attended by several hundred people. He worked for British Petroleum at the time. When the time came for questions, Scott Chubb went to the microphone and urged Koonin to come to Chubb's "cold fusion" session at the APS meeting. Here's what happened:

> Steven Koonin: BP pays me to pay attention to promising technologies, and we have money to invest. They pay me to make judgments about that. For better or worse, I've decided that that's not something I want to invest in right now —
>
> Scott Chubb: You might want to look at the proceedings of ICCF-11 and ICCF-10.
>
> Steven Koonin: — and if I make the wrong call, maybe I'll lose my job.
>
> Scott Chubb: And I hope you do.

Moments later, while still at the microphone, Chubb publicly apologized to Koonin.

As the years went by, the scientists who believed in room-temperature fusion didn't seem to recognize, or want to recognize, that their insistence on the increasingly untenable idea reinforced the increasingly insurmountable stigma against them and their research.

Four months before the February 2011 ICCF-16 meeting in

Chennai, India, two of India's pioneers in helping their country gain nuclear weapons capabilities in 1974 took opposing positions on "cold fusion." One was Rajagopala Chidambaram, who, in 2011, was the principal scientific adviser to the government of India. The other was Mahadeva Srinivasan, who was retired and the chairman of ICCF-16. Their disagreements went back 20 years to when they both worked at the Bhabha Atomic Research Centre (BARC).

From 1989 through the end of 1990, Srinivasan had been the director of the Neutron Physics Division and the assistant director of the Physics Group. Chidambaram had been the director of the Physics Group. As discussed in Chapter 7, both Srinivasan and Padmanabha Krishnagopala Iyengar, the director of BARC at the time, enthusiastically supported a dozen teams at BARC to explore the work. Most of the teams found clear nuclear evidence (neutrons and tritium) in palladium-deuterium electrolytic cells as well as in deuterium gas cells.

But Chidambaram rejected all the data and the entire subject. The day he took Iyengar's place as the director of BARC, Chidambaram terminated all funding for the research. Srinivasan, as he told me, continued with his work. But in order to present his results at a conference, he had to get Chidambaram's permission.

During a staff meeting with a dozen associate directors of BARC, Srinivasan presented his results. When Srinivasan was finished, Chidambaram gave his approval but insulted him in front of the group: "Srinivasan, you are a loner."

Fast-forward to October 2010. Widom, Larsen and their Indian colleague Yogendra N. Srivastava had just published a (non-fusion) theory paper in *Pramana — Journal of Physics*, which is associated with the Indian National Science Academy and Indian Physics Association. Chidambaram must have helped the authors because they thanked him in the acknowledgements.

On Oct. 12, 2010, Kumar Chellappan wrote an article in the *Deccan Chronicle* about India's fast-breeder technology and LENRs. When Chellappan asked Chidambaram about the research, calling it "cold fusion," Chidambaram dismissed it with contempt. "I'm not convinced about the feasibility of the cold fusion. I do not intend to take up cold fusion experiments. Anybody interested in pursuing the same is welcome to do it on their own," Chidambaram said. He undoubtedly understood the distinction between LENRs and "cold fusion."

On Jan. 25, 2011, with Srinivasan's assistance, Balasubramanian Sivakumar wrote a story in the *Times of India* about the coming ICCF-16 conference in Chennai. The story did not mention the term "LENR" or the recent paper in the prestigious *Pramana* journal.

On Friday, Feb. 4, 2011, Srinivasan stopped by the Gandhinagar Guest House in Chennai, India, which he had graciously arranged for me. I had arrived from the U.S. very early that morning. While we sipped tea in the courtyard, Srinivasan told me that a major news story would appear in the *Times of India* about him and the conference. I told him that it was too bad that Chidambaram seemed so negative in the *Deccan Chronicle* story. "Maybe Chidambaram would be supportive if you discussed the research without the presumption of fusion," I said.

In the years I had known Srinivasan, I couldn't remember a time when the easy-going, gentle scientist didn't have a smile on his face. In an instant, Srinivasan was a different man. I saw the tension in his face and heard the anger in his voice. He told me how Chidambaram had shamed him. Srinivasan told me that it would make no difference what he called the research. The past was very much present with Srinivasan, and he could not see how promoting his conference as "cold fusion" to the news media wasn't useful.

On Feb. 7, 2011, Srinivasan held a press conference at ICCF-16. Joining him was S. Kailas, president of the Indian Physics Association, McKubre and Nagel. Their primary message was that the stigma attached to the subject was the biggest impediment to progress and was responsible for its perception as pathological science. Yet they didn't see that their continued promotion of "cold fusion" as the underlying mechanism for LENRs and their continued insistence on the use of that term were directly responsible for perpetuating the stigma.

Nevertheless, they flew the flag of "cold fusion" boldly. On Feb. 9, 2011, Srinivasan, Melich and Nagel presented "A Colloquium on Cold Fusion" at a nearby university, the Indian Institute of Technology at Madras. Nagel's talk was "Cold Fusion: An Emerging Energy Technology?" Melich moderated a panel discussion called "Cold Fusion:

Future Prospects."

When Nagel displayed a slide showing more than a dozen proposed theories, the Widom-Larsen theory was conspicuously absent. During the question-and-answer session, I asked Nagel why he had excluded the Widom-Larsen theory from his and Melich's presentations. "They did not come to our [ICCF-14] conference," Nagel said. "That's why they don't appear on that slide." Nagel also complained about Robert Park. "Park still won't acknowledge there is new science," Nagel said. In fact, Park had acknowledged LENRs three times by then, once in 2006 while Nagel was in the room at the Defense Threat Reduction Agency meeting.

In 2006, Nagel had pushed me hard to look at the Widom-Larsen theory. Now he had come full circle. After Nagel took on Melich as his co-organizer for the ICCF-14 meeting in 2008, Nagel shifted his priorities and reverted to "cold fusion."

After the colloquium, I talked with Srinivasan. He was disappointed and angry. "Not a single professor from the Physics Department came. I invited them all."

The Never-Ending Wait for Vindication

By 2011, LENR scientists had begun to talk openly about the field dying. But the field wasn't dying; the scientists and their colleagues were, and so were their opportunities.

Notable members of the field who had died in 2008-2011 included James Patterson, Andrei Lipson, John Alfred Thompson, Leslie Case, Scott Chubb, P.K. Iyengar, Yan Kucherov, and Talbot Chubb. Notable members who died in 2012-2016 included Martin Fleischmann, Hal Fox, John Bockris, Sergio Focardi, Robert W. Bass, Emilio del Giudice, and John Dash.

The battle faced by these unsung researchers was larger than most of them recognized. Their failure to recognize the distinction between "cold fusion" and LENRs was only a part of the problem. The opposition they faced was not just against reluctant colleagues in the scientific community; it was bigger. They were the first to challenge the dominant paradigm, which held as one of its foundations that nuclear reactions could not take place under these moderate conditions.

During off-hours and in back rooms, many of these scientists worked diligently and quietly to collect key experimental evidence despite condemnation and mockery from their scientific peers and a general lack of funding. Many of them had made tremendous personal and professional sacrifices. Despite their fixation on fusion, they had performed heroic research.

For years, critics had demanded a viable theory before they would accept the experimental data as valid. This demand was unfair and inconsistent with the scientific method. Widom and Larsen were able to develop their theory of LENRs, thanks to the excellent data collected by the experimentalists, but the cold fusioneers rejected the newcomers' theoretical ideas.

Letting go of "cold fusion," conceptually or even just in name, was not negotiable for them. Retention of the term was directly tied to what they wanted most: vindication. As door after door closed, they began blaming their problem on what they called a reputation trap. LENRs were not a reputation trap; "cold fusion" was. Now things were about to get worse.

Around the time of the ICCF-16 meeting in February 2011, well-known scientists in the LENR field, including Nobel Laureate Brian Josephson, 71, embraced Andrea Rossi, a convicted scam artist who claimed he had developed a working 1,000 Watt thermal LENR reactor. In 22 years, nobody had developed the science to demonstrate, with 100% reproducibility, 1 Watt of excess heat in the laboratory. The jump was nearly beyond belief for many people, but not for key scientists in the field who believed in fusion.

Melich, 71, Storms, 80, McKubre, 63, Nagel, 73, and Srinivasan, 74, all qualified scientists, lent their names and the reputation of the field to Rossi, a man who was dishonest and disreputable and who had already been convicted of fraud.

CHAPTER 35

The Big Con

M ichael McKubre gave a lecture at SRI International, in Menlo Park, California, on Oct. 11, 2011, and described Andrea Rossi as "a dodgy character who has had trouble with the law." Despite McKubre's knowledge of Rossi's past, McKubre suggested that Rossi had created a machine that produced heat at a level 1 million times greater than typical LENR experiments.

"People that I know and trust," McKubre said, "have stood in front of Rossi's reactor and come away convinced that it really is doing, more or less, what Rossi claims. This includes my ex-program manager [Bob Nowak] at DARPA ... [and another] good friend of mine."

McKubre displayed a graph of heat measurements made from Rossi's apparatus. What McKubre didn't disclose was that the graph was created by Horace Heffner, a LENR enthusiast who identifies himself as an "unqualified amateur." Heffner, in turn, got the data from a news story written by journalist Mats Lewan for the prominent Swedish technology newspaper *Ny Teknik*. Lewan, in turn, obtained the data by measuring it himself, by hand, on behalf of Rossi.

The story began on Sept. 25, 2009, in Bedford, New Hampshire. Rossi, an Italian national with a master's degree in philosophy and a criminal record for fraud and other crimes in Italy, showed Michael Melich, one of McKubre's close friends, a demonstration of his nickel-hydrogen gas-based apparatus.

According to a story pitch sent to the *Christian Science Monitor* newspaper by Marianne Macy, Melich's wife, Melich had "been involved with testing the Rossi reactor ... for the U.S. government since 2008." Melich had indeed been involved but not on behalf of the government.

On a tip from Pamela Mosier-Boss on June 29, 2011, I contacted Tony Tether, a former director of DARPA, who confirmed that he had also seen a Rossi demonstration in 2009. "If it was a hoax," Tether said, "it was a damn good one."

I saw one of Rossi's demonstrations on June 14, 2011, when I travelled to Bologna, Italy, to interview Rossi. Rossi had hired a plumber, Carlo Leonardi, to solder some copper pipes together. The apparatus was about 3 feet long and consisted of one horizontal section and one vertical section. At one end, Rossi connected a small peristaltic pump. The pump drew water from a small plastic jug and fed it into the assembly. An electric heater was attached to the outside of the center portion of the assembly. The key to his unprecedented results, Rossi said, was a nuclear reaction between hydrogen gas and a secret nickel-based catalyst that he put in the center of the apparatus.

The proof for his claim of kilowatts of excess heat, he said, was the steam it produced. He asserted that his apparatus created so much steam that all of the liquid water in a jug sitting on the floor had been totally vaporized into steam. He claimed that he obtained a rate of several thousand Watts from electrical input power of just a few hundred Watts coming from the wall plug.

Had all that water been truly vaporized, his calculations would, indeed, have supported his claim. Rossi told people, including me, that we could take his word that all the water was being vaporized into steam, that the steam was completely dry, free of water droplets, and that he had hired an expert who used the correct instrument to confirm the dryness of the steam. After I filmed his demonstration and uploaded my video to YouTube, *New Energy Times* readers, several of whom had degrees in mechanical engineering, figured out Rossi's sleight of hand. I published their analyses in a lengthy report on July 30, 2011.

Mechanical engineers could see the glaring problem that the physicists and chemists who looked at Rossi's device had failed to notice. When water vaporizes to steam, it expands 1,600 times in volume. For example, on July 8, 2011, I spoke with Hanno Essén, a Swedish theoretical physicist at the Royal Institute of Technology, and the president of the Swedish Skeptics Society, an organization that is supposed to debunk pseudoscience. He had seen a demonstration of

Rossi's apparatus on March 29, 2011. Here's part of our conversation:

> KRIVIT: Are you aware that many people have calculated that the exit speed [of the steam] should be 50 to 100 km/h?
>
> ESSÉN: Sounds like a lot. No, I wasn't aware of that. There was not a huge water flow going in, so, intuitively, it was consistent with the amount of steam coming out.
>
> KRIVIT: People who have made the speed calculations have written that the speed coming out of a 10mm hose of that length should be really fast; it should have [audible] sound and [visible] momentum. It doesn't seem like you saw anything like that.
>
> ESSÉN: So where did it go then?
>
> KRIVIT: Probably down the hose into the drain.
>
> ESSÉN: Down the drain? There was no drain.
>
> KRIVIT: Where was the hose going when you saw it?
>
> ESSÉN: In the room next [door], going into a hole in the wall.

When I was in Bologna, Rossi showed me his apparatus, but he didn't volunteer to show me any steam coming from it. I had to ask him three times. Finally, when he pulled the hose out of the wall, rather than massive amounts of steam coming out at high speed, my video showed tiny, gentle puffs, no greater than what would be produced by an electric tea kettle. None of the scientists who went to see Rossi's apparatus saw hissing clouds of superheated steam coming out of the apparatus; they had no experience with steam and didn't know what to look for.

Initially, I didn't know anything about steam, either. I didn't figure out what Rossi was doing. My readers told me what to look for, and they figured it out based on my video. Rossi had an opaque black rubber hose attached to the outlet of the apparatus. He was feeding a mixture of water and steam into the hose, through a passageway to the adjacent bathroom, and down a drain hole. It would've taken only seconds to undo the hose clamp and visually confirm that no water appeared at the apparatus outlet. None of the visiting scientists who endorsed Rossi's claim performed that check.

As public confidence in Rossi's claims grew, he attracted the interest of a few wealthy investors and companies. One potential investor was Robert King. Brian Ahern, who had earned his Ph.D. in materials science from MIT was slated to perform the due diligence testing for King. Ahern explained to me what happened. "Rossi agreed to perform a demonstration for King," Ahern wrote, "if King placed $1.5 million in escrow and released the funds to Rossi after a successful test. However, after King agreed to the terms, Rossi increased the price to $15 million. King walked away."

Investor John Preston, president and chief executive officer of Continuum Energy Technologies LLC, also came to Bologna, Italy, to see a Rossi demonstration. He brought with him Michael Nelson, a propulsion engineer from NASA's Marshall Space Flight Center, and Jim Dunn, a past director of the NASA Northeast Regional Technology Transfer Center.

The people who went to Bologna with Preston were not permitted to tell me details of what happened because they had signed nondisclosure agreements. However, one of them spoke with me under the condition that I not identify him. The group saw problems from the moment they arrived. "Rossi changed the game totally, the test plan, the device, everything," the witness said. "There was nothing there that we had agreed on. He had a 30-liter reservoir in there, and he wouldn't even let us see what was in the box or weigh the box." By September, Rossi had changed his output measurement from a steam-based calculation to water-based mass-flow calorimetry.

The Sept. 5, 2011, demonstration was inconclusive; Rossi's device sprang a leak. The team came back the next day to try again. Throughout both days, Rossi was making demands for his first progress payment. The deal had been that, if Rossi could demonstrate a device producing energy with no input for two hours, Preston would pay Rossi $500,000. Preston gave him nothing.

The Sept. 6, 2011, demonstration was inconclusive, again because there was no outflow of steam or water. When Dunn asked Rossi whether his device had an internal reservoir, Rossi became enraged. Preston's team decided they had had seen enough and left. Dunn offered to come back in a few days to give Rossi time to fix the problem. Rossi, however, declined Dunn's offer and said he was "too busy."

A day later, Lewan showed up and filmed the operation of Rossi's device. As Lewan told me later, he had no idea what had transpired the day before. But Lewan saw an outflow and, he said, excess heat.

For several years, Rossi repeatedly claimed to have a working one megawatt LENR thermal generator, courted new suitors, and dropped their names to Lewan and other excited bloggers. Rossi asked for huge payouts for the privilege of performing due diligence on his apparatus. He failed to deliver tangible evidence of any such system.

Rossi couldn't lose: Either he got his money, or the mark in his confidence game walked away silently, bound by the nondisclosure agreement. Rossi had burned through at least 10 opportunities with investors, companies or institutions by the time he made a deal with Thomas Francis Darden II, the founder and chief executive officer of Cherokee Investment Partners, LLC, in Raleigh, North Carolina.

In August 2013, after performing a cursory test, Darden and 13 investors put a total of $11,555,050 into developing Rossi's pipe assembly for what would have been a multibillion-dollar energy technology. That same month, according to public records, Rossi purchased 10 residential real estate properties in Miami Beach, Florida, for a total of $2,172,000. Within two months, Rossi purchased another $893,000 worth of Florida property.

On April 7, 2016, after Rossi conducted a year-long test, Darden conceded in a press release that he had "worked for over three years to substantiate the results claimed by Mr. Rossi from the E-Cat technology — all without success." Two days earlier, Rossi had sued Industrial Heat and other parties in federal court in Miami, accusing the defendants of fraud and attempting to steal his intellectual property.

Behind the Curtain and in Plain Sight

Rossi had two key collaborators whom few people seemed to notice. The first was Melich, a member of Rossi's Board of Advisers who was also employed by the U.S. Navy. On Jan. 17, 2011, after Rossi gave his first public demonstration, Melich wrote the first public technical report on Rossi's demonstration. In his brief report, Melich claimed that Rossi's apparatus produced 12 kilowatts of excess heat. Melich convinced Jed Rothwell to publish his report on Rothwell's Web site without attribution, showing only Rothwell's name as editor.

On Feb. 6, 2011, Melich gave a 10-minute presentation at the Sixteenth International Conference on Cold Fusion (ICCF-16) on the Rossi demonstration. Melich told the attendees that he had "responsibilities as a federal employee" to keep an eye on the LENR field.

Melich also said that he knew "a fair amount about what Dr. Rossi has done." Melich disregarded all of the technical problems that Francesco Celani had just explained to the audience. Instead, Melich told the audience that he thought there was a 10% chance that Rossi's claims were valid and that the LENR community should help Rossi.

Rossi's second collaborator was technology reporter Mats Lewan, who circumvented journalistic protocol and journalism ethics on multiple occasions. He recorded data for Rossi on April 19 and 28, and Oct. 6, 2011, wrote technical reports for Rossi based on those data, then wrote news stories based on his technical reports.

When Lewan reported on Rossi's Oct. 6, 2011, test, he wrote that Rossi's device "ran in a completely stable self-sustained mode for over three hours." Lewan explained that Rossi had pre-heated the device at a rate of 2.7 kilowatts. Lewan failed to mention the large diesel generator outside, as shown in images from a similar test that took place on Oct. 28, 2011. Lewan explained that the heating power was cut after four hours. Lewan wrote that, by "putting [his] hand on the insulated enclosure, [he] could clearly feel the water boiling." Based on his sense of touch, along with the instruments Lewan brought with him and used, he calculated that the power output rate for the next three hours was 2 to 3 kilowatts. This is the same rate as the power input and suggests that, rather than inventing the world's most innovative energy device, Rossi had designed an efficient heat-storage process. In his calculations, Lewan ignored the energy input during the pre-heating phase.

Lewan consistently turned a blind eye to Rossi's many factual inconsistencies. The best evidence comes from one of the last posts from former Rossi supporter Paul Burns, who wrote under the pen name Paul Story, on his blog eCatNews.

"October is dead," Burns wrote on Nov. 2, 2012, "taking with it any hope I had that Andrea Rossi would deliver on yet another claim. For some time now, it has been obvious to most [people] that Rossi cannot be trusted. Even his supporters admit he lies [while] they excuse him for one reason or another. ... I can no longer ignore the fact that Andrea Rossi is acting like a fraudster."

A Collective Hallucination

The red flags were not obvious to some of the most well-known LENR scientists. The responses from them indicate that further credible, significant LENR research from them is unlikely.

On Feb. 11, 2011, Robert Duncan, then the vice chancellor for research at the University of Missouri, announced at a lecture his intention to purchase two of Andrea Rossi's Energy Catalyzers. "When they show up at Home Depot," Duncan said, "I'm going to go out there with my credit card." Duncan was not a LENR researcher. He gained instant visibility in the field in 2009, however, when he endorsed the heat claims of Energetics Technologies in a CBS-TV *60 Minutes* program about LENRs, calling it "cold fusion."

On or about May 15, 2011, David Nagel gave a presentation at the 15th International Conference on Emerging Nuclear Energy Systems, in San Francisco, California. He said that Rossi's apparatus had heated a factory in 2007 for 24 hours per day for six months and, as a result, had reduced the factory's electricity bill by 90%. Nagel also said that Rossi had 97 devices operating in four countries. There was no evidence whatsoever for this.

When McKubre, Peter Hagelstein, Nagel and George Miley were interviewed in the fall of 2011 for a case study of emerging scientific fields for the U.S. Intelligence Advanced Research Projects Activity, they told the government that Rossi's device was a real, working, commercial LENR energy device, according to documents I received from the Office of the Director of National Intelligence under the Freedom of Information Act.

On Aug. 2, 2011, Edmund Storms, one of the most prolific Internet "cold fusion" promoters, wrote an e-mail to colleagues and enthusiasts, anticipating the commercial viability of Rossi's apparatus.

"The entire field of cold fusion will grow rapidly and provide jobs for those of us who have slaved in the dark for so long," Storms wrote. "Personally, I hope [Rossi's apparatus] works as claimed, and I will do everything I can to promote the idea and work to make it better. Kicking Rossi at this stage just because the claim is not fully proven seems counter-productive to everyone. We desperately need the Rossi claim to be real."

Perhaps the most illuminating statement came from Mahadeva Srinivasan, the chairman of the ICCF-16 conference, on Dec. 21, 2011, after his longtime friend, colleague, and former chairman of the Indian Atomic Energy Commission P.K. Iyengar died.

"In the recent months, [Iyengar] was very happy to learn of the development of the nickel-hydrogen Rossi reactor and the imminent commercialization of cold fusion/low energy nuclear reactions," Srinivasan said. "Iyengar's bold and far-sighted stand on this controversial subject stands vindicated."

Desperation, Despair and Dismay

On Oct. 20, 2011, at the World Green Energy Symposium, in Philadelphia, Pennsylvania, George Miley, 78, started taking liberties uncharacteristic of someone with a long history of respectable scientific research. A person who attended the symposium filmed Miley's presentation and put it on the Internet. "At the moment," Miley said, "we can run continuously at levels of a few hundred watts." On Oct. 24, 2011, Miley sent an e-mail to Dennis Cravens, confirming his claim. "Yes," Miley wrote, "we are getting some good gas-loading results at the 100s of Watt level!!"

As with Rossi's claim, Miley's claim, if true, would have turned the world's energy industry on its head. Miley was a real scientist. I had no choice but to investigate. Soon after I began my investigation, I called "Sam," my anonymous sponsor, to ask for another check. He had heard about my investigation and told me that he wasn't sure if he could continue funding me. "Steven," Sam said, "do you really have to do this investigation?" I knew that Sam and George went back many years. I also knew that Sam had funded Miley's LENR research. For all I knew, he was still funding him.

I remember the phone call taking place on a Friday night sometime in November. Sam had been very patient for several years while some of the most well-known American LENR researchers had put pressure on him to stop funding me. I was expecting his support to end at some point, but I wasn't sure if I wanted it to end now. He told me I needed to make a choice. I asked whether I could sleep on it.

Saturday morning, I called Sam and told him that I would go through with the investigation and I would give Miley full opportunity to present his side. Sam grumbled and wasn't pleased with my decision. Nevertheless, I thanked him for his years of support. I knew then that it was time to begin writing this book.

After I exchanged many e-mails with Miley, we got to his key experimental data. It showed a meager level of 8 Watts of excess heat for only 100 seconds. I felt discouraged. As I struggled to understand, I asked him why he thought it was acceptable to make unsupported claims at a science conference. "What you do in a meeting," Miley said, "is different from what you do in a peer-reviewed journal publication."

Science Shutdown at SPAWAR

On Oct. 28, 2011, Rossi demonstrated his device to curious onlookers and a few journalists in his rented garage in Bologna, Italy. Peter Svensson, a reporter with the Associated Press (AP), was there, as he told me, on his own dime. The AP didn't let him run a story on Rossi.

A Utah man named Sterling Allan, who ran a "free-energy" Web site, joined Svensson and Lewan in Bologna. On Nov. 2, 2011, reporter John Brandon published a brief story on the Rossi demonstration in *Fox News*. Brandon obtained much of his information from Allan, and it contained many errors. Brandon also spoke with Rossi, who told him that Paul Swanson, a researcher at SPAWAR, had observed a demonstration. That part was true. But Rossi twisted it and told Brandon that Swanson could "vouch for the demo," which was not true.

Four hours after the *Fox News* story published, Admiral Patrick H. Brady, commander of SPAWAR headquarters, sent an e-mail to the top leadership at SPAWAR-Pacific. The subject line encapsulated the body of the message "Foxnews.com / cold fusion / SPAWAR SSC PAC." When Brady's orders were passed down to Mosier-Boss the following day, according to documents obtained by FOIA requests, she received the following e-mail from her commanding officer, Stephen Russell:

> You are directed to cease all work related to Low Energy Nuclear Reactions (LENR) as part of your SSC Pacific employment. This includes the continued prohibition of experimentation that has the potential for ionizing radiation or use of radiological materials, any related LENR control experimentation, and any LENR theoretical or simulation work. You should also recall any pending proposals from potential sponsors, as this work will no longer be pursued at SSC Pacific. Furthermore, this direction also includes no further publishing of any results from your LENR work as a SSC Pacific employee.

Russell also wrote that he would contact Mosier-Boss' Defense Threat Reduction Agency (DTRA) sponsor and explain why she would be unable to complete her work and that SPAWAR would seek to return any remaining funds. On Nov. 21, Beel sent a letter to Larry Forsley, and advised his company, JWK International Corp, that SPAWAR was unilaterally terminating its cooperative research and development agreement with JWK and returning unused funds.

The SPAWAR group had published 25 LENR papers in peer-reviewed journals, more than any other individual or group in the U.S. Worldwide, only one group, in Japan, even came close and tied their publishing record in LENRs. In two decades, the SPAWAR researchers made a wide variety of contributions to the body of experimental research in LENRs. As a result of the media attention about the illegitimate Rossi claims, promoted in large part by certain fusion believers, legitimate LENR research in the U.S. ground to a halt.

CHAPTER 36

From Science to Engineering

Many experimental reports described in this book show that LENRs suggest the possibility of providing a practical energy source. The Widom-Larsen theory also supports this possibility. This chapter provides technical details to understand the potential energy capacity of LENRs. Readers who do not want these details may want to skip to the next chapter.

A common misconception of the Widom-Larsen theory is that, because the reactions begin with "weak nuclear interactions," LENRs are intrinsically a poor source of energy. However, Widom and Larsen's 2007 paper "Theoretical Standard Model Rates of Proton-to-Neutron Conversions Near Metallic Hydride Surfaces" included calculations that show LENRs can potentially produce substantial amounts of energy. (Widom, 2007).

In a 2008 article, Larsen began with the basic fuel required (either hydrogen or deuterium) to produce the neutrons: A heavy-mass surface plasmon electron of at least 0.78 MeV must be created. (Larsen, Oct. 12, 2008)

Only one heavy-mass surface plasmon electron is required to produce a neutron from each proton. If the fuel is deuterium (or tritium), then the heavy-mass electron reacts directly with a deuteron (or triton) and two (or three) neutrons are produced, respectively. Therefore, in the case of deuterium, the energy cost to create neutrons is reduced to 0.39 MeV per neutron, and for tritium, it is 0.26 MeV per neutron.

Once a LENR reaction network starts, a variety of nuclear processes, as discussed in Chapter 26, releases nuclear binding energy. Each beta decay, for example, may release from a few keVs up to about 20 MeV of heat.

"In numerous experiments involving well-performing electrolytic LENR cells," Larsen wrote, "with either light or heavy water, ultralow-energy neutron production rates on the order of 1×10^{11} to as high as 1×10^{16} per second have been measured with reasonable precision. These values for reaction rates hold true whether the LENR transmutation products are in the form of helium-4 or the complex arrays of different-mass isotopic products such as those found in the experiments of [George] Miley and [Tadahiko] Mizuno."

Larsen's simplified calculations assume an idealized system with perfect efficiency and establish a theoretical upper bound on a potential energy release from a deuterium-based LENR device with a working surface area of 1 cm^2. In the example, Larsen's calculations were based on an assumed rate of about 10^{14} reactions/cm^2/sec, each reaction releasing 26.9 MeV of heat. After subtracting energy needed to produce neutrons, the theoretical upper limit on total heat production is 34 times the input power. Larsen calculated the surface power density of this hypothetical LENR-based device as 428 Watts per cm^2.

Scaling up total surface area of a LENR-based thermal source from a square centimeter to a square meter — a 10,000-fold increase — would result in a 4.28 MW power source. This is about 4,000 times the amount of power generated by an ideal solar panel of the same size.

Although total power output is important, the real measure of value is energy density. The energy density of LENRs is unknown at this point. However, being nuclear, LENR processes have intrinsic energy densities that are at least a million times larger than chemical energy sources.

Self-Sustaining Thermal Feedback

In 1993, Fleischmann and Pons reported a phenomenon in their experiments that they called a positive thermal feedback effect. (Fleischmann and Pons, 1993) They found that, as the temperature of the experiments got hotter, the rate of temperature increase accelerated.

Fleischmann told Melvin Miles, one of his close collaborators, that the cell temperature needs to be above 60° C in order for the positive thermal feedback to produce large excess-heat effects.

According to Miles, few researchers in the field seemed aware of the benefit of operating with cell temperatures above 60° C. Most LENR researchers performing electrolytic experiments use mass-flow calorimetry rather than isoperibolic calorimetry, which Fleischmann and Pons used. Although the calculations for mass-flow calorimetry are much simpler and easier than for isoperibolic calorimetry, the downside is that mass-flow calorimetry typically maintains cell temperature near room temperature, thereby cooling the cell. For this reason, Fleischmann and Pons avoided mass-flow calorimetry. Giuliano Mengoli, an Italian physicist, also reported great success with the electrolytic experiments close to the boiling temperature of water. (Mengoli, 1998)

Although running cells at higher temperatures appears to have helped these researchers, the more fundamental issue has been the precise and as-yet-unidentified material and metallurgical properties that are essential to reliably produce substantial amounts of excess heat.

After Fleischmann and Pons left Utah, they went to work in the Toyota-sponsored laboratory Institut Minoru de Recherche Avancée (IMRA), near Nice, France. Few LENR researchers since Fleischmann and Pons have achieved the level of control and magnitude of excess heat that the pair did while they worked at IMRA.

However, by 1996, Pons and Fleischmann appear to have ended their collaboration. In a paper that year, perhaps his last at the IMRA lab, Pons lists two colleagues as co-authors, Thierry Roulette and Jeanne Roulette, but not Fleischmann.

Roulette, Roulette and Pons reported seven runs in a series of palladium-deuterium electrolysis experiments; two of them were exemplary. Run #3 produced 101 Watts of excess heat for 30 days, giving 294 MJ of excess energy. Run #4 produced 17 Watts of excess heat for 70 days, giving 102 MJ of excess energy. (Roulette, 1996)

The power, energy, and duration of these experiments fall within the same order of magnitude as the Francesco Piantelli group's nickel-hydrogen gas experiments. As discussed in Chapter 13, their two best experiments produced 18 and 72 Watts of excess heat and 600 and 900 MJ of integrated energy, over 319 and 278 days, respectively.

As discussed in earlier chapters of this book, many other successful nickel-hydrogen LENR experiments have been reported, mostly in the 1990s. An excellent review of this work is provided in a 1998 paper by Mengoli and his colleagues. (Mengoli, 1998)

By comparison, the last set of Michael McKubre's palladium-deuterium electrolytic experiments, performed in 2006-07 at SRI International, produced only a meager 0.7 Watts of excess heat. McKubre resigned from SRI a decade later, in 2016. (McKubre, 2008)

According to Larsen, the Piantelli experiments provide an easy way to understand the natural ability of gas-based LENR systems to support thermal energy feedback loops. Getting the system to start requires the addition of enough input energy from an external source to create the first neutrons. The Piantelli design does this with an electric heater. Once the system gets going, it starts emitting infrared heat, as well as gamma radiation, which is directly converted to infrared radiation at LENR-active sites. That infrared energy radiates back from the cell walls, is absorbed on the LENR-active surface, and thus feeds the production of the neutrons. In theory, once the cycle starts and reaches a net positive energy balance, the system doesn't need the electric heater, and it can run until the fuel targets and/or hydrogen or deuterium sources are exhausted.

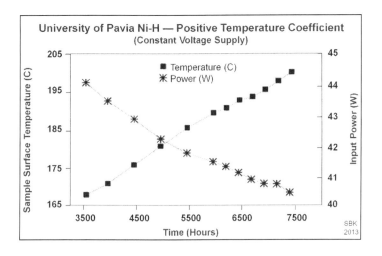

A replication of the Piantelli experiment (see preceding graph) performed by Luigi Nosenzo (University of Pavia) and Luigi Cattaneo, of Consiglio Nazionale Ricerche (National Research Council), at the University of Pavia illustrates the thermal feedback phenomenon. (Piantelli, 2012)

Another graph, from one of Piantelli's experiments, also displays the typical inverse relationship between input power and cell temperature. According to Piantelli, the pressure and voltage remain constant. After a period during which the hydrogen is loaded into the nickel, and power and temperature remain constant, some kind of activation phenomenon occurs. In the graph below, this point occurs at 3.2 hours. The temperature begins to rise sharply as the power drops. (Piantelli, 2010)

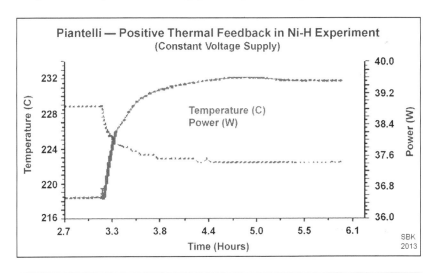

Thermal Runaway

In the history of the field, there have been a few rare but significant thermal runaway events. There are two ways to look at these events. One is to assume that they are fictional or the result of self-delusion. The other way is to consider that they are early indicators of a future energy source with vast potential. As Larsen wrote in his Nov. 29, 2015, slides, the thermal output of the first nuclear fission reactor, constructed in 1942 at the University of Chicago, was only 0.5 watts. That's about typical for LENR experiments. The difference is that LENR experiments weigh less than 5 pounds. The primitive Chicago reactor weighed more than 400 tons.

In March 1991, Tadahiko Mizuno began a new electrolysis experiment. The only active material in the cell was 100 grams of palladium. In early April, he began to see a small but significant amount of excess heat. On April 22, Mizuno stopped the electrolysis. Three days later, he noticed that the cell was still emitting heat even though there was no input power. The next day, he submerged the cell in a 15-liter bucket. Within a day, most of the water in the bucket evaporated. The cell remained hot. Mizuno continued replenishing the water until the cell finally cooled down on May 7. During those final eight days, the cell vaporized 17.5 liters of water, accounting for 85 megajoules of energy. Calorimetry doesn't get much simpler and more foolproof than this. (Mizuno, 1997)

Around September 1993, Piantelli was running a nickel-hydrogen gas experiment in his lab in Siena, Italy. Around 7 in the evening, he looked at the temperature reading. Something didn't look normal: The cell had passed the activation period, but he didn't see the typical plateau of the temperature rise. Instead, the temperature continued increasing rapidly. He had already turned off the heater input power.

A rapidly increasing temperature in an enclosed steel container was a serious problem. He wasn't sure what to do. Should he terminate the experiment, and if so, how would he stop it? He wondered whether he should just leave the building. Instead, he called his colleague Sergio Focardi in Milan — at 2 in the morning — and asked for advice.

This was before Piantelli learned about the poisoning effect of using different isotopes of hydrogen. Admixtures of deuterium and hydrogen in LENR experiments don't work, according to Larsen, because each isotope oscillates at different frequencies on the surfaces of metal hydrides, and the different frequency oscillations inhibit the coherent collective quantum effects that are required to create LENR-active sites.

Focardi came up with a solution: Feed nitrogen into the cell. And it worked. It stopped the uncontrolled temperature rise and terminated the experiment before anything damaging occurred. Piantelli typically ran his experiments at an internal temperature of 200-400 °C. But

Piantelli didn't know the temperature this time because the metal thermocouples inside the cell had melted. That in itself, like Mizuno's vaporized water, was direct data; it meant that the temperature — with no input power — exceeded 1,450 °C.

Clues on the Road From Science to Engineering

For many years, researchers have known the key experimental parameters to produce LENR phenomena. Fleischmann and Pons, for example, knew as early as 1989 that an unusually high loading ratio of deuterium to palladium was required. They also knew that some sort of sharp trigger, for example an abrupt change in input current, was required to initiate the reactions. Other key variables that have been identified include a high rate of deuterium (or hydrogen) flux into or through the reactant material. In the electrolytic systems, a high current density is also required.

But even given all of these correct conditions, experiments do not always consistently produce excess heat. With rare exceptions, the ability to demonstrate measurable excess-heat on demand has been elusive. Countless times, researchers or companies in the field have claimed that they possessed the knowledge of how to reliably produce excess heat or that they knew how to make palladium cathodes that reliably did so.

Invariably, for one reason or another, a year or two later, the same researchers or companies were no longer making such claims. The remaining challenges are to identify the correct reactant materials and to identify how to properly fabricate those materials.

On the other hand, researchers who have sought evidence of isotopic shifts and heavy-element transmutations such as those at Mitsubishi Heavy Industries, Osaka University and Toyota, appear to be able to do so reliably.

Larsen believes that he understands how to fabricate LENR-active sites capable of producing excess heat, based on proprietary extensions of his and Widom's theory. If true, this knowledge could form the basis of new industries. Since 2006, he's been invited to participate in numerous U.S. government and corporate briefings to discuss his theory and LENRs in general.

After I reported, on Dec. 16, 2012, that researchers from Royal Dutch Shell plc, one of the largest energy companies in the world, were interested in exploring LENRs, Larsen told me that he had already been approached by Shell. Larsen visited Houston, Texas, to meet with members of the Emerging Technologies Department at Shell Global Solutions International. He also participated in extensive telephone and videoconference discussions with members of Shell's Europe research and development team. The technical people Larsen spoke with at Shell, he said, were enthusiastic; however, upper-level management decided that, rather than try to move forward in 2012, they would watch the technology develop for a while longer.

In another encounter, Larsen was approached by Dennis Bushnell and Joseph Zawodny, at NASA, who led him to believe that NASA might provide two things for his company. One, as Bushnell suggested to Larsen, was that NASA would give Larsen's company a public endorsement, what Bushnell called the technology equivalent of the "*Good Housekeeping* seal of approval." The second incentive they suggested was NASA funding, as Larsen explained:

> In a series of telephone calls I had during the spring and summer of 2008 with Zawodny and Bushnell, they dangled a carrot — the possibly of significant funding from NASA. I told them that I was willing to teach them the basic physics but I would not transfer Lattice's proprietary knowledge about how to use nanotechnology to improve the reliability of LENRs without having a contract. I told them, "Under contract, I will show you how to make transmutations every time, but I will not show you how to reliably make large amounts of heat."
>
> In January 2009, after an internal NASA meeting, Bushnell and Zawodny informed Lattice that they would not be funding us but they would welcome any free advice we wanted to offer NASA. We declined.
>
> About six months later, Zawodny called me up and said,

"Lew! You didn't tell us everything we need to implement your theory."

I told him, "What do you think I am, stupid?"

Larsen has been careful to avoid giving engineering-related technical information in his papers; however, one of Widom and Larsen's papers gives a hint. "Successful fabrication and operation of long-lasting energy-producing devices with high percentages of nuclear-active surface areas will require nanoscale control over surface composition, geometry and local field strengths," they wrote. (Widom and Larsen, Sept. 25, 2007)

All indicators suggest that nanotechnology is a key element in the development of LENRs. The International Center for Technology Assessment says, "Nanotechnology is the science of engineering on a molecular scale, in effect building matter atom by atom from the bottom up. The prefix nano denotes a fraction of one-one billionth, and nanotechnology involves the construction of matter a billionth of a meter in size: roughly the size of several atoms."

Nanotechnology also includes the ability to produce and make productive use of very tiny particles. It is the extraordinary surface area and perhaps the special surface characterization of these tiny particles that could produce useful LENR reaction rates.

The transmutations on the surfaces of cathodes as shown by many researchers demonstrate that the LENR reaction sites are not uniformly distributed across their surfaces. Instead, large numbers of tiny, short-lived hot spots comprising micron-scale LENR reaction sites are randomly scattered across an active cathode's working surface. This phenomenon is clearly visible in the SPAWAR infrared video. If the structure or surface characterization of LENR reaction sites can be predictably fabricated, perhaps by engineered reactant materials (cathodes in electrolytic or glow discharge experiments; nanopowders in gas permeation experiments) using precise surface geometries, researchers will make a major step toward practical technology.

Toward a New Energy Technology

The prospect of a new clean energy source from hydrogen as well as a method of making rare elements from common elements are mindboggling. As exciting as the possibilities are, by far the biggest question is timing: When will the science develop into technologies, or, alternatively, when will sufficient exploration reveal that the science is destined to be only a laboratory curiosity?

The latter is unlikely; every known aspect of the experimental and theoretical research indicates that extremely compact, long-lived, high-energy-density power sources should be possible with LENRs.

No deadly radiation or radioactive waste has been observed during thousands of experiments. Neither scarce or exotic reactant materials nor controlled nuclear materials are required.

Neither greenhouse gas emissions nor nuclear chain reactions are observed. Therefore, all of this indicates that commercialization of LENR technology would enable new types of nuclear power systems that would not need complex containment or disposal systems.

Virtually any energy technology can and will be put to destructive purposes; that is a matter not of science but of sociology. However, the absence of chain reactions in the neutron-catalyzed transmutation process suggests that it is unsuitable for use as a weapon.

Instead, the evidence described in this book shows that the reactions take place in highly localized, LENR hot spots. These hot spots, according to Larsen, are born and die within 200-400 nanoseconds because the quantum-mechanical effects required to produce neutrons and, according to Larsen, are locally destroyed by the heat arising from the reaction site itself. There is thus no evidence that the reaction sites trigger chain reactions or provide a mechanism for exponentially accelerated energy release. Once the production of ultra-low-momentum neutrons, whether externally triggered or from an internal thermal feedback loop, stops, so do the neutron-capture reactions.

Nevertheless, we cannot rule out the possibility that some means of simultaneous triggering of widespread LENR activity could be devised, thus providing a means of high-rate energy release.

CHAPTER 37

The End of the Beginning

A century ago, from 1912 to 1927, scientists around the world observed atomic transmutations that contradicted the prevailing scientific paradigm. Volume 3 in this series, *Lost History*, tells this story, which foreshadowed modern low-energy nuclear reaction research.

In 1989, Martin Fleischmann and Stanley Pons claimed they had achieved sustained nuclear fusion in a room-temperature experiment. They correctly measured evidence of excess heat that suggested nuclear reactions were taking place, but they were greatly mistaken in their proposed explanation. Their error impeded acceptance of the research for many years. Volume 2 in the series, *Fusion Fiasco*, tells the story of the divisive 1989-to-1990 fusion conflict.

I conducted my last interview with Fleischmann on June 3, 2009. He had appeared on CBS-TV's *60 Minutes* program in March that year, and in that telecast, he had said that he regretted calling his work "fusion." I asked him to say more.

"Well, fusion has a special meaning in the scientific literature — hot fusion — and perhaps it was a mistake to call this process fusion," Fleischmann said. "It should have been called a nuclear effect, you see."

On Aug. 3, 2012, Fleischmann, the figurehead of the field, died from age-related health problems. His death coincided with a general lack of scientific progress by the first generation of LENR researchers. These events marked the end of the beginning of a new field of science.

This book, *Hacking the Atom,* the first volume in the series, explains the science of low-energy nuclear reaction research. It also documents the history of the field in its formative years, from 1990 to 2015.

What Happened and Why

Why, after nearly three decades, is more progress in LENRs not evident? There are three reasons.

First, from the broadest perspective, LENRs represent a potential example of a scientific paradigm shift. Stakeholders in the current scientific paradigm, whether consciously or unconsciously, demonstrably delayed the development of the science, as explained in *Fusion Fiasco*. Some paradigm shifts in science require a generation (25-30 years) to occur.

The second reason for the lack of progress is revealed in the events that took place from 1990 to 2015. Politically dominant scientists in the LENR field took actions and maintained attitudes that delayed its development because of their unshakable ideological attachment to the idea of room-temperature fusion. They controlled the selection of presentations at the international conference series. They influenced interested government agencies to the extent they could. They coordinated their messaging in presentations and to news media, which, as time went on, comprised increasingly less-credible media outlets.

When given the chance to interest governments and industrialists in LENRs, researchers like Michael McKubre and Peter Hagelstein in the U.S., Akito Takahashi in Japan, and Vittorio Violante in Italy pushed fusion. They made little headway. When industrialist Bill Gates expressed an interest in LENRs, he went to these scientists.

Aside from Mitsubishi Heavy Industries, few large industrial companies have openly acknowledged their research in LENRs. Other large industrial companies likely are involved but have kept a low profile because, as of 2015, the incentive to be publicly affiliated with the research is not as strong as the risk of being associated with science that is perceived as illegitimate.

When two newcomers to the field — Lewis Larsen and Allan Widom — published the first potentially viable theory to explain LENRs, and revealed how weak interactions and neutron-capture processes were a much better explanation for the phenomena than fusion, history repeated itself.

Back in 1989, scientists working at traditional fusion laboratories lashed out and politically attacked Martin Fleischmann and Stanley Pons. In 2006, the recognized LENR authorities responded much the same way to Widom and Larsen. This behavior impeded scientific progress.

The third reason for the lack of progress is that those scientists — primarily Americans — were stuck in a rut. They could have searched for nuclear transmutation products, isotopic shifts, alpha particles and neutrons. Instead, most of them focused their efforts on looking for evidence of excess heat and, to a lesser extent, helium-4. On June 25, 2009, Larsen published slides with his critique of this approach:

> Achieving success with such an approach is, at best, a random proposition. It is a bit like trying to fabricate modern microprocessor chips with submicron feature sizes on silicon dies using machinists' T-squares, rulers and scribes rather than using advanced lithography and CMOS process technologies.
>
> Even when substantial macroscopic excess heat is achieved in a 1 cm^2 device, heat as the sole metric of success provides little or no insight into the underlying mechanisms of heat production or what a researcher might do to improve the quantity and duration of heat output in future devices. For example, exhaustive detection and identification of all nuclear reaction products, to whatever extent possible, provide crucial technical information.
>
> Unguided, random, Edisonian exploration of LENR's vast physics and materials parameter space likely is responsible for the lack of readily reproducible experimental results and limited research and development progress that have characterized the field of LENRs for the past 20 years.

Despite the appeal of the idea of amateurs in their garages making revolutionary changes, the science is unlikely to progress meaningfully in those circumstances. Rather, progress will require extremely sophisticated nanotechnology expertise and machines. Real progress is unlikely to occur on a shoestring budget. Research efforts funded at a minimum of $100 million probably will be necessary. To bring the science to a level at which a practical technology may be delivered likely will require financial investment on the order of $1 billion.

Many questions remain. Among them are the following: Can the output of small devices be reliably scaled from milliwatts to many Watts? If so, can total system power outputs be scaled up efficiently? Can researchers learn the key contributing materials science and operating parameters? Can the risk of thermal-runaway-induced explosions or fires, as have occurred in lithium-ion batteries, be prevented or managed? Would any eventual engineered device be cost-effective? Can government regulatory hurdles be overcome? What is the potential impact of material pairing (for example, deuterium-palladium systems and nickel-hydrogen systems)? Are electrolytic-based systems obsolete? Are gas-based systems the future? What can be learned from experiments performed in the first few years of the field? What will be the impact of investments in and development of LENRs in industry, technology, finance, and education?

Perhaps most significantly for a field that has been on the periphery of science for nearly three decades, can the scientific community accept LENRs sufficiently to make available the extensive exploration capacities of traditional academic and industrial research institutions?

Voices From the Past

It is unlikely that the tumultuous and, at times, tortuous path of the introduction of the new idea of room-temperature nuclear reactions could have gone much differently. Simply, the conflict has revealed the intrinsic nature of human beings and the nature of science. German philosopher Arthur Schopenhauer (1788-1860) is proven correct, again:

> When a new truth enters the world, the first stage of reaction to it is ridicule, the second stage is violent opposition, and in the third stage, that truth comes to be regarded as self-evident.

The new idea of LENRs, when distinguished from the mistaken idea of "cold fusion," has been quietly and slowly gaining acceptance. If German theoretical physicist Max Karl Ernst Ludwig Planck (1858-1947) is still correct, which he probably is, broader acceptance of LENRs is just a couple of years away, because of the age of the original participants in and former opponents of the field. He explained:

> A new scientific truth does not triumph by convincing its opponents and making them see the light, but rather because its opponents eventually die, and a new generation grows up that is familiar with it.

Asking when broader acceptance of LENRs by the scientific community will occur is a fair question. In fact, in the last decade, there has been no overt rejection of LENRs. Visible signs of acceptance may appear when more widespread funding opportunities occur.

British geneticist and evolutionary biologist John Burdon Sanderson Haldane (1892-1964) offered a framework in which to evaluate and to predict acceptance of new ideas. Haldane wrote that the process of acceptance passes through four usual stages:

(i) This is worthless nonsense.
(ii) This is an interesting but perverse point of view.
(iii) This is true but quite unimportant.
(iv) I always said so.

A Moment of Confusion

The potential implications of LENRs are surprising, in fact, difficult to believe. This is particularly true for people who have been trained in physics or chemistry. One such advance reader of this book explained to me why LENRs create such intellectual turmoil for him:

> Something has been bothering me the last couple of weeks. I've gone through a mini-crisis of faith, to use a poor

descriptive term. I don't mean faith, per se, because faith is not a suitable term for discussion of science. Science is supposed to rest solely on facts, experiments, data, hypothesis, etc. But it captures the flavor of my thoughts. My crisis is more accurately described by the fact that this subject presents a challenge to my existing interpretive structure of the physical world.

The science you are writing about is inconsistent with everything I have learned about science in school and in my self-education about physics and chemistry, and in all my reading, and what I know about the scientific field. There isn't yet an established scientific structure within which to explore the topic of LENRs. Usually, when scientists explore new topics, it is within a framework that permits uncertainty within certain areas; but in this case, the activity is wholly outside of any established framework. I have enormous respect for scientists because they rely not at all on belief and entirely on evidence and the scientific method.

LENR science, in particular the work on transmutations, makes me think that the assumptions I am familiar with are wrong. Fundamental to my assumptions is the idea that nuclear reactions can take place only by fusion or fission and only in reactors or in high-energy accelerators but not from tabletop environments. This is a very difficult thing to accept, because where it leads I know not.

The profound turmoil that gripped the advance reader signals the presence of a scientific paradigm shift. Beyond that, LENRs offer great promise as a game-changing energy technology and may lead to other novel applications. A quotation from the character Neo in the movie *The Matrix* is apropos:

I don't know the future. I didn't come here to tell you how this is going to end. I came here to tell you how it's going to begin.

Glossary of Scientific Terms

Absorber: *See* neutron absorber.

Accelerator: *See* particle accelerator.

Activation: A process in which a non-radioactive material is subjected to nuclear radiation and becomes radioactive.

Activity: A measure of the level of radioactivity of a material. Measured by the number of spontaneous nuclear disintegrations in a specific amount of material during a specific interval.

Alchemy: Primarily a reference to ancient methods and practices intended to effect elemental or personal transformation.

Alpha (particle, emission): A Greek letter used to describe one of the first types of radioactive emissions. It is emitted during a nuclear reaction and was later identified as a helium-4 nucleus. (*See also*: Appendix C)

Alpha decay: Radioactive decay in which an alpha particle is emitted. Each emitted alpha lowers the atomic number of the nucleus by two and its atomic mass by four.

AMU: Atomic Mass Unit (*See* atomic mass)

Anode: In electrolysis, the metal contact point of the electrical circuit that attracts the flow of electrons.

Atom (atomic): Basic building block of all matter. Atoms comprise three elementary particles: protons, neutrons and electrons. Each atom has one nucleus in its center containing the protons and neutrons. The nucleus is surrounded by electrons, normally equal in number to the number of protons in the nucleus of a neutral atom.

Atomic energy: *See* nuclear energy.

Atomic mass: Effectively measured by the total number of protons and neutrons in an atom's nucleus; determines the type of isotope within a specific range of possible isotopes. *See also* atomic number.

Atomic number: Measured by the total number of protons in an atom's nucleus; determines the type of chemical element. *See also* atomic mass.

Atomic transformation: *See* transmutation.

Beta (particle, emission): A Greek letter used to describe one of the first types of radioactive emissions. It is emitted during a nuclear reaction and was later identified as an electron. (*See also*: Appendix C)

Beta decay: A weak interaction in which a neutron inside a nucleus decays into a proton, an energetic electron and a neutrino. The energetic electron released in a beta decay exits the nucleus as a beta particle.

BF3 detector (counter): Boron tetrafluoride detector; used to measure neutron emissions. Detector consists of a cylindrical tube filled with boron trifluoride gas, used to detect low energy "thermal" neutrons. With the addition of a neutron moderator surrounding the detector (to bring down neutron energy), the detector can also be used to detect higher-energy "fast neutrons."

Binding energy: a) For a particle in a system, the net energy required to remove it from the system; b) for a system, the net energy required to decompose it into its constituent particles.*

Branching ratio (D+D fusion): According to the well-understood theory of deuterium-deuterium nuclear fusion, the reaction paths occur through one of three possible branches. The first branch produces a neutron. The second branch produces tritium. The third branch produces helium-4. In D+D fusion reactions, on average, a neutron is produced almost 50% of the time, tritium is produced almost 50% of the time, and helium-4 is produced less than 1% of the time. Since the discovery of D+D fusion, these ratios have always been consistent.

Bulk effect: In LENRs, the initial mistaken concept that effects are proportional to the mass or volume of the reactant material, rather than proportional to the surface area.

Cathode: In electrolysis, the metal contact point of the electrical circuit that emits the flow of electrons and attracts positive ions or protons.

Charged particle: A fundamental particle such as an electron, proton or positron, or a compound particle that carries a net positive or negative electrical charge.

Chemistry: An area of science primarily involved with interactions between atoms and electrons, their structures and properties. There are two historical exceptions. The first was in the early 20th century, when chemists were as involved in nuclear research as physicists were. The second period began in 1989, with the introduction of what was later called LENRs.

Collective effects: Describes the interaction of many-body groups of essentially identical items, such as elementary particles. When the items interact as a group, they create different effects than they would produce either alone or with a few others. The concept can apply to many-body physics, such as electrons oscillating together, or to a flock of birds flying in formation and thus creating lift efficiencies that none of the birds could create individually.

Coulomb barrier: An electrostatic barrier surrounding positively charged nuclei that, under normal temperatures and pressures, prevents nuclei from interacting with each other.

Cross-section: a measure of the probability of a specified interaction between an incident photon or particle radiation and a target particle or system of particles. It is the reaction rate per target particle for a specified process divided by the flux density of the incident radiation.*

Decay: *See* radioactive decay.

Deuterium: A stable isotope of hydrogen that has one proton and one neutron in its nucleus. Also known as heavy hydrogen.

Deuteron: The nucleus of a deuterium atom, comprising one proton and one neutron.

Disintegration: A process in which constituent parts of the nucleus of an atom separate from the nucleus and fly off, leaving a smaller atom in place.

Electrochemistry: The study of electricity and how it relates to chemical reactions. In electrochemistry, electricity can be generated by movements of electrons from one element to another in a reaction known as redox reaction, or oxidation-reduction reaction. (U.C. Davis)

Electrode: In electrochemistry, the metal contact point of the electrical circuit that conducts the flow of electrons.

Electrolysis: Chemical decomposition by an electric current of a liquid, or solution containing ions, into constituent elements.

Electrolytic fusion: The idea of creating nuclear fusion by electrolysis.

Electromagnetic force: One of the four fundamental physics forces. Repels protons from one another and keeps atomic nuclei separate from one another. *See also* strong force.

Electron: A stable elementary particle that is a component of an atom. It possesses a negative electrical charge and exists outside and orbits around the nucleus of an atom.

Electron-volt (MeV, Mega-electron-volt): A unit of energy equal to the change in energy of one electron in passing through a voltage difference of 1 volt.*

Electroweak interaction: The unified description of two of the four known fundamental interactions of nature: electromagnetism and the weak interaction.

Element: Designates a form of matter that is distinguished by a unique number of protons in its nucleus and unique chemical properties.

Emissions: *See* alpha, beta and gamma rays.

Energy: Power during a given period.

Fission: *See* nuclear fission.

Fusion: *See* nuclear fusion.

Gamma rays (gamma radiation, gamma emission): Gamma rays are highly penetrating forms of electromagnetic radiation emitted from nuclear transitions. Gamma rays are a class of photons (a larger group of massless entities) that, according to quantum mechanics, behave both as waves and as particles. On Earth, they are encountered from radioactive material decays and a few rare terrestrial events. Gamma rays are identified by their energy from the so-called photo-peak. A range of various-energy gamma ray interaction-related peaks and continua are depicted in a typical gamma spectrum ranging from the photo-peak at the upper end of the energy scale down to zero. (*See also*: Appendix C)

Gamow factor: The probability that two nuclear particles will overcome the Coulomb barrier and undergo nuclear fusion reactions.

Gas-loading: An experimental method in which molecules of a gas, typically deuterium or hydrogen, dissociate, ionize and then move into hydride-forming sites in a host metal or metal-oxide structure.

Glow discharge: A plasma formed by the passage of electric current between metal electrodes through a low-pressure gas.

Half-life: The time required for half of a given quantity of a radioactive material to decay.

Heavy electrons (in LENRs): Electrons that possess higher effective mass than electrons at rest in free space. In particle physics, the term used for this concept is mass renormalization of dressed electrons. In the context of the Widom-Larsen theory, this increase in effective mass comes directly from local nuclear-strength electric fields that confer additional mass through quantum electrodynamics. These are not the same as "heavy electrons" or "heavy fermions" found in certain intermetallic compounds such as UPt_3.

Heavy hydrogen: *See* deuterium.

Heavy water (D_2O): Water molecules made from deuterium instead of hydrogen. *See also* light water.

Hydride (Deuteride): Compounds that hydrogen (or deuterium) form with other chemical elements, typically within metals or alloys. Some metals can absorb hundreds of times their own volume of hydrogen or deuterium.

Hydrogen: A chemical element with a single proton in its nucleus. Its normal isotope, known as protium, is stable. Its second isotope, known as deuterium, is also stable. Its third isotope, known as tritium is unstable.

Ion: An atom that has either an excess or shortage of an electron or electrons. Ordinarily, neutral atoms have an equal number of electrons to their protons.

Ionization: A process by which an atom gains or loses an electron or electrons.

Isotope: A variation of an element that contains the same number of protons but a different number of neutrons from the most abundant version of that element. Isotopes have the same atomic number but a different atomic mass.

Isotope, stable: An isotope that is not undergoing radioactive decay or emitting gamma radiation.

Isotope, unstable: An isotope that is undergoing radioactive decay or emitting gamma radiation.

Isotopic abundance: The relative number of atoms of a specific isotope among all the isotopes of a given element, expressed as a fraction of all the isotopes of that element.

Isotopic shift: A change in the ratios among isotopes of one species of elements away from the isotopic abundance of the same species that exists in nature.

LENR, LENRs: *See* low-energy nuclear reactions.

Light water (H_2O) (normal water): Water composed of the normal hydrogen isotope, which contains one proton and no neutron. The term "normal water" is used sometimes synonymously; however, one of every 6,000 molecules of normal water is a molecule of heavy water.

Loading: The process of placing atoms of deuterium (or hydrogen) interstitially into vacant spaces within the crystalline lattice of metallic elements.

Loading ratio: The ratio between the number of atoms of deuterium (or hydrogen) and the number of atoms of the host metal into which they have been loaded.

Low-energy nuclear reactions (LENRs): A class of nuclear reactions — based on Standard Model physics — that can occur in condensed matter under mild macrophysical conditions. Key steps in LENR processes, unlike nuclear fusion or fission, are based primarily on electroweak interactions rather than strong-force interactions. Unlike fission reactions, low-energy nuclear reactions do not produce nuclear chain reactions. (*See also* Appendix A)

Metal hydrides or deuterides: Metals that have absorbed hydrogen or deuterium in their atomic structure, or lattice. *See also* loading.

Moderator: *See* neutron moderator.

Neutrino: An elementary particle having virtually no mass. Like a neutron, it has no electrical charge; it barely interacts with ordinary matter.

Neutron: An unstable (when outside a nucleus) elementary particle that is a component of an atom and exists inside the nucleus of an atom. It has no electrical charge. A free neutron outside of a nucleus has a half-life of approximately 10.3 minutes before it decays into a proton, an electron, and an electron antineutrino.

Neutron absorber: A material or object with which neutrons interact, resulting in their disappearance as free particles without production of other neutrons.

Neutron capture: A nuclear reaction in which an atomic nucleus and one or more neutrons collide and merge to form a heavier nucleus.

Neutron, cold: Neutrons of kinetic energy on the order of 1 milli-electron-volt or less (0.001 eV).

Neutron, fast: Neutrons having kinetic energy between 1 MeV and 20 MeV.

Neutron moderator: Material used to reduce the speed of neutrons, without absorbing them into the moderator material.

Neutron, prompt: Neutrons emitted from a nuclear process, at the time of the reaction, without measurable delay.

Neutron, slow: Neutrons having kinetic energy between 1 eV and 10 eV.

Neutron, thermal: A free neutron that has been slowed down by a moderator, is in equilibrium with its surroundings, has an energy between 0.025 eV and 0.2 eV.

Neutron, ultra-low-momentum (ULMN): A neutron with kinetic energies that are effectively zero, on the order of 10^{-12} eV or less — that is, .000000000001 eV. The kinetic energy of ULMNs is an estimated value because it has never been measured. ULMNs are extremely slow neutrons with extremely low kinetic energies and commensurately large DeBroglie wavelengths because they are created through a many-body collective process (as opposed to being produced by a two-body nuclear reaction that occurs inside a star). ULMNs are thus orders of magnitude slower than so-called "ultra-cold" neutrons, which are typically produced for experiments aiming to better measure the lifetime of free neutrons located outside nuclei. (Courtesy: Lewis Larsen)

Neutron, virtual: *See* virtual neutron

Neutronization: The process in a collapsing star in which protons and electrons directly react to form neutrons and release neutrinos.

Nuclear: Activity or properties having to do with characteristics of or changes in an atomic nucleus.

Nuclear capture: A nuclear process by which a nucleus acquires an additional particle.

Nuclear chemistry: Chemistry-related aspects of nuclear and atomic research.

Nuclear energy: Energy that is released during a nuclear reaction, such as nuclear fission, nuclear fusion, radioactive decay or a variety of nuclear processes that capture nuclear particles.

Nuclear fission: The process in which a larger nucleus is split into two (or, rarely, more) parts. The process is usually accompanied by the emission of neutrons, gamma radiation and, rarely, small charged nuclear fragments.

Nuclear fission, spontaneous: Nuclear fission that occurs spontaneously, without the addition of particles or energy to the nucleus.*

Nuclear fusion: The process in which two light nuclei overcome electrostatic repulsion and form one newer, heavier atom.

Nuclear physics: Approaches to and studies of nuclear science and technology based on principles, processes and devices common to physics.

Nuclear process: A mechanism, such as fission, fusion, or radioactive decay, which changes the energy, form, or structure of the nucleus.

Nuclear reaction: An event, occurring from a nuclear process, in which the energy, form, or structure of the nucleus is changed.

Nuclear science: The study of nuclear processes and reactions.

Nuclear transformation: *See* nuclear transmutation.

Nuclear transmutation: A nuclear process in which an element changes into another element by the increase or decrease in the number of protons in its nucleus.

Nuclei: *See* nucleus.

Nucleosynthesis: The formation of new nuclides by any number of nuclear processes, including nuclear decay.

Nucleus (nuclei, pl.): Center part of an atom that contains protons and neutrons. Comprises nearly the entire mass of the atom but only a tiny part of its total volume.

Nuclide: A distinct species of an atom identified by the number of protons and neutrons in its nucleus and its nuclear energy state.

Particle: *See* charged particle, neutrino and neutron.

Particle accelerator: A device for imparting kinetic energy to charged particles.*

Photon: A massless elementary particle that is a unit of light and other forms of electromagnetic radiation. It can have properties of both waves and particles.

Physics: A field of science and technology that measures, studies and influences matter, motion and energy.

Plasmonics: The study of the interaction between the electromagnetic field and free electrons in a metal.

Power: The ability to perform work; measured by a specific rate.

Protium: The common isotope of hydrogen; distinct from deuterium and tritium.

Proton: Stable elementary particle that is a component of atoms. The proton has a positive electrical charge and exists in the nucleus of an atom.

Quantum electrodynamics: The study of the properties of electromagnetic radiation and the way in which it interacts with charged matter in terms of quantum mechanics. (Source: Oxford Physics Dictionary)

Quantum mechanics: The branch of physics that deals with the mathematical description of the motion and interaction of subatomic particles, incorporating the concepts of quantization of energy, wave-particle duality, the uncertainty principle, and the correspondence principle. (Source: Oxford Dictionary)

Radiation, nuclear (Radioactive Emission): Emission of charged or uncharged particles or electromagnetic rays, including alphas, betas, neutrons, and gamma-rays.

Radiation, prompt: Prompt radiation is produced and emitted from its source immediately. When the reaction stops, so does the prompt radiation. *See also* radioactive decay.

Radioactive: The property of an unstable material that spontaneously emits particles or gamma rays.

Radioactive decay: A form of nuclear radiation that emits alpha and/or beta particles from radioactive materials. The emissions may take place during nuclear reactions as well as after the reactions stop. The decay causes the radioactive interior to lose some of its constituent material. An element that undergoes radioactive decay will change into a new element or a new isotope.

Radioactive half-life: For a single radioactive decay process, the time required for the activity to decrease to half its value by that process.*

Radioactivity: A naturally occurring process in which unstable elements spontaneously emit particles or gamma rays. In addition to naturally occurring radiation, man-made nuclear processes can cause some non-radioactive elements to become radioactive.

Radiochemist: Person who works in the field of radiochemistry.

Radiochemistry: The part of chemistry that deals with radioactive materials.*

Radionuclide: A radioactive nuclide.

Radium: An unstable radioactive chemical element with a half-life of about 1,600 years, and 88 protons in its nucleus.

Radon: An unstable chemical element and radioactive gas with a half-life of 3.8 days, and 86 protons in its nucleus. It is produced from the decay of uranium or thorium.

Scattering: A process in which a change in direction or energy of an incident particle or incident radiation is caused by a collision with a particle or a system of particles.*

Scattering, elastic: Scattering in which the total kinetic energy is unchanged.*

Sonic implantation: An experimental method that uses acoustic cavitation to stimulate activity on metal surfaces and induce low-energy nuclear reactions.

Spectral lines: Bright and dark lines — seen in spectra of photon-emitting items, such as candle flames, glowing gas, or stars — that are characteristic of a given atom or molecule.

Standard Model: The Standard Model of particle physics is a theory that explains the physics of the world and what holds it together. It encompasses the behavior of fundamental particles across three of the four fundamental forces in physics: electromagnetic, weak interactions, and strong interactions.

Strong force: One of the four fundamental physics forces; works only at very short distances within nuclei. Keeps protons and neutrons bound together inside atomic nuclei. *See also* electromagnetic force.

Surface plasmon electrons (surface plasmons): A collective many-body effect; coherent oscillations of entangled electrons that take place at the surface of metals and at other interfaces.

Transmutation, biological: A speculative process that indicates the increase or decrease of elements in human and other living matter that cannot easily be explained environmentally or by conventional biology.

Transmutation, man-made: Human-triggered nuclear transmutation; traditionally occurs by exposure to radioactive sources. Can also occur by nontraditional processes, specifically LENRs.

Transmutation, natural: Spontaneous nuclear transmutation that occurs by the natural activity of a radioactive element.

Transmutation, nuclear: The changing of one element to another by a change in the number of protons in its nucleus.

Tritium: An unstable isotope of hydrogen. A chemical element with a single proton in its nucleus and two neutrons. It is radioactive with a half-life of 12.3 years.

Virtual neutron: The idea of a virtual particle such as a virtual neutron is an explanatory conceptual entity. It is employed in mathematical calculations by some theoretical physicists to help analyze and calculate subatomic processes, especially in the context of quantum field theory. Some physicists avoid using the virtual particle idea because they believe it is confusing and misleading. Nevertheless, hidden interaction of virtual particles is sometimes used to explain observable physical phenomena. Several key attributes characterize virtual particles: they do not appear directly among truly observable and detectable input and output quantities of calculations, are typically very short-lived and transient on subatomic time scales (attoseconds), may temporarily violate known laws of physics, and may possess a different mass than the corresponding real particle. A hypothetical "virtual neutron" that appears in a theoretical calculation differs greatly from a free real neutron that has a mass of 1.008 amu and half-life to beta decay of about 15 minutes when it is situated outside an atomic nucleus, and beyond the short-range influence of the strong force inside atomic nuclei. When a real neutron is captured by a nucleus, total atomic weight of that particular nucleus will, at least temporarily, increase by about 1 amu which can be measured; by contrast, a virtual neutron by itself would not leave any detectable traces signaling its fleeting existence in the observable world. (Definition: L. Larsen)

Weak force: A fundamental force of physics that produces weak interactions.

Weak interaction: An elementary particle interaction that is involved in many forms of nuclear decay (radioactivity), for example, a beta decay process. In all such interactions, various types of neutrinos are emitted or absorbed. Weak interactions are distinct from strong-force interactions because, at low average particle energies, weak-interaction cross-sections are vastly lower than strong-force interactions. Weak interactions are not necessarily weak energetically, and some can involve very large releases of energy. For example, beta decays of some extremely neutron-rich nitrogen isotopes can release more than 20 MeV of nuclear binding energy. For comparison, the strong-force deuterium-tritium fusion reaction releases 17.6 MeV.

* Source: Glossary of Terms in Nuclear Science and Technology, American Nuclear Society, ISBN 0894485539

Appendix A — Definition* of Low-Energy Nuclear Reactions

Low-energy nuclear reactions (LENRs) are a class of nuclear reactions — based on Standard Model physics — that can occur in condensed matter under mild macrophysical conditions. Key steps in LENR processes, unlike nuclear fusion or fission, are based primarily on electroweak interactions rather than strong-force interactions. Unlike fission reactions, low-energy nuclear reactions do not produce nuclear chain reactions.

LENRs involve a broad set of nuclear phenomena spanning many length-scales that have two characteristics in common: a) production of neutrons from electroweak reactions; and b) many-body collective effects between oppositely charged particles. (In condensed-matter systems, these particles are typically quantum mechanically entangled).

LENRs take place in three realms: a) electrically dominated reactions, in which nuclear-strength local electric fields on micron scales in condensed matter enable electroweak neutron production; b) magnetically dominated reactions, in which many-body collective magnetic-field effects directly accelerate charged particles in plasmas; and c) mixed reactions, in which components of dusty plasmas behave in ways characteristic of the electrically dominated reactions and the magnetically dominated reactions.

The word "low" in "low-energy nuclear reactions" refers to the magnitude of input energies that are required to trigger LENR reactions; the magnitude of output energies released after triggering may be either low or high. Researchers chose this term to distinguish it from the field of high-energy particle physics, which uses very high temperatures or particle accelerators to trigger nuclear reactions.

The two most unusual characteristics of LENRs are that neutron-catalyzed transmutation reactions, which typically occur only in stars, fission reactors, or high-energy particle accelerators, can be initiated in tabletop condensed-matter experimental systems without releasing biologically dangerous amounts of energetic neutron or gamma radiation.

Electrically Dominated LENR Reactions

Electrically dominated reactions take place in condensed matter. These

LENRs take place under relatively mild conditions — that is, without the requirement of using large nuclear fission reactors, extremely high temperatures, or high-energy particle accelerators.

Given proper types and amounts of input energy, these LENRs take place when specific conditions are present on the surfaces of metals or at metal-oxide interfaces, in the presence of hydrogen or ones of its isotopes, deuterium or tritium. No radioactive seed elements are required. Neutrons produced in an electroweak reaction at micron-scale LENR-active sites on surfaces or at interfaces are subsequently captured by nearby atoms; these energy-releasing captures induce nuclear transmutations. Neutrons produced in LENRs have ultra-low energy, so almost all of them are captured locally; externally detectable emissions of deadly energetic neutrons are thus also avoided.

LENR experiments typically produce a variety of nuclear transmutation products and various types of effects and may produce macroscopically measurable excess heat. A variety of elements may be synthesized from one another, and isotopic shifts may occur; these transmutation products are generally stable elements produced by beta decays of short-lived, neutron-rich unstable isotopes created by previous neutron captures.

According to the Widom-Larsen theory, LENRs in the electrically dominated realm have two unique characteristics: a) produced neutrons have ultra-low-momentum and b) unreacted heavy electrons present in LENR-active sites suppress dangerous energetic gamma emission by locally converting incident gamma radiation from any source directly into infrared radiation (heat). (Widom and Larsen, 2006; Srivastava et al., 2010)

Magnetically Dominated LENR Reactions

Magnetically dominated reactions take place in plasmas. These LENRs can, for example, occur in magnetic flux tubes of solar flares; these processes may produce GeV neutrons, other elementary particles, and energetic gamma rays that are not suppressed.

Mixed LENR Reactions

Mixed reactions take place in dusty plasmas. These LENRs can occur in organized magnetic fields present in a plasma as well as on solid surfaces of micron- to nanometer-sized dust particles of condensed matter, which are embedded in such plasmas. Examples include exploding wire experiments and in natural lightning. *(See also concise LENRs definition in Glossary)*

Appendix B — Timeline of Related 20th Century Events

1920

Charles Galton Darwin describes collective many-body excitations of electrons. His ideas lay the foundation for the understanding of the collective behavior of electrons. (Darwin, 1920)

1923 (November)

Robert Millikan speculates that transmutations going from lighter to heavier elements might occur in the stars. The idea of nuclear disintegration going from heavy to light elements is known at this time. (Millikan, 1923)

1929

Fritz Houtermans and Robert Atkinson propose that thermal kinetic energies inside stars are high enough to allow nuclei of light elements to overcome the Coulomb barrier and form heavier elements. The concept is later identified as thermonuclear fusion. In 1933, Oliphant experimentally confirms nuclear fusion. (Atkinson and Houtermans, 1929)

1932

James Chadwick experimentally confirms the existence of the neutron. In 1910, Rutherford had theorized its existence. (Chadwick, 1932)

1932

John Cockcroft and Ernest Walton, at the Cavendish laboratory, build the first apparatus for accelerating atomic particles to high energies. They report the first man-made transmutation by artificially accelerated particles: the disintegration of lithium by fast hydrogen protons. (Cockcroft and Walton, 1932)

1933

Leó Szilárd conceives the idea of the nuclear chain reaction. (Rhodes, 1986)

1933

Mark Oliphant experimentally confirms fusion of deuterons into various targets to create helium-3 and tritium, using a particle accelerator at Cavendish. Concurrently, he observes the liberation of excess nuclear binding energy, which prompts him to speculate that fusion is the process that powers the sun. (Oliphant, Harteck and Rutherford, 1934)

1934 (January)

Frédéric and Irène Joliot-Curie create artificial radioactivity in previously stable elements. (Joliot and Joliot-Curie, 1934)

1934 (March 3)

Gian-Carlo Wick proposes the concept of electron capture. (Wick, 1934)

1935

Harold John Taylor reports the first neutron-capture-based transmutations. He experimentally demonstrates that boron-10 nuclei capture thermal neutrons and fission into helium-4 and lithium-7. (Taylor, 1935)

1938 (Nov. 18)

Otto Hahn and Fritz Strassmann bombard uranium with neutrons and transmute uranium into smaller atoms. In 1939, Lise Meitner and Otto Frisch identify the effect as fission. (Hahn and Strassmann, 1938)

1939 (Feb. 11)

Lise Meitner and Otto Frisch propose the concept of neutron-induced uranium fission. (Meitner and Frisch, 1939)

1939 (March 1)

Hans Bethe proposes that 1) hydrogen-hydrogen fusion is the process that powers the stars, 2) no elements heavier than helium-4 could have been formed in stars, and 3) the production of neutrons in stars is negligible. His first proposal was correct; the second and third were not. (Bethe, 1939)

1946 (March 14)

Fred Hoyle makes an early contribution to the theory of supernovae-exploding stars. He proposes that neutron creation in the hot cores of collapsing stars can be explained by the reaction of an electron with a proton ($e + p \rightarrow n + \nu$). (Hoyle, 1946)

1951

Ernest Sternglass observes neutron production in keV-energy (low-energy) electric discharge experiments in a hydrogen-filled X-ray tube directed at targets of silver and indium. *See* Darwin 1920. Albert Einstein suggests that collective effects may explain the results. (Sternglass, 1997; Trost, 2013)

1957

Geoffrey and Margaret Burbidge, William Fowler and Fred Hoyle propose what is later regarded as the modern concept of nucleosynthesis of elements in stars. They theorize that fusion reactions create elements up to the atomic mass of iron and that neutron capture processes and decays create heavier elements beyond iron. Prior to their work fusion-based concepts alone were unable to fully explain the production of heavier elements The group later propose that nucleosynthesis also occurred outside the cores of stars. (Burbidge, Burbidge, Fowler and Hoyle, 1957)

1960

Corentin Louis Keruran (1901-1983) begins publicly discussing his research in biological transmutation. (*See also:* 1971)

1968

Sheldon Glashow, Abdus Salam, and Steven Weinberg develop modern electroweak theory. (Nobelprize.org)

1971

Keruran publishes a biological transmutation book. (Keruran, 1971)

1973

The Gargamelle collaboration at the European Organization for Nuclear Research performs the first stage of experimental confirmation of electroweak interactions. (CERN)

1983

The "UA1" and the "UA2" collaborations of the European Organization for Nuclear Research perform the second stage of experimental confirmation of electroweak interactions. (CERN)

Appendix C — Basic Types of Radioactive Emissions

Type	Nature of Radiation	Penetrating Power[1]	Ionizing Power [2]
Alpha α $^4_2 He$	A helium nucleus of 2 protons and 2 neutrons. Mass = 4 Charge = +2	**Low** Particles are stopped by a few cm of air or a thin sheet of paper.	**Very High** The biggest mass and charge of the three. Packs the biggest punch.
Beta β $^0_{-1} e$	High kinetic energy electrons. Mass = 1/1850 Charge = -1	**Moderate** Most particles are stopped by a few mm of metals like aluminum.	**Moderate** Less than the alpha particle.
Gamma γ $^0_0 \gamma$	Very high frequency electromagnetic radiation. Mass = 0 Charge = 0	**Very High** Most, but not all, gamma rays are stopped by a thick layer of steel or concrete, or a few cm of dense lead.	**Lowest** Carries no electric charge and has no mass, so it has very little punch when it collides with an atom.

1. When penetrating denser material, more radiation is absorbed and stopped than when penetrating less-dense material. However, as mass or charge decreases, the penetrating power increases.
2. Ionizing power is the ability to remove electrons from atoms and form positive ions. Ionizing radiation is harmful to living cells. Courtesy Georgia State University, adapted by *New Energy Times*.

Appendix D — Helium Permeation in Metals Analysis

Gas Behavior in Hydride-Forming Metals at or Near Standard Temperature and Pressure	Hydrogen	Helium
Readily permeates hydride-forming metals	Yes	No
Diffuses through defects, cracks or grain boundaries in metals	Yes	Yes
Soluble (dissolves) in hydride-forming metals	Yes	No

Bowman Jr., Robert C. (Feb. 7, 2007) "NMR Studies of 3He Retention and Release in Metal Tritides — A Review," Hydrogen & Helium Isotopes in Materials Conference, Albuquerque, N.M. *[Helium does not outgas from metals easily or quickly.]*

Chien, Chun-Ching, Hodko, Dalibor, Minevski, Zoran and Bockris, John O'M. (April 1992) "On an Electrode Producing Massive Quantities of Tritium and Helium," *Journal of Electroanalytical Chemistry*, **338**, 189-212 *[Helium on near-surface areas on cathode can be retained if quickly immersed in liquid nitrogen.]*

Gozzi, D., Cellucci, F., Cignini, P.L., Gigli, G., Tomellini, M., Cisbani, E., Frullani, S., Urciuoli, G.M. (1998) "X-Ray, Heat Excess and 4He in the D:Pd System," *Journal of Electroanalytical Chemistry*, **452**, 253, and Erratum, **452**, 251-71 *[Helium does not show up in the bulk if the cathode is vaporized.]*

McKubre, Michael, et al., (June 1998) "Development of Energy Production Systems from Heat Produced in Deuterated Metals, Volume 1," Electric Power Research Institute, TR-107843 *[Researchers hypothesized, but did not test, that helium was retained (occluded) in metal during experiment.]*

Ramsay, W., and Travers, M.W. (January 1897) "An Attempt to Cause Helium or Argon to Pass Through Red-Hot Palladium, Platinum, or Iron." *Proceedings of the Royal Society of London* (1854-1905), 61(-1), 266-7 *[Helium won't dissolve in metal even at high temperature.]*

Schultheis, D. (2007) "Permeation Barrier for Lightweight Liquid Hydrogen Tanks," Ph.D. dissertation, University of Augsburg *[Defect-free metal will not allow helium to pass through.]*

Xia, Ji-xing, Hu, Wang-yu, Yang, Jian-yu, and Ao, Bing-yun (2006) "Diffusion Behaviors of Helium Atoms at Two Pd Grain Boundaries," *Transactions of Nonferrous Metals Society of China*, 16, S804-7 *[Helium has low solubility in metals, grain boundaries support permeation.]*

Bibliography

Apicella, M.L., Castagna, Emanuele, Hubler, Graham, McKubre, Michael, Sarto, Francesca, Sibilia, Concita, Rosada, A., Santoro, E., Tanzella, Francis, and Violante, Vittorio (slides presented Nov. 2004 at ICCF-11) "Progress on the Study of Isotopic Composition in Metallic Thin Films Undergone to Electrochemical Loading of Hydrogen," *Procs. of the 12th International Conference on Cold Fusion*, 264-271 (published paper, 2006)

Apicella, M.L., Castagna, E., Lecci, S., Sansovini, M., Sarto, F., Violante, Vittorio (presented 2009) "Mass Spectrometry: Critical Aspects Related to the Particles Detection in the Condensed Matter Nuclear Science," Slide presentation given at the 15th International Conference on Condensed Matter Nuclear Science

Atkinson, Robert and Houtermans, Fritz (1929) "Zur Frage Aufbaumöglichkeit in Sternen," *Z. für Physik*, **54**, 656-65

Baard, Erik (Oct. 6, 1999) "Researcher Claims Power Tech That Defies Quantum Theory," *Dow Jones NewsWires*; (Dec. 21, 1999) "Quantum Leap: Doctor Randell Mills Says He Can Change the Face of Physics. The Scientific Establishment Thinks He's Nuts" *The Village Voice*

Bass, Robert (April 1999) "Cold Fusion: The First 10 Years," *Infinite Energy*, **24**, 18

Bazhutov, Yuri, Chertov, Y., Krivoshein, A., Skuratnik, Y., and Khokhlov, N., (Dec. 1993) "Excess Heat Observation During Electrolysis of $CsCO_3$ Solution in Light Water," *Procs. of the 4th Intl. Conf. on Cold Fusion*, 24-1-24-4

BBC Horizon (March 21, 1994) "Too Close to the Sun"

Bethe, Hans (March 1, 1939) "Energy Production in Stars," *Physical Review*, **55**, 434

Biberian, Jean-Paul, ed. (Feb. 2011) *Journal of Condensed Matter Nuclear Science*, **4**, http://www.iscmns.org

Bishop, Jerry (Aug. 1993) "It Ain't Over Till It's Over - Cold Fusion," *Popular Science*, 47-82

Bockris, John (Nov. 1993) "Summary Of Results For Anomalous Phenomena In Experiments On Solid State Reactions On Noble Metal Content," available at *New Energy Times* Web site

Bockris, John and Sundaresan, Raghavan (Nov. 1994) "Anomalous Reactions during Arcing between Carbon Rods in Water," *Fusion Technology*, **26**, 261-266

Bockris, John (Aug. 1996) "Summary Report on the Second International Low Energy Nuclear Reactions Conference (ILENR2)," *Infinite Energy*, **9**, 11

Bockris, John (Aug. 1996) "Letter to Emile Schweikert - Meeting on Ultra-Low-Energy Nuclear Reaction," *Infinite Energy*, **9**, 17-18

Bockris, John (1999) "Early Contributions From Workers At Texas A&M University To (So-Called) Low Energy Nuclear Reactions," *Journal of New Energy*, **4**(2), 1999, 40

Bockris, John (2000) "Accountability and Academic Freedom — The Battle Concerning Research on Cold Fusion at Texas A&M University,"

Accountability in Research, **8**(2), 103-19

Bockris, John (2004) "History of the Discovery of Transmutation at Texas A&M University," *Procs. of the 11th Intl. Conf. on Cold Fusion*, 562-586

Bockris, John (Sept. 2011) "Letter to Joe Champion"

Bressani, Tulio (presented Oct. 1996) "Nuclear Products in Cold Fusion Experiments Comments and Remarks after ICCF-6," *Procs. of the 6th Intl. Conf. on Cold Fusion*, 703-709

Bressani, Tulio (presented April 1998) "Nuclear Physics Aspects of Cold Fusion Experiments - Scientific Summary After ICCF-7," *Procs. of the 7th Intl. Conf. on Cold Fusion*, 32-37

Burbidge, E. Margaret, Burbidge, Geoffrey R., Fowler, William A., and Hoyle, Fred (1957) "Synthesis of the Elements in the Stars," *Reviews of Modern Physics*, **29**(4), 547-650

Bush, Benjamin F., Lagowski, Joseph J., Miles, Melvin M., Ostrom, Greg S. (1991) "Helium Production During the Electrolysis of D_2O in Cold Fusion Experiments," *Journal of Electroanalytical Chemistry*, **304,** 271-278

Bush, Robert T. (Sept. 1992) "A Light Water Excess Heat Reaction Suggests That 'Cold Fusion' May Be 'Alkali Hydrogen Fusion'," *Fusion Technology*, 301-332

Bush, Robert and Eagleton, Robert (presented Dec. 1993) "Calorimetric studies for several light-water electrolytic cells with nickel fibrex cathodes and electrolytes with alkali salts of potassium, rubidium, and cesium," *Procs. of the 4th Intl. Conf. on Cold Fusion*, **2**, 13-1-13-22

Campari, E., Focardi, S., Gabbani, V., Montalbano, V., Piantelli, F., and Veronesi, S., (presented 2004) "Overview of H-Ni Systems: Old Experiments and New Setup," 5th Asti Workshop on Anomalies in Hydrogen/Deuterium-Loaded Metals, Asti, Italy

Case, Leslie C. (1998) "Catalytic Fusion of Deuterium into Helium-4," *Procs. of the 7th Intl. Conf. on Cold Fusion*, 48-50

Case, Leslie C. (July, 1999) "Progress in Les Case's Catalytic Fusion," *Infinite Energy*, **23**, 9-15

Castagna, Emanuele, Sibilia, Concita, Paoloni, S., Violante, Vittorio, and Sarto, Francesca (presented Nov. 2004 at ICCF-11) "Surface Plasmons and Low-Energy Nuclear Reactions Triggering," *Procs. of the 12th Intl. Conf. on Cold Fusion*, 156-162

Celani, Francesco, Spallone, Antonio, Marini, P., Di Stefano, V., Nakamura, M., Mancini, A., D'Agostaro, G., Righi, E., Trenta, G., Quercia, P., Catena, C., Andreassi, V., Fontana, F., Garbelli, D., Gamberale, Luca, Azzarone, D., Celia, E., Falcioni, F., Marchesini, M., and Novaro, E., (Oct. 2002) "Unexpected Detection of New Elements in Electrolytic Experiments with Deuterated Ethyl-Alcohol, Pd Wire, Sr and Hg Salts," *Procs. of the 4th Japan CF Conference*, 17-21

Celani, Francesco, Spallone, Antonio, Righi, E., Trenta, G., Catena, C., D'Agostaro, G., Quercia, Andreassi, V., Marini, Di Stefano, V., Nakamura, M., Mancini, A., Sona, G., Fontana, F., Gamberale, Luca, Garbelli, D., Falcioni, F., Marchesini, M., Novaro, E. and Mastromatteo, Ubaldo (Aug. 2003) "Thermal and Isotopic Anomalies When Pd Cathodes Are Electrolyzed in Electrolytes Containing Th-Hg Salts Dissolved at Micromolar Concentration in

C2H5OD/D2O Mixtures," *Procs. of the 10th Intl. Conf. on Cold Fusion,*" 379-397

CERN (retrieved Dec. 1, 2014) (web.cern.ch)

Chadwick, James (1932) "The Existence of a Neutron," *Procs. of the Royal Society of London,* **A136,** 692

Chadwick, James (1932) "Possible Existence of a Neutron," *Nature,* **129,** 312

Chien, Chun-Ching, Hodko, Dalibor, Minevski, Zoran, and Bockris, John O'Mara (April 1992) "On an Electrode Producing Massive Quantities of Tritium and Helium," *Journal of Electroanalytical Chemistry,* **338,** 189–212

Chubb, Scott (April, 2010) "At 21, Cold Fusion Is Still in Its Infancy," *Infinite Energy,* **90,** 8-11

Cirillo, Domenico, Dattilo, Alessandro, and Iorio, Vincenzo (presented Nov. 2004) "Transmutation of Metal to Low Energy in Confined Plasma in the Water," *Procs. of the 11th Intl. Conf. on Cold Fusion,* 492-504

Cirillo, Domenico, Germano, Roberto, Tontodonato, V., Widom, Allan, Srivastava, Yogendra, Del Giudice, Emilio, and Vitiello, G. (online Nov. 2011) "Experimental Evidence of a Neutron Flux Generation in a Plasma Discharge Electrolytic Cell," *Key Engineering Materials,* **495,** 104-107, (2012)

Cirillo, Domenico, Germano, Roberto, Tontodonato, Valentino, Widom, Allan, Srivastava, Yogendra N., Del Giudice, Emilio, Vitiello, Giuseppe (2012) "Experimental Evidence of a Neutron Flux Generation in a Plasma Discharge Electrolytic Cell," *Key Engineering Materials,* **495** 104-107

Ciuchi, S., Maiani, L., Polosa, AD, Riquer, V., Ruocco, G., Vignati, M. (arXiv Sept. 28, 2012) "Low Energy Neutron Production by Inverse Beta Decay in Metallic Hydride Surfaces," *The European Physical Journal C,* **72,** 2193-6 (online Oct. 26, 2012)

Claytor, Thomas N., Jackson, Damon D., and Tuggle, Dale G. (1996) "Tritium Production from a Low Voltage Deuterium Discharge on Palladium and Other Metals," Los Alamos report LAUR#95-2687

Cockcroft, John and Walton, Ernest (1932) "The Disintegration of Elements by High-Velocity Protons," *Procs. of the Royal Society of London,* **137,** 229-42; Cockcroft and Walton (1932) *Nature* **129,** p. 242

Conte, Elio, (April, 1999) "Theoretical Indications of the Possibility of Nuclear Reactions at Low Energy," *Infinite Energy,* **24,** 49-55

Coupland, D. R., Doyle, M. L., Jenkins, J. W., Notton, J. H. F., Potter, R. J., and Thompson, D. T. (1993) "Some Observations Related to the Presence of Hydrogen and Deuterium in Palladium," *Procs. of the 3rd Intl. Conf. on Cold Fusion,* Nagoya, Japan, Oct. 21-26, 1992, 265-284

Daddi, Lino (Oct. 1995) "On the Possible Role of Virtual Neutrons in Cold Fusion," *Infinite Energy,* **35,** 58-62

Daddi, Lino (May 27, 2005) "Hydrogen Miniatoms," Coherence 2005 Seminar, Rome, Italy

Darwin, Charles Galton (1920) "Motion of Charged Particles," *Philosophical Magazine,* Series 6, **39,** 537-51

Dash, John, Nobel, Grant, and Diman, D. (presented in 1993) "Surface Morphology and Microcomposition or Palladium Cathodes after Electrolysis in Acidified Light and Heavy Water: Correlation with Excess Heat," *Procs. of the 4th Intl. Conf. on Cold Fusion,* 25-1-25-11

Dash, John (April 1995) "Letter to the Editor," *Infinite Energy*, 1(1), 5

Dash, John (April 27, 2001) "Interaction of Titanium with Hydrogen Isotopes-Final Progress Report " US Army Research Office Grant #DAAAG55-97-1-0357

Dominguez, Dawn, Hagans, Patrick, L., and Imam, M. Ashraf (Jan. 9, 1996) "A Summary of NRL Research on Anomalous Effects in Deuterated Palladium Electrochemical Systems," *Naval Research Laboratory*, NRL/MR/6170-96-7803

Enyo, Michio (February 24, 2000) "Open Minded Attitudes to the Science," Portland State University Web site, retrieved Feb. 6, 2016

Faccini, R., Pilloni, A., Polosa, A.D., Angelone, M., Castagna, E., Lecci, S., Pietropaolo, A., Pillon, M., Sansovini, M., Sarto, F., Violante, V., Bedogni, R., and Esposito, A. (Oct. 17, 2013) "Search for Neutron Flux Generation in a Plasma Discharge Electrolytic Cell," http://arxiv.org/abs/1310.4749

Faccini, R., Pilloni, A., Polosa, A.D., Angelone, M., Castagna, E., Lecci, S., Pietropaolo, A., Pillon, M., Sansovini, M., Sarto, F., Violante, V., Bedogni, R., and Esposito, A. (Jan. 30, 2014) "Further Investigations on the Neutron Flux Generation in a Plasma Discharge Electrolytic Cell," http://arxiv.org/abs/1401.8218

Fisher, John Crocker (April 1974) *Energy Crises in Perspective*, John Wiley & Sons Inc.

Flatow, Ira (March 26, 2010) "Scientific Disciplines Mix At Chemistry Meeting," National Public Radio's *Science Friday*

Fleischmann, Martin and Pons, Stanley (April 10, 1989) "Electrochemically Induced Nuclear Fusion of Deuterium," *Journal of Electroanalytical Chemistry*, 261(2), Part 1, 301-308; Errata: Fleischmann, Martin, Pons, Stanley, and Hawkins, Marvin (May 10, 1989) 263, 187-188

Fleischmann, Martin (presented March 1990) "An Overview of Cold Fusion Phenomena," *Procs. of the 1st Annual Conference on Cold Fusion*, 344-350

Fleischmann, Martin and Pons, Stanley (May 3, 1993) "Calorimetry of the Pd-D_2O System; From Simplicity Via Complications to Simplicity," *Physics Letters A*, 176, 118

Fleischmann, Martin (2000) "Reflections on the Sociology of Science and Social Responsibility in Science, in Relationship to Cold Fusion," *Accountability in Research*, 8, 19

Freeman, Leslie J. ed. (1981) *Nuclear Witnesses: Insiders Speak Out*, Norton & Co., 54

Focardi, Sergio, Gabbani, V., Montalbano, V., Piantelli, Francesco, and Veronesi, S. (1998) "Large Excess Heat Production in Ni-H Systems," *Nuovo Cimento*, 111A, 1233-1242

Focardi, S., Gabbani, V., Montalbano, V., Piantelli, F., and Veronesi, S. (presented Nov. 2004) "Evidence of Electromagnetic Radiation From Ni-H Systems," *Procs. of the 11th Intl. Conf. on Condensed Matter Nuclear Science*, 70-80

Fox, Hal (Aug. 1995) "Does Low-Temperature Nuclear Change Occur in Solids? A Report on the Low-Energy Transmutation Conference held at Texas A&M University, June 19, 1995," *Infinite Energy*, 1(3), 8-11

Fox, Hal, ed. (Jan. 1996) *Journal of New Energy*, 1(1), 1-138

Gerischer, Heinz, (1991) "Is Cold Fusion a Reality? The Impressions of a Critical Observer," *Procs. of the 2nd Annual Conference on Cold Fusion*, 465-474

Giaever, Ivar (Oct. 2007) "The Discovery of Superconducting Tunneling," slide presentation at BCS Conference, University of Illinois

Goodstein, David, (Dec. 17, 2006) "Conduct and Misconduct in Science," *Annals of the New York Academy of Sciences*, **775**, 31-38

Hagelstein, Peter (1990) "Status of Coherent Fusion Theory," *Procs. of the 1st Annual Conference on Cold Fusion*, 99-118

Hagelstein, Peter, L. (Feb. 16, 1993) "Summary Of The Third Annual Conference on Cold Fusion," www.lenr-canr.org, 15-20

Hagelstein, Peter, (1996) "Anomalous Energy Transfer between Nuclei and the Lattice," *Procs. of the 6th Intl. Conf. on Cold Fusion,* 382-386

Hagelstein, Peter, (presented April 1998) " Anomalous Energy Transfer," *Procs. of the 7th Intl. Conf. on Cold Fusion*, 140-146

Hagelstein, Peter, McKubre, Michael, Nagel, David, Chubb, Talbot, and Hekman, Randy (Aug. 2004) "New Physical Effects In Metal Deuterides," U.S. Department of Energy Review of Low-Energy Energy Nuclear Reactions

Hagelstein, Peter and Chaudhary, Irfan (online Jan. 24, 2008) "Electron Mass Shift in Nonthermal Systems," *Journal of Physics B: Atomic, Molecular and Optical Physics*, **41**(12), 125001-125010 (June 6, 2008)

Hagelstein, Peter L. (Feb. 9, 2010) "Constraints on Energetic Particles in the Fleischmann–Pons Experiment," *Naturwissenschaften*, **97**(4), 345-352

Hahn, Otto and Strassmann, Fritz (Nov. 18, 1938) "Über die Entstehung von Radiumisotopen aus Uran durch Bestrahlen mit Schnellen und Verlangsamten Neutronen," *Naturwissenschaften*, **26**(46), 755-6

Hansen, Wilford, N. and Melich, Michael, E. (1993) "Pd/D Calorimetry - The Key to the F/P Effect and a Challenge to Science," *Procs. of the 4th Intl. Conf. on Cold Fusion*, 11-1-11-12 and Transactions of Fusion Technology, 26(4T), Part 2, (Dec. 1994), 355

Higashiyama, Taichi, Sakano, Mitsuru, Miyamaru, Hiroyuki, and Takahashi, Akito (presented August 2003) "Replication of Mitsubishi Heavy Industries Transmutation Experiment By D_2 Gas Permeation Through Pd Complex," *Procs. of the 10th Intl. Conf. on Cold Fusion*, 447-454

Hioki, Tatsumi, Takahashi, Naoko, Kosaka, Satoru, Nishi, Teppei, Azuma, Hirozumi, Hibi, Shogo, Higuchi, Yuki, Murase, Atsushi, and Motohiro, Tomoyoshi (October 4, 2013) "Inductively-Coupled Plasma Mass Spectrometry Study on the Increase in the Amount of Pr Atoms for Cs-Ion-Implanted Pd/CaO Multilayer Complex with Deuterium Permeation," *Japanese Journal of Applied Physics*, **52**(10R), 107301-1-8

Hoyle, Fred (1946) "The Synthesis of the Elements from Hydrogen," *The Monthly Notices of the Royal Astronomical Society*, **106**(5), 343-383

Iyengar, Padmanabha Krishnagopala (presented July 1989) "Cold Fusion Results in BARC Experiments," *Procs. of the 5th Intl. Conf. on Emerging Nuclear Energy Systems*, Karlsruhe, Germany, World Scientific, 291-295 (1989)

Iwamura, Yasuhiro, Itoh, Takehiko, and Toyoda, Ichiro (presented Dec. 1993) "Observation of Anomalous Nuclear Effects in D2-Pd System," *Procs. of the 4th Intl. Conf. on Cold Fusion*, 12-1-12-9

Iwamura, Yasuhiro, Gotoh, Nobuaki, Itoh, Takehiko, and Toyoda, Ichiro (presented April 1995) "Characteristic X-ray and Neutron Emissions from Electrochemically Deuterated Palladium," *Procs. of the 5th Intl. Conf. on Cold Fusion*, 197-200

Iwamura, Yasuhiro, Itoh, Takehiko, Gotoh, Nobuaki, and Toyoda, Ichiro (July 1998) "Detection of Anomalous Elements, X-ray and Excess Heat in a D2-Pd System and its Interpretation by the Electron-Induced Nuclear Reaction Model," *Fusion Technology*, **33**, 476-492

Iwamura, Yasuhiro, Itoh, Takehiko, Gotoh, Nobuaki, Sakano, Mitsuru, Toyoda, Ichiro, and Sakata, Hiroshi (1998) "Detection of Anomalous Elements, X-Ray and Excess Heat Induced by Continuous Diffusion of Deuterium through Multilayer Cathode (Pd/CaO/Pd)," *Procs. of the 7th Intl. Conf. on Cold Fusion*, 167-171

Iwamura, Yasuhiro, Sakano, Mitsuru, and Itoh, Takehiko (July 2002) "Elemental Analysis of Pd Complexes: Effects of D2 Gas Permeation," *Japanese Journal of Applied Physics A*, **41**, 4642-4650

Iwamura, Yasuhiro, Itoh, Takehiko, and Tsuruga, Shigenori, (Feb. 2015) "Transmutation Reactions Induced by Deuterium Permeation through Nano-structured Palladium Multilayer Thin-Film," *Current Science*, **108**(4), 628-32

Jevtic, Nada, Pronko, John, McKubre, Michael, Crouch-Baker, Steven, Tanzella, Francis L., Bush, Benjamin, Williams, Mark, Wing, Sharon (Nov. 1999) "Development of Energy Production Systems from Heat Produced in Deuterated Metals, Volume 2, Electric Power Research Institute TR-107843

Jiang, X. L., Han, L. J., and Kang, W. (April 1998) "Anomalous Element Production Induced by Carbon Arcing Under Water" *Procs. of the 7th Intl. Conf. on Cold Fusion*, 172-174

Joliot, Frédéric and Joliot-Curie, Irène (1934) "Artificial Production of a New Kind of Radioelement," *Nature*, **133**, 201

Karabut, Alexander B., Kucherov, Yan R., and Savvatimova, Irina B. (June 26, 1990) "Nuclear Reactions at the Cathode in a Gas Discharge," *Soviet Technical Physics Letters*, **16**(6), 463-464

Karabut, Alexander B., Kucherov, Yan R., and Savvatimova, Irina B. (Nov. 9, 1992) "Nuclear Product Ratio for Glow Discharge in Deuterium," *Physics Letters A*, **170**(4), 265-272

Kervran, Louis C. (1971) *Biological Transmutations*, (Swan House Pub.)

Kim, Yeong (Dec. 1990) "Neutron Burst from a High-Voltage Discharge between Palladium Electrodes in D_2 Gas," *Fusion Technology*, **18**, 680-682

Knies, David, Violante, Vittorio, Grabowski, Kenneth, Hu, J.Z., Dominguez, Dawn, He, J.H., Quadri, S.B., and Hubler, Graham (Oct. 18, 2012) "In Situ Synchrotron Energy Dispersive X-Ray Diffraction Study of Thin Pd Foils With Pd: D and Pd: H Concentrations up to 1:1," *Journal of Applied Physics*, **112**, 083510, 1-6

Kopecek, Radovan and Dash, John (presented Sept. 1996) "Excess Heat and Unexpected Elements from Electrolysis of Acidified Heavy Water with Titanium Cathodes," *Procs. of the 2nd International Low Energy Nuclear Reactions Conference*, 46-53

Krivit, Steven B. (Sept. 2008) "Review of 'Voodoo Science: The Road From

Foolishness to Fraud'" by Robert L. Park, Oxford University Press, USA (Nov. 15, 2001), *Journal of Scientific Exploration*, **22**(3), 428-433

Krivit, Steven B. and Marwan, Jan (online Sept. 3, 2009) "A New Look at Low-Energy Nuclear Reaction Research," Journal of Environmental Monitoring, Royal Society of Chemistry, (Oct. 2009) **11**(10), 1731-46

Krivit, Steven. B. (Aug. 15, 2013) "Nuclear Phenomena in Low-Energy Nuclear Reaction Research," *Naturwissenschaften*, **100**(9) 899-900 (print, Sept. 2013)

Kunznetsov, Vladimir D., Mishinsky, Gennady V., Penkov, Fedor M., Arbuzov, Vladimir, I., and Zhemenik, Valentin, I. (2003) "Low Energy Transmutation of Atomic Nuclei of Chemical Elements," *Annales Fondation de Louis de Broglie*, **28**(2), 173-213

Laing, Jonathan (Sept. 1, 1986) "Shining Prophecy: The Coming Renaissance of U.S. Industry," *Barron's*, **66**(35), 13-28

Laing, Jonathan (Feb. 1, 1988) "Back to the Futurist: Lewis Larsen Says We Ain't Seen Nothing Yet," *Barron's*, **68**(5), 58-59

Laing, Jonathan (Jan. 11, 1999) "Are We Headed for a New Age?" *Barron's*, **79**(2), 27-30

Larsen, Lewis (Oct. 12, 2008) "Portable and Distributed Power Generation from LENRs," *Institute of Science in Society*, http://www.i-sis.org.uk/

Larsen, Lewis (Feb. 7, 2009) "Weak Interaction LENR Transmutation Reactions on Earth versus Nucleosynthesis in Stars," slideshare.com

Larsen, Lewis and Widom, Allan (Feb. 11, 2011) "Apparatus and Method for Absorption of Incident Gamma Radiation and its Conversion to Outgoing Radiation at Less Penetrating, Lower Energies and Frequencies," U.S. Patent #7,893,414

Lochak, Georges and Urutskoev, Leonid, I. (presented Nov. 2004) "Low-Energy Nuclear Reactions and the Leptonic Monopole," *Procs. of the 11th Intl. Conf. on Cold Fusion*," 421-437

Los Angeles Times (Oct. 26, 1992) "Japan Keeps Working on Cold Fusion: Technology: a Senior Researcher at NTT Now Claims to Have Evidence of the Controversial Phenomenon," D3

Machiels, Albert and Passell, Thomas (Nov. 1999) "Trace Elements Added to Palladium by Electrolysis in Heavy Water," EPRI TP-108743

Mallove, Eugene, (May 1994) "An Italian Cold Fusion Hot Potato," *Cold Fusion*, **1**(1), 44-45

Mallove, Eugene, (April 1999) "Cold Fusion: The First 10 Years," *Infinite Energy*, **24**, 7-63

Mallove, Eugene (May 1999) "Professor George H. Miley of the University of Illinois Awarded a DOE Contract for Low Energy Nuclear Reactions (LENR) Study," *Infinite Energy*, **26**, 43-44

Mallove, Eugene (Dec. 1999) "Critics Kill Professor George Miley's Historic U. S. DOE Low-Energy Nuclear Reactions Contract," *Infinite Energy*, **28**, 44

Mallove, Eugene (2000) "Breaking Through: Welcome ICCF8 - Liberate Science," *Infinite Energy*, **31**, 4

Mallove, Eugene (2003) "Review of the 10th International Conference on Cold Fusion (ICCF10)" *Infinite Energy*, **52**, 6-8

Mallove, Eugene (2004) "Breaking Through: Vindication!?" *Infinite Energy*, **55**, 7

Marwan, Jan and Krivit, Steven B., eds. (Aug. 2008) *Low-Energy Nuclear Reactions Sourcebook*, American Chemical Society/Oxford University Press

Marwan, Jan and Krivit, Steven B., eds. (Dec. 2009) *Low-Energy Nuclear Reactions and New Energy Technologies Sourcebook (Vol. 2)* American Chemical Society/Oxford University Press

McKubre, Michael, Crouch-Baker, Steven, Tanzella, Francis L., Smedley, Stuart I., Williams, Mark, Wing, Sharon, Maly-Schreiber, Maria, Rocha-Filho, Romeu, Searson, Peter C., Pronko, John G., Kohler, Donald A. (August 1994) "Development of Advanced Concepts for Nuclear Processes in Deuterated Metals, Electric Power Research Institute, TR-104195, Research Project 3170-01

McKubre, Michael, Crouch-Baker, Steven, Hauser, Alan, Jevtic, Nada, Smedley, Stuart I., Tanzella, Francis L., Williams, Mark, Wing, Sharon, Bush, Benjamin, McMahon, Fiona, Srinivasan, Mahadeva, Wark, Alister, Warren, Derek (June 1998) "Development of Energy Production Systems from Heat Produced in Deuterated Metals, Volume 1, Electric Power Research Institute, TR-107843

McKubre, Michael, (presented 1998) "Michael McKubre's (SRI) Summary During ICCF-7 Closing Remarks," *Infinite Energy*, **20**, 34-35 (Transcribed by Jed Rothwell)

McKubre, Michael, Tanzella, Francis, Tripodi, Paolo, Hagelstein, Peter (presented May 2000) "The Emergence of a Coherent Explanation for Anomalies Observed in D/Pd and H/Pd System: Evidence for 4-He and 3-He Production," *Procs. of the 8th Intl. Conf. on Cold Fusion*, 3–16

McKubre, Michael (slides presented 2004) "DoE Review: New Physical Effects in Metal Deuterides," 11th International Conference on Cold Fusion

McKubre, Michael, Tanzella, Francis, Dardik, Irving, El-Boher, Arik, Zilov, Tonya, Greenspan, Ehud, Sibilia, Concita, and Violante, Vittorio (August 2008) "Replication of Condensed Matter Heat Production," in *Low-Energy Nuclear Reactions Sourcebook,* Marwan, Jan and Krivit, Steven B., eds., American Chemical Society/Oxford University Press

McKubre, Michael (slides presented 2009) "Cold Fusion, LENR, the Fleischmann-Pons Effect; One Perspective on the State of the Science," 15th International Conference on Condensed Matter Nuclear Science

Meitner, Lise and Frisch, Otto (Feb. 11, 1939) "Disintegration of Uranium by Neutrons: A New Type of Nuclear Reaction," *Nature,* **143**, 239-40

Melich, Michael E. and Hansen, Wilford N. (1992) "Some Lessons from 3 Years of Electrochemical Calorimetry," *Procs. of the 3rd Intl. Conf. on Cold Fusion*, 397-400

Melich, Michael E. and Hansen, Wilford N. (1993) "Back to the Future, The Fleischmann-Pons Effect in 1994," *Procs. of the 4th Intl. Conf. on Cold Fusion*, 10-1-10-10

Mengoli, Giuliano, Bernardini, M., Manducchi, C., and Zannoni, G. (1998) "Anomalous Heat Effects Correlated With Electrochemical Hydriding of Nickel," *Il Nuovo Cimento*, **20-D**, 331-352

Mengoli, Giuliano, Bernardini, M., Manducchi, C., and Zannoni, G. (1998)

"Calorimetry Close to the Boiling Temperature of the Pd-D$_2$O Electrolytic System," *Journal of Electroanalytical Chemistry*, **444**, 155

Miles, Melvin M., Bush, Ben F., Ostrom, Greg S., and Lagowski, Joseph J., (1991) "Heat and Helium Production in Cold Fusion Experiments," *Procs. of the 2nd Annual Conference on Cold Fusion*, 363-372

Miles, Melvin, Bush, Benjamin F., and Johnson, Kendall B. (Sept. 1996) "Anomalous Effects in Deuterated Systems, Final Report," Naval Air Warfare Center Weapons Division, NAWCWPNS-TP-8302

Miles, Melvin, Fleischmann, Martin, and Imam, Ashraf (March 26, 2001) "Calorimetric Analysis of a Heavy Water Electrolysis Experiment Using a Pd-B Alloy Cathode," Naval Research Laboratory, NRL/MR/6320-01-8526

Miles, Melvin (2015) "Excerpts From Martin Fleischmann Letters," presented at the 17th *Intl. Conf. on Cold Fusion*

Miley, George (Aug. 1996) "Summary Report on the 2nd International Low Energy Nuclear Reactions Conference (ILENR2)," *Infinite Energy*, **9**, 11

Miley, George H. and Patterson, James A. (August 1996) "Nuclear Transmutations in Thin-Film Nickel Coatings Undergoing Electrolysis," *Infinite Energy*, **9**, 19-32; also nearly identically published, Fall 1996, *Journal of New Energy*, 1(3): 5-30

Miley, George H., Narne, Gokul, Williams, M.J., Patterson, James A., Nix, J., Cravens, Dennis, Hora, Heinrich (presented Oct. 1996) "Quantitative Observation of Transmutation Products Occurring in Thin-Film Coated Microspheres During Electrolysis," *Procs. of the 6th Intl. Conf. on Cold Fusion*, 629-644

Miley, George H. (presented May 2000) "On the Reaction Product and Heat Correlation for LENRs," *Procs. of the 8th Intl. Conf. on Cold Fusion*, 419-24

Millikan, Robert A. (Nov. 1923) "Gulliver's Travels in Science," *Scribner's Magazine*, **74**(5), 577-85

Minevski, Z. and Bockris, John, (Feb. 1996) "Two Zones of 'Impurities' Observed After Prolonged Electrolysis of Deuterium on Palladium," *Infinite Energy*, **6**, 67-68

Mizuno, Tadahiko, Ohmori, Tadayoshi, and Michio Enyo (April 1996) Anomalous Isotopic Distribution in Palladium Cathode After Electrolysis," *Infinite Energy*, **7**, 10-13, also nearly identically published, Summer 1996, in *Journal of New Energy*, 1(2), 37-44

Mizuno, Tadahiko, Ohmori, Tadayoshi, Enyo, Michio (presented Sept. 1996) " Isotopic Changes of the Reaction Products Induced by Cathodic Electrolysis in Pd," *Procs. of the 2nd International Low Energy Nuclear Reactions Conference*, 31-45

Mizuno, Tadahiko, Ohmori, Tadayoshi, Akimoto, Tadashi, Kurokawa, Kazuya, Kitaichi, Masatoshi, Inoda, Koichi, Azumi, Kazuhisa, Simokawa, Shigezo, and Enyo, Michio (presented 1996) "Isotopic Distribution for the Elements Evolved in Palladium Cathode After Electrolysis in D$_2$O Solution," *Procs. of the 6th Intl. Conf. on Cold Fusion*, 665-69; also *Denki Kagaku (Japanese Journal of Electro-Chemistry,)* **64** (11), 1160-1165, (1996)

Mizuno, Tadahiko (1997) *Nuclear Transmutation: the Reality of Cold Fusion*, Infinite Energy Press

Mizuno, Tadahiko, Ohmori, Tadayoshi, Azumi, Kazuhisa, Akimoto, Tadashi, and Takahashi, Akito (presented May 2000) "Confirmation of Heat Generation and Anomalous Element Caused by Plasma Electrolysis in the Liquid," *Procs. of the 8th Intl. Conf. on Cold Fusion*, 75-80

Mizuno, Tadahiko (Feb. 2001) "A Letter from Dr. Tadahiko Mizuno to Jed Rothwell," *Infinite Energy*, **35**, 14

Mizuno, Tadahiko (2005) "Accident Report," www.lenr-canr.org

Mizuno, Tadahiko and Toriyabe, Yu (2005) "Anomalous Energy Generation during Conventional Electrolysis," *Procs. of the 12th Intl. Conf. on Cold Fusion*, 65-74

Mizuno, Tadahiko (March 2009) "Isotopic Changes of Elements Caused by Various Conditions of Electrolysis," presented at the 237th American Chemical Society National Meeting, Salt Lake City, Utah

Mosier-Boss, Pamela, Szpak, Stanislaw, Gordon, Frank E., and Forsley, Larry (2008) "Reply to Comment on 'The use of CR-39 In Pd/D Co-deposition Experiments': A Response to Kowalski," *European Physics Journal of Applied Physics*, **44**, 291-295

Mosier-Boss, Pamela, Szpak, Stanislaw, Gordon, Frank E., and Forsley, Larry (2009) "Triple Tracks in CR-39 as the Result of Pd/D Co-deposition: Evidence of Energetic Neutrons," *Naturwissenschaften*, **96**, 135-142

Nagel, David and Melich, Michael (2008) "Preface," *Procs. of the 14th Intl. Conf. on Condensed Matter Nuclear Science and the 14th Intl. Conf. on Cold Fusion*, i-xii

Naval Postgraduate School (Dec. 2006) "Summary of Research 2005," NPS-09-06-010

Naval Postgraduate School (Dec. 2010) "Draft Summary of Research 2009," NPS-04-10-005

Niedra, Janis M., Myers, Ira T., Fralick, Gustave C., Baldwin, Richard S. (1996) "Replication of the Apparent Excess Heat Effect in a Light Water-Potassium Carbonate-Nickel Electrolytic Cell", NASA TM-107167

Noninski, Vesselin C. (March 1992) "Excess Heat During the Electrolysis of a Light-Water Solution of K_2CO_3 With a Nickel Cathode," *Fusion Technology*, **21**, 163-7

Notoya, Reiko and Enyo, Michio (presented Oct. 1992) "Excess Heat Production in Electrolysis of Potassium Carbonate Solution with Nickel Electrodes," *Procs. of the 3rd Intl. Conf. on Cold Fusion*, 421-426

Ohmori, Tadayoshi and Enyo, Michio (presented June 1995) "Iron Formation in Gold and Palladium Cathodes," *Procs. of the [1st] Low-Energy Nuclear Reactions Conference*, 15-19

Ohmori, Tadayoshi, Mizuno, Tadahiko, Enyo, Michio (presented Sept. 1996) "Isotopic Distributions of Heavy Metal Elements Produced During the Light Water Electrolysis on Au Electrode," *Procs. of the 2nd International Low Energy Nuclear Reactions Conference*, 90-99

Ohmori, Tadayoshi, Mizuno, Tadahiko, Enyo, Michio (presented April 1998) "Strong Excess Energy Evolution, New Element Production, and Electromagnetic Wave and/or Neutron Emission in the Light Water Electrolysis with a Tungsten Cathode," *Procs. of the 7th Intl. Conf. on Cold*

Fusion, 279-284

Oliphant, Mark, Harteck, Paul, and Rutherford, Ernest (1934) "Transmutation Effects Observed With Heavy Hydrogen," *Procs. of the Royal Society of London A*, **144**, 692-703

Oriani, Richard A., Nelson, John C., Lee, Sung-Kyu, and Broadhurst, J. H. (1990) "Calorimetric Measurements of Excess Power Output During the Cathodic Charging of Deuterium Into Palladium," *Fusion Technology*, **18**, 652

Park, Robert L. (May 2000) *Voodoo Science: The Road from Foolishness to Fraud*, Oxford University Press

Park, Robert (June 6, 2008) "Hydrinos: How Long Can a Really Dumb Idea Survive?" What's New?, University of Maryland

Passell, Thomas, (1994) "Foreword," *Procs. of the 4th Intl. Conf. on Cold Fusion*, (no page number)

Passell, Thomas O. (presented April 1995) "Charting the Way Forward in the EPRI Research Program on Deuterated Metals," *Procs. of the 5th Intl. Conf. on Cold Fusion*," 603-18

Passell, Thomas (1996) "Search for Nuclear Reaction Products in Heat-Producing Palladium," *Procs. of the 6th Intl. Conf. on Cold Fusion*, 282-290

Passell, Thomas O. and George, Russell (presented May 2000) "Trace Elements Added to Palladium by Exposure to Gaseous Deuterium," *Procs. of the 8th Intl. Conf. on Cold Fusion*, 129-133

Passell, Thomas (2002) "Evidence for Lithium-6 Depletion in Pd Exposed to Gaseous Deuterium and Hydrogen," *Procs. of the 9th Intl. Conf. on Cold Fusion*, 299-304

Passell, Thomas O. (Presented Aug. 2003) "Pd-110/Pd-108 Ratios and Trace Element Changes in Particulate Palladium Exposed to Deuterium Gas," *Procs. of the 10th Intl. Conf. on Cold Fusion*, 399-403

Piantelli, Francesco (Sept. 2010) "Proton Reactor," Presented at the 9th International Workshop on Anomalies in Hydrogen/Deuterium Gas Loaded Metals, Siena, Italy

Piantelli, Francesco and Collis, William (April 2012) "Some Results from the Nichenergy Laboratory," Presented at the 10th International Workshop on Anomalies in Hydrogen Loaded Metals, Siena, Italy

Pons, Stanley and Fleischmann, Martin (July 1990) "Calorimetric Measurements of the Palladium/Deuterium System: Fact and Fiction," *Fusion Technology*, **17**, 669-679

Ramamurthy, Halasyam, Srinivasan, Mahadeva, Mukherjee, U.K., and Adibabu, (presented Dec. 1993) "Further Studies on Excess Heat Generation in Ni-H$_2$O Electrolytic Cells," *Procs. of the 4th Intl. Conf. on Cold Fusion*, **2**, 15-1-15-31

Rhodes, Richard (1986) *The Making of the Atomic Bomb*, Simon and Schuster

Rolison, Debra R. and O'Grady, William E. (presented Oct. 1989) "Mass/Charge Anomalies in Pd After Electrochemical Loading With Deuterium," Procs. of NSF/EPRI Workshop on Anomalous Effects in Deuterided Metals, Washington, D.C., August 1993

Rolison, Debra R., O'Grady, William E., Doyle, Jr., Robert J., and Trzaskoma, Patricia (presented March 1990) "Anomalies in the Surface Analysis of Deuterated Palladium," *Procs. of the 1st Annual Conference on Cold Fusion*,

272-280

Rolison, Debra R. and O'Grady, William E. (Sept. 1, 1991) "Observation of Elemental Anomalies at the Surface of Palladium after Electrochemical Loading of Deuterium or Hydrogen," *Analytical Chemistry*, (**63**)17, 1697-1702

Rosada, A, Santoro, E., Sarto, Francesca, Violante, Vittorio, Avino, Pasquale (presented 2009) "Impurity Measurements by Instrumental Neutron Activation Analysis on Palladium, Nickel and Copper Thin Films," *Procs. of the 15th Intl. Conf. on Condensed Matter Nuclear Science*, 221-226

Rothwell, Jed (Aug. 1996) "Summary Report on the Second International Low Energy Nuclear Reactions Conference (ILENR2)," *Infinite Energy*, **9**, 10-17

Rothwell, Jed (Oct. 1996) "Review of the 6th International Conference on Cold Fusion ICCF-6," *Infinite Energy*, **10**, 13-21

Rothwell, Jed (Feb. 1999) "The American Chemical Society Conference Cold Fusion Sessions," *Infinite Energy*, **29**, 18-25

Rothwell, Jed and Mallove, Eugene (August 2000), "Summary Report on ICCF8: The Eighth International Conference on Cold Fusion," *Infinite Energy*, **32**, 25-68

Rothwell, Jed (Feb. 2001) "Report on the Second Annual Japan Cold Fusion Society Conference (JCF-2)," *Infinite Energy*, **35**, 9-17

Roulette, Thierry, Roulette, Jeanne, and Pons, Stanley (1996) "Results of Icarus-9 Experiments Run at IMRA Europe," *Procs. of the 6th Intl. Conf. on Cold Fusion*, 85-92

Sanderson, Katharine (March 22, 2010) "ACS: Cold Fusion Calorimeter Confusion," *Nature* blog, http://blogs.nature.com/

Sankaranarayanan, Thevarmadhom Krishna, Srinivasan, Mahadeva, Bajpai, M., and Gupta, D. (Presented Dec. 1993) "Investigation of Low Level Tritium Generation in Ni-H_2O Electrolytic Cells," *Procs. of the 4th Intl. Conf. on Cold Fusion*, (published July 1994), **3**, 3-1-3-14

Savvatimova, Irina B., Kucherov Yan, and Karabut, Alexander B. (Presented Dec. 1993) "Cathode Material Change after Deuterium Glow Discharge Experiments," *Procs. of the 4th Intl. Conf. on Cold Fusion*, (published July 1994)

Scaramuzzi, Francesco (2000) Foreword to the *Procs. of the 8th Intl. Conf. on Cold Fusion*, xii-xiv

Science (June 18, 1999) "Letters to the Editor," **284**, 1929

Science (July 23, 1999) "DOE to Review Nuclear Grant," **285**, 505-506

Shkedi, Zvi, Mcdonald, Robert C., Breen, John J., Maguire, Stephen J., and Veranth, Joe (Nov. 1995) "Calorimetry, Excess Heat and Faraday Efficiency in Ni-H_2O Electrolytic Cells," *Fusion Technology*, **28**, 1720-1731; also Shkedi, Zvi, (Sept. 1996) "Response to 'Comments on Calorimetry, Excess Heat and Faraday Efficiency in Ni-H_2O Electrolytic Cells'," *Fusion Technology*, **30**, 133

Singh, Mahavir, Saksena, M., Dixit, V., and Kartha, V. (Nov. 1994) "Verification of the George Oshawa Experiment for Anomalous Production of Iron from Experiment Carbon Arc in Water," *Fusion Technology*, **26**, 266-270

Srinivasan, Mahadeva, Shyam, A., Sankaranarayanan, T. K., Bajpai, M. B., Ramamurthy, H., Mukherjee, U. K., Krishnan, M. S., Nayar, M. G., and Naik, Y. (presented Oct. 1992) "Tritium and Excess Heat Generation During

Electrolysis of Aqueous Solutions of Alkali Salts with Nickel Cathode," *Procs. of the 3rd Intl. Conf. on Cold Fusion,* 123-132

Srinivasan, Mahadeva and McKubre, Michael (May 1994) "Two Balance Method of Faraday Efficiency Measurement with External Recombiner and Open Cell Calorimetry for Identifying Origin of Excess Heat in NiH_2O Electrolytic Cells," 1(1), 73-4

Srivastava, Yogendra N., Widom, Allan, and Larsen, Lewis (Oct. 2010) "A Primer for Electro-Weak Induced Low Energy Nuclear Reactions," *Pramana - Journal of Physics,* 75(4) 617-637

Sternglass, Ernest (July 7-8, 1951) Laboratory Notebook, July 1951-Nov. 1951, Cornell University, Kroch Library, Rare & Manuscript Collections

Sternglass, Ernest J. (Aug. 26, 1951) "Letter to Albert Einstein," The Albert Einstein Archives, The Hebrew University of Jerusalem, Archival Call Number: 22-247

Sternglass, Ernest (Nov. 16, 1951) Laboratory Notebook, July 1951-Nov. 1951, Cornell University, Kroch Library, Rare & Manuscript Collections

Sternglass, Ernest (Jan. 12-13, 1953) Laboratory Notebook, Nov. 1951-Aug. 1953, Cornell University, Kroch Library, Rare & Manuscript Collections

Sternglass, Ernest (1997) *Before the Big Bang - the Origin of the Universe,* Four Walls Eight Windows, New York

Storms, Edmund, (Sept. 2010) "Status of Cold Fusion (2010)," *Naturwissenschaften,* 97, 861–881

Szpak, Stanislaw and Mosier-Boss, Pamela, eds., (Feb. 2002) "Thermal And Nuclear Aspects Of The Pd/D_2O System Vol. 1: A Decade Of Research At Navy Laboratories, SPAWAR Systems Center, San Diego, TR 1862, Vol. 1

Szpak, Stanislaw and Mosier-Boss, Pamela, eds., (Feb. 2002) "Thermal And Nuclear Aspects Of The Pd/ D_2O System Vol. 2: Simulation Of The Electrochemical Cell (Icarus) Calorimetry, SPAWAR Systems Center, San Diego, TR 1862, Vol. 2

Szpak, Stanislaw, Mosier-Boss, Pamela, and Gordon, Frank E. (2007) "Further Evidence of Nuclear Reactions in the Pd/D Lattice: Emission of Charged Particles," *Naturwissenschaften,* 94, 511-514; see also erratum: DOI 10.1007/s00114-007-0247-x

Takahashi, Akito (May 1989) "Opening Possibility of Deuteron-Catalyzed Cascade Fusion Channel in PdD under D_2O Electrolysis," *Journal of Nuclear Science and Technology,* 26(5), 558-560

Tanzella, Francis, Earle, Ben, and McKubre, Michael (Oct. 2007) "The Search for Nuclear Particles in the Pd-D Co-Deposition Experiment," Presented at the *8th* International Workshop on Anomalies in Hydrogen/ Deuterium Loaded Metals, Catania, Italy

Taubes, Gary (June 15, 1990) "Cold Fusion Conundrum at Texas A&M," *Science,* 248(4961), 1299-1304

Taylor, Harold John (1935) "The Disintegration of Boron by Neutrons," *Procs. of the Physical Society,* 47(5), 873-76

Teller, Edward (Presented Oct. 16-18, 1989) "Anomalous Effects on Deuterided Metal," *Procs. of NSF/EPRI Workshop on Anomalous Effects in Deuterided Metals,* Washington, D.C., (pub. August 1993)

Trost, Hans Jochen (Nov. 28, 2013), translation of Aug. 3, 1951 letter in "Einstein's Lost Hypothesis," *Nautilus* magazine, Winter 2014, 21-29

Urutskoev, Leonid, Liksonov, Vladimir I., Tsinoev, Vladlen G. (2002) "Observation of Transformation of Chemical Elements," *Annales Fondation Louis de Broglie*, **27**(4), 701-726

Urutskoev, Leonid, I., Filippov, Dmitri, Rukhadze, Anri A., Biryukov, A.O., Markolia, A.A., Alabin, Kirill A., Shpakovsky, T.V., Steshenko, G.K., Levanov, A.A., Belous, V. (2012) "The Development of Research Methodology of Gas Phase Formed After Electrical Explosion of Conductors," *Prikladnaya Fizika [Applied Physics]*, **4**, 60-68

Urutskoev, Leonid, (2015) "The Chernobyl Memorial," in *La Saga Nucleaire*, Colas-Linhart, Petiet, Anne, eds., L'Harmattan, Paris, France

Violante, Vittorio, Tripodi, Paolo, Di Gioacchino, D., Borelli, R., Bettinali, L., Santoro, E., Rosada, A., Sarto, Francesca, Pizzuto, A., McKubre, Michael, and Tanzella, Francis (presented May 2002) "X-Ray Emission During Electrolysis of Light Water on Palladium and Nickel Thin Films," *Procs. of the 9th Intl. Conf. on Cold Fusion*, 376-382

Violante, Vittorio, Apicella, M.L., Capobianco, L., Sarto, Francesca, Rosada, A, Santoro, E., McKubre, Michael, Tanzella, Francis, and Sibilia, Concita (presented Aug. 2003) "Search For Nuclear Ashes In Electrochemical Experiments," *Procs. of the 10th Intl. Conf. on Cold Fusion*, 405-420

Violante, Vittorio, Castagna, Emanuele, Sibilia, Concita, Paoloni, S., Sarto Francesca (presented August 2003) "Analysis Of Ni-Hydride Thin Film After Surface Plasmons Generation by Laser Technique," *Procs. of the 10th Intl. Conf. on Cold Fusion*, 421-434

Violante, Vittorio, Mazzitelli, G, Capobianco, L., Sarto, Francesca, Santoro, E., McKubre, Michael, Tanzella, Francis, Miley, George H., Luo, Nie, Shrestha, Prajakti Joshi, Sibilia, Concita (presented Aug. 2003) "Study of Lattice Potentials on Low-Energy Nuclear Processes In Condensed Matter," *Procs. of the 10th Intl. Conf. on Cold Fusion*, 667-680

Violante, Vittorio, Sarto, Francesca, Castagna, Emanuele, Lecci, Stefano, Sansovini, M., Torre, A., Hubler, Graham, Knies, David, Grabowski, Kenneth, McKubre, Michael, Tanzella, Francis, Sibilia, Concita, Del Prete, Z., and Zilov, Tanya (presented 2009) "Evolution and Progress in Material Science for Studying the Fleischmann and Pons Effect," *Procs. of the 15th Intl. Conf. on Cold Fusion*, 1-4

Wakefield, Dawn Lee, v. Texas A&M Development Foundation and Texas A&M University (1994) Case No. 40,518-CV, Brazos County Court, Texas

Walling, Cheves and Simons, John (June 15, 1989) "Two Innocent Chemists Look at Cold Fusion," *Journal of Physical Chemistry*, **93**(12), 4693-6

Wendt, Gerald L. and Irion, Clarence E. (Sept. 1922) "The Decomposition of Tungsten to Helium," *Journal of the American Chemical Society*, **44**(9), 1887-1894

White, Carol (Dec. 11, 1992) "Japan Cold Fusion Conference Sets New Direction For Science," *Executive Intelligence Review*, **19**(49), 20-24

Wick, Gian-Carlo (March 3, 1934) "Sugli Elementi Radioattivi di F. Joliot e I. Curie," Rendiconti Accademia, Lincei, Italy, **19**, 319-24

Widom, Allan and Larsen, Lewis (pre-print May 2, 2005) "Ultra Low Momentum Neutron Catalyzed Nuclear Reactions on Metallic Hydride Surfaces," *European Physical Journal C - Particles and Fields*, **46**(1), 107-110, (publication March 9, 2006)

Widom, Allan and Larsen, Lewis (Sept. 10, 2005) "Absorption of Nuclear Gamma Radiation by Heavy Electrons on Metallic Hydride Surfaces," http://arxiv.org/abs/cond-mat/0509269

Widom, Allan and Larsen, Lewis (Feb. 20, 2006) "Nuclear Abundances in Metallic Hydride Electrodes of Electrolytic Chemical Cells," http://arxiv.org/abs/cond-mat/0602472

Widom, Allan, Larsen, Lewis, and Srivastava, Yogendra N. (Sept. 8, 2007) "Energetic Electrons and Nuclear Transmutations in Exploding Wires," http://arxiv.org/abs/0709.1222

Widom, Allan and Larsen, Lewis (Sept. 25, 2007) "Theoretical Standard Model Rates of Proton to Neutron Conversions Near Metallic Hydride Surfaces," http://arxiv.org/abs/nucl-th/0608059v2

Widom, Allan, Srivastava, Yogendra N., and Larsen, Lewis (Feb. 5, 2008) "Errors in the Quantum Electrodynamic Mass Analysis of Hagelstein and Chaudhary," http://arxiv.org/abs/0802.0466

Widom, Allan, Larsen, Lewis, and Srivastava, Yogendra. N. (April 16, 2008) "High Energy Particles in the Solar Corona," http://arxiv.org/abs/0804.2647

Widom, Allan, Srivastava, Yogendra N., and Larsen, Lewis (Oct. 17, 2012) "Erroneous Wave Functions of Ciuchi et al. for Collective Modes in Neutron Production on Metallic Hydride Cathodes," http://arxiv.org/abs/1210.5212

Widom, Allan, Swain, John, Srivastava, Yogendra, and Cirillo, Domenico, (Nov. 11, 2013) "Analysis of an Attempt at Detection of Neutrons Produced in a Plasma Discharge Electrolytic Cell," http://arxiv.org/abs/1311.2447

Worledge, David H. (1990) "Technical Status of Cold Fusion Results," *Procs. of the 1st Annual Conference on Cold Fusion*," 252-260

Yamaguchi, Eiichi and Nishioka, Takahashi (April 1990) "Cold Fusion Induced by Controlled out Diffusion of Deuterons in Palladium," *Japanese Journal of Applied Physics*, **29**(4) L666-L669

Zhang, Wushou and Dash, John (2007) "Excess Heat Reproducibility And Evidence Of Anomalous Elements After Electrolysis In Pd/D_2O+H_2SO_4 Electrolytic Cells," *Procs. of the 13th Intl. Conf. on Condensed Matter Nuclear Science*, 202-16

Zimmerman, Peter (May 12, 1989) "Fusion Flap Proved an Elementary Axiom: Glitter Isn't Gold," *Los Angeles Times*, 11

Zimmerman, Peter (Oct. 2009) "Transcript of Presentation at the American Physical Society Meeting," *Procs. of the 3rd Intl. Conf. on Future Energy*, 91-94

Index

About the Author

Steven B. Krivit lives in San Rafael, California, and is an investigative science journalist and international speaker. He studied industrial design at the University of Bridgeport (Connecticut) and completed his bachelor's degree in business administration and information technology at National University (Los Angeles). He was a computer network systems engineer until 2000, when he became curious about low-energy nuclear reaction (LENR) research. He founded the *New Energy Times* Web site and online news service to share what he learned. By 2016, he had spoken with nearly all the scientists who were involved in the field. He has lectured nationally and international to scientific as well as lay audiences. He has advised the U.S. intelligence community, the U.S. Library of Congress, members of the Indian Atomic Energy Commission and the interim executive director of the American Nuclear Society. He is the leading author of review articles and chapters about LENRs, including invited papers for the Royal Society of Chemistry (2009), Elsevier (2009 and 2013) and John Wiley & Sons (2011). He was an editor for the American Chemical Society 2008 and 2009 technical reference books on LENRs and editor-in-chief for the 2011 Wiley *Nuclear Energy Encyclopedia*.

Krivit was the first science journalist to publicly identify and teach the distinctions between the unproven theory of "cold fusion" and the experimentally confirmed neutron-catalyzed LENRs. He did so in 2008 at the 236th national meeting of the American Chemical Society. His chapters in the *Elsevier Encyclopedia of Electrochemical Power Sources* were the first chapters on LENRs in a print encyclopedia.

Other Volumes in This Series

Fusion Fiasco: Explorations in Nuclear Research, Vol. 2

This book tells the behind-the-scenes story of the 1989-1990 fusion fiasco, one of the most divisive scientific controversies in recent history. It explains how credible experimental low-energy nuclear reactions research emerged from the erroneous idea of "cold fusion."

Lost History: Explorations in Nuclear Research, Vol. 3

This book explores the story of forgotten chemical transmutation research during the 1910s and 1920s, a precursor to modern low-energy nuclear reactions research. This work has been obscured and absent in the dialogue of the scientific community for a century.

For More Information
www.stevenbkrivit.com